TWO SCIENCES OF MIND

ADVANCES IN CONSCIOUSNESS RESEARCH

ADVANCES IN CONSCIOUSNESS RESEARCH provides a forum for scholars from different scientific disciplines and fields of knowledge who study consciousness in its multifaceted aspects. Thus the Series will include (but not be limited to) the various areas of cognitive science, including cognitive psychology, linguistics, brain science and philosophy. The orientation of the Series is toward developing new interdisciplinary and integrative approaches for the investigation, description and theory of consciousness, as well as the practical consequences of this research for the individual and society.

Volume 9

Seán Ó Nualláin, Paul Mc Kevitt and Eoghan Mac Aogáin (eds)

Two Sciences of Mind
Readings in cognitive science and consciousness

TWO SCIENCES OF MIND

READINGS IN COGNITIVE SCIENCE AND CONSCIOUSNESS

Edited by

SEÁN Ó NUALLÁIN
National Research Council, Ottawa
Dublin City University

PAUL MC KEVITT
Aalborg University

EOGHAN MAC AOGÁIN
Linguistics Institute of Ireland, Dublin

JOHN BENJAMINS PUBLISHING COMPANY
AMSTERDAM/PHILADELPHIA

TM The paper used in this publication meets the minimum requirements of American National Standard for Information Sciences — Permanence of Paper for Printed Library Materials, ANSI Z39.48-1984.

Library of Congress Cataloging-in-Publication Data

Two sciences of mind : readings in cognitive science and consciousness / edited by Seán Ó Nualláin, Paul Mc Kevitt, Eoghan Mac Aogáin.
p. cm. -- (Advances in consciousness research, ISSN 1381-589X ; v. 9)
Papers originally presented at a workshop on "Reaching for Mind."
Includes bibliographical references.
1. Cognitive science--Congresses. 2. Consciousness--Congresses. I. Ó Nualláin, Seán. II. Mc Kevitt, Paul. III. Mac Aogáin, Eoghan. IV. Series.
BF311.T87 1997
153--dc21 96-52164
ISBN 90 272 5129 0 (Eur.) / 1-55619-189-8 (US) (Pb; alk. paper) CIP

John Benjamins Publishing Co. • P.O.Box 75577 • 1070 AN Amsterdam • The Netherlands
John Benjamins North America • P.O.Box 27519 • Philadelphia PA 19118-0519 • USA

Table of Contents

About the Editors

Seán Ó Nualláin holds an M.Sc. in Psychology from University College and a Ph.D. in Computer Science from Trinity College, both in Dublin, Ireland. He has just returned from sabbatical leave at the National Research Council (NRC), Canada to his lecturing post at Dublin City University, Ireland where he initiated and directed the B.Sc. in Applied Computational Linguistics. He is the author of a book on the foundations of Cognitive Science: *The Search for Mind* (Ablex, 1995). He chaired the first workshop on which this book is based (co-chair Paul Mc Kevitt) at Sheffield in April, 1995.

Paul Mc Kevitt is a Visiting Professor of Intelligent Multimedia Computing at Aalborg University in Denmark, and a British EPSRC (Engineering and Physical Sciences Research Council) Advanced Fellow in the Department of Computer Science at the University of Sheffield, England. The Fellowship releases him from his Associate Professorship for five years to conduct full-time research on the integration of natural language, speech and vision processing. He is currently pursuing a Master's Degree in Education at the University of Sheffield. He completed his Ph.D. in Computer Science at the University of Exeter, England in 1991. His Master's Degree in Computer Science was obtained from New Mexico State University in 1988 and his Bachelor's Degree in Computer Science from University College Dublin in 1985. His primary research interests are in Natural Language Processing including the processing of pragmatics, beliefs and intentions in dialogue. He is also interested in philosophy, Multimedia and the general area of Artificial Intelligence.

Eoghan Mac Aogáin is Director of Instidiud Teangeolaiochta (Linguistics Institute) of Ireland. He studied philosophy at University College Dublin, Ireland and psychology at the Centre for Advanced Study in Theoretical Psychology at the University of Alberta, Canada. He has participated in a number of EU Natural Language Processing (NLP) projects, and is currently involved in PAROLE, which will provide computational lexica and corpora for 13 Union languages to a common standard. His interests include corpus-based approaches to grammar writing and the design of language learning programs. He is the founder editor of the international journal: Language, Culture and Curriculum.

List of Contributors

Kenneth Aizawa
Dept. of Philosophy
Centenary College
2911 Centenary Blvd
SHREVEPORT, LA 71134-1188, USA
kaizawa@beta.centenary.edu

Bernard J. Baars
The Wright Institute
2728 Durant Avenue
BERKELEY CA 94704, USA
baars@cogsci.berkeley.edu

John A. Barnden
Computing Research Lab &
Comp. Science Dept.
New Mexico State Univ.
LAS CRUCES, NM 88003-8001, USA
jbarnden@crl.nmsu.edu

István S.N. Berkeley
Dept. of Philosophy
Univ. of Southwestern
Louisiana, P.O. Box 43770
LAFAYETTE, LA 705-3770, USA
istvan@usl.edu

Mark H. Bickhard
Dept. of Psychology
Lehigh University
17, Memorial Drive East
BETHLEHEM, PA 18015-3068, USA
mhb0@lehigh.edu

Andrew Brook
2217 Dunton Tower
Carleton University
1125 Col. By Drive
OTTAWA, ON K1S 5B6, Canada
abrook@ccs.carleton.ca

Sung-Bae Cho
Dept. of Computer Science
Yonsei University
SEOUL, Korea

Dr Thomas Christaller
German National Center for
Information Technology GMD
AI Research Division
Sloß Birlinghoven
D 53754 Sankt Augustin, Germany
christaller@gmd.de

Terry Dartnall
Griffith University
Fac. Science & Technology
Nathan Campus, Kessels Rd
NATHAN, BRISBANE QUE 4111,
Australia

Kerstin Dautenhahn
VUB-AI Lab.
Autonomous Agents Group
Bldg G10, Room 725
Pleinlaan 2
1050 BRUSSELS, Belgium
Belgiumkerstin@arti9.vub.ac.be

Craig DeLancey
Indiana University
Dept. Philosophy
Sycamore Hall 026
BLOOMINGTON IN 47405-2601, USA

Chenming Guo
Computers Science Dept.
Tsinghua University
BEIJING 100084, China
cmguo@sun.ihep.ac.cn

William C. Hoffmann
2591 W Camino Llano
TUSCON AZ 85742, USA

B.F. Katz
School of Cognitive and
Computing Sciences
University of Sussex
BRIGHTON BN1 9QH, UK

Phil Kime
Dept. of Philosophy
Univ. of Edinburg, David
Hume Tower, George Sq.
EDINBURGH EH8 9JX, UK
Phil.kime@ed.ac.uk

Eoghan Mac Aogáin
Inst. Teangeolaiochta
(Irish Linguistics Inst.)
31, Plás Mac Liam
Dublin 2, Ireland
Eoghan.MacAogain.M248@eurokom.ie

K.F. MacDorman
Computer Laboratory
Pembroke Street
CAMBRIDGE CB2 3QG, UK
karl.macdorman@cl.cam.ac.uk

Paul Mc Kevitt
Center for Pesonkommunikation
Fredrik Bajers Vej 7-A2
Institute of Electric Systems
Aalborg University
DK 9220 AALBORG, Denmark
pmck@cpk.auc.dk

Ajit Narayanan
University of Exeter
Dept. of Comp. Science
Old Library
EXETER EX4 4PT, UK
Ajit@dcs.exeter.ac.uk

Prof.dr. James Newman
Colorado Neurological
Institute
740 Clarkson Street
DENVER, CO 80806, USA
newmanjb@aol.com

Seán Ó Nualláin
Dublin City University
Baile Átha Cliath 9
EIRE/IRELAND
sean@CompApp.DCU.IE
sean@ai.iiT.nrc.ca

N.C. Riley
School of Cognitive and
Computing Sciences
University of Sussex
BRIGHTON BN1 9QH, UK

Gerard Sabah
Language and Cognition
Group (LIMSI), CNRS
B.P. 133
F 91403 ORSAY Cedex, France
gs@limsi.fr

Dr P.D. Scott
Dept. of Computer Science
University of Essex
COLCHESTER CO4 3SQ, UK
scott@essex.ac.uk

J.G. Taylor
Centre for Neural Netwoks
Dept. of Mathematics
Kings College, Strand
LONDON WC2R 2LS, UK
udah057@bay.cc.kcl.ac.uk

Stuart Watt
Department of Psychology
The Open University
MILTON KEYNES, MK7 6AA, UK
S.N.K.Watt.@open.ac.uk

Introduction

Seán Ó Nualláin
Dublin City University and NRC, Canada

The "Reaching for Mind": Foundations of Cognitive Science (CS) workshop was announced over the Internet as follows:

1. Workshop Description

The assumption underlying this workshop is that Cognitive Science (CS) is in crisis. The crisis manifests itself, as exemplified by the recent Buffalo summer institute, in a complete lack of consensus among even the biggest names in the field on whether CS has or indeed should have a clearly identifiable focus of study; the issue of identifying this focus is a separate and more difficult one. Though academic programs in CS have in general settled into a pattern compatible with classical computationalist CS (Pylyshyn 1984, Von Eckardt 1993), including the relegation from focal consideration of consciousness, affect and social factors, two fronts have been opened on this classical position.

The first front is well-publicized and highly visible. Both Searle (1992) and Edelman refuse to grant any special status to information-processing in explanation of mental process. In contrast, they argue, we should focus on Neuroscience on the one hand and Consciousness on the other. The other front is ultimately the more compelling one. It consists of those researchers from inside CS who are currently working on consciousness, affect and social factors and do not see any incompatibility between this research and their vision of CS, which is that of a Science of Mind.

2. Workshop Issues

The tension which riddles current CS can therefore be stated thus: CS, which gained its initial capital by adopting the computational metaphor, is being constrained by this metaphor as it attempts to become an encompassing Science of Mind. Papers are invited for this workshop which:

1. Address this central tension.
2. Propose an overall framework for CS (as attempted, *inter alia*, by Ó Nualláin (1995)).
3. Explicate the relations between the disciplines which comprise CS.
4. Relate educational experiences in the field.
5. Describe research outside the framework of classical computationalist CS in the context of an alternative framework.
6. Promotes a single logico-mathematical formalism as a theory of Mind (as attempted by Harmony theory and using category theory).
7. Moderately or indeed violently disagree with the premise of the workshop.

Ó Nualláin, S. 1995. *The Search for Mind: A New Foundation for CS.* Norwood: Ablex.
Pylyshyn, Z. 1984. *Computation and Cognition.* Cambridge, MA: MIT Press.
Searle, J. 1992. *The Rediscovery of the Mind.* Cambridge, MA: MIT Press.
Von Eckardt 1993. *What is Cognitive Science?* Cambridge, MA: MIT Press.

The Workshop Committee was as follows:

John Barnden	(New Mexico State University, NM, USA & University of Reading, England)
Istvan Berkeley	(University of Alberta, Canada)
Mike Brady	(Oxford, England)
Harry Bunt	(ITK, Tilburg, The Netherlands)
Daniel Dennett	(Tufts University, USA)
Eric Dietrich	(SUNY Binghamton, NY, USA)
Jerry Feldman	(ICSI, UC Berkeley, USA)
Stevan Harnad	(University of Southampton, England)
James Martin	(University of Colorado at Boulder, CO, USA)
Eoghan MacAogain	(Instidiud Teangeolaiochta/Irish Linguistics Institute, Dublin, Ireland)
John Macnamara†	(McGill University, Canada)
Mike McTear	(Universities of Ulster and Koblenz, Germany)
Ryuichi Oka	(RWC P, Tsukuba, Japan)
Jordan Pollack	(Ohio State University, OH, USA)

Zenon Pylyshyn	(Rutgers University, USA)
Ronan Reilly	(University College, Dublin)
Roger Schank	(ILS, Northwestern, Illinois, USA)
Walther v.Hahn	(University of Hamburg, Germany)
Yorick Wilks	(University of Sheffield, England)

† We regret to say that John Macnamara passed away in January, 1996

A remarkable 40 papers had been submitted within two months of the announcement. The fact that so many distinguished members of the CS community agreed to act as members of a committee discussing a crisis in the foundations of their discipline also was telling. The workshop itself was an extremely lively affair, as may be inferred from the diversity of approaches manifest in the papers in each of the three separate parts of this book. The final discussion focussed on two issues:

- the relation between epistemology and ontology,
- what is semantics?

The latter issue recurs throughout this book; the former is implicit in, inter alia, the debate on what precisely to do with consciousness. I have found myself forced to conclude that it is necessary to found a separate science of consciousness, combining normal "science" with phenomenal analysis, alongside a CS based (less controversially) on information. Hopefully, after reading part 3, readers will be likewise convinced.

Part 1 ("CS in Crisis? Cognition and Mind") features papers mainly addressing workshop issues 1, 2 and 7; Part 2 concerns itself more with 4. In Part 3, we will encounter a series of *Weltanschauungen* (all-encompassing views) of great consequentiality which confront all the workshop issues, and more. I hope this book will convey some of the excitement of the event. Some papers which are accepted by the committee but not presented at the workshop due to valid reasons also are included.

3. Acknowledgments

As often happens, Paul Mc Kevitt woke me from my slumbers to suggest I give a public expression to what had been half-formulated thoughts by writing the workshop spec. His work as publicist and general scrummager was invaluable. Eoghan Mac Aogáin's contribution as a reviewer was such that this book would not otherwise have been produced. I wish to thank those who provided extra reviews, in particular the Canadian trio of Arnold Smith, Rob Stainton, Peter

Turney. Any errors that may remain are my responsibility, and remain there even after Noel Evan's best efforts to rein me in. Finally, I dedicate this book to all researchers and teachers in this area who are open-mindedly trying to do justice to the bewildering range of data obtained.

Seán Ó Nualláin

Part I: Cognitive Science in Crisis?

Cognition and Mind

Seán Ó Nualláin

Phil Kime's lucidly-written paper provides an excellent introduction to this part. Kime agrees with the premise of the workshop, but for reasons other than those given in its description. His argument is two-pronged. First of all, the "semantic" formalisms used in CS, particularly where its concerns converge with those of linguistics, are borrowings from the work of logicians like Tarski and Frege who warned of their inappropriateness for natural language. Secondly, different CS camps have wildly different notions of evidence and explanation.

One particular issue straddles the two pillars of the argument; the relation between formalism and *datum*. Since they normally are *a priori* constructions which are not built on any systematic correspondence with the data, there is no difficulty in extending formalisms to handle any given new *datum*. The force of their claim does not diminish in this case because it is essentially an artifact of their internal logical consistency, which can be preserved. Much of cognition is, Kime continues (in logical terms) contingent rather than necessary and will be captured, if at all, by non-Euclidean geometric systems and the like. Indeed, these might yet supply a counter-argument to the systematicity objection (see Aizawa's paper in Part 2). Almost as a parting shot, Kime takes the sacred cow of native speaker intuitions to task with respect to the evidential and explanatory schisms he sees.

Narayanan agrees with the premise of the workshop. He begins by citing some objections to classical CS and AI, i.e., rule-following is not enough for intelligent behavior and in any case falls prey to Gödelian arguments. However, his major concerns are other; he claims that even the eliminative materialists' bet that cognition will eventually be explained with respect to neural activity is too sure to be interesting. He wishes to examine cognition at a finer level of granularity, i.e. that of biochemistry. Emboldened by the example of Penrose, who with Hameroff suggests that the cytoskeleton of cells will provide answers about the seat of consciousness, Narayanan takes us on a biological tour. There is an implication that CS is in some way scientifically obliged to ground its descriptions at this level. Granted, DNA has been already speculated about as a computational device (e.g. solving the travelling salesman's problem) and protein-folding may in general be computationally exploited. However, several good reasons exist for not journeying in this particular direction.

The first is, quite simply, that reductionism is insatiable. Having rejected the Churchlands' granularity as too coarse, we are logically obliged to end at the subatomic level. Secondly, we lack biomolecular explanatory mechanisms for any aspects of cognition. Thirdly, if as Watt later argues, eliminative materialism gives free rein to the mind to settle on (perhaps outlandish) metaphors to understand itself, biomolecular CS will encourage even more promiscuous behavior.

Unsurprisingly, my paper addresses itself to practically all of the workshop issues, with varying degrees of coverage. The notion of "crisis" is interpreted as "opportunity" as well as "quagmire." Von Eckardt's recent characterization of CS is used to focus on the tension that inevitably arises when an essentially informational notion of cognition confronts the phenomenon of consciousness. Several positive recommendations are issued; the domain of CS is to be those aspects of mind which can be informationally characterized, with obvious consequences for its hitherto nebulous academic domain. Several synthetic themes arising from the disciplines which inform CS are then discussed, and the sources of evidence for "egocentric" cognition proposed as a proof of concept for argument from synthesis. Finally, it is argued that a neuroscientific "invasion" may be salutary for, rather than destructive of CS.

For the next two writers CS is not in crisis. Watt argues that metaphors are the stuff of science; somewhat controversially, he states that even the "substantive assumptions" of CS as described by Von Eckardt are metaphorical. We cannot but be anthropomorphic; we project ourselves onto every phenomenon, and Searle's Chinese Room argument gains its force from this. (Barnden extends this type of reasoning in Part 3). Scott's approach is much harder-nosed; scientific theories are communicable and thus inevitably Turing-computable. Cognitivism, the notion that mentation is computation, must be scientifically correct; any theory of mind is essentially a program. However, there are two flaws in Scott's argument. The first is that communication can occur between initiates on matters (like art) for which it is excessive to predicate "computability" as he wishes to use the term. Secondly, his treatment of consciousness is quite simply incorrect; as the papers in Part 3 demonstrate, there is a consensus that it has a functional purpose. However, Scott's paper alerts us to the necessity of finding ways to handle social factors, emotion and consciousness informationally and our inability to handle subjectivity.

Dartnall's substantial contribution revives an old issue. Our minds somehow seem to cope with logical systems employing notions like necessity and certainty. There are two fallacious explanations for this, each flawed in its way. The first, logicism (or reverse psychologism) insists that the laws of logic are the laws of thought. Recent culprits, Dartnall argues, include Noam Chomsky and John MacNamara. The second fallacy, psychologism, attempts to reduce all acts of logical inference to purely psychological processes, which cannot have the characteristic of logical necessity.

Dartnall follows Kime's anti-logicism with a broad historical range of reference. He argues that the central dichotomy is "act versus content" i.e. confusion arises only when we fail to distinguish psychological acts and their intentional objects. The PSSH (physical symbols hypothesis) in AI tries to pack in both act and content; this is precisely the source of the Chinese Room argument's force. An appropriate alternative contrast, Dartnall concludes, is state versus content.

Finally, Bickhard eases the transition into the next part with a sustained attack on the standard story about representation. He argues that our current notion of cognition is parasitic on this story; its putative nature as an autonomous subsystem is equally misconceived.

Representation, (R) as classically conceived, involves "encodingism"; the external world is encoded in some form, which contains a Cartesian flaw. Lacking a homunculus, we need to ground representations otherwise. Some solace may be found in the Piagetian notion that R is an internalization of interactions midway between subject and object. Drescher has extended this work but a better model is one in which representation, action and motivation are seen as manifestations of a single underlying ontology.

It is indeed true that a set of concepts (like egocentric/intersubjective) may have to be superimposed on our notion of R; what is however more urgent for CS is an explicit realist stance. Bickhard points out the problems with R with considerable skill; my own guess is that the notion of context and the role of self in delimiting context will eventually be seen as crucial. However, this leaves us with the problem of how some of our most abstract constructions (like Riemann geometry) refer to anything; what Wigner encapsulates as "the unreasonable effectiveness of mathematics."

Reinventing the Square Wheel

The Nature of the Crisis in Cognitive Science

Phil Kime
Centre for Cognitive Science
University of Edinburgh

1. Preliminaries

Before bemoaning a crisis, it is always best to look around to see if there really is one. Ever since Montague, the momentum in semantics carrying it towards a formal theory of meaning for natural languages has been a significant spoke in the wheel of Cognitive Science. We learnt how to deal with the intensional identity problems that Montague left us with; we learnt how to construct detailed and computationally tractable models of compositionality and how to approach the incorporation of notions of tense, modality and quantification. For all the progress that the field boasts, you would hardly think there was a crisis at all.

However, this is really an illusion. Semantics and traditional theories of meaning are not in a good way and this is, in my opinion, central to the problem that besets Cognitive Science as it underlies the problem with the entire computational program. The crisis as I see it has two main threads: the first being that the computationalist program is wedded to formalisms and ideas that were imported wholesale from people who had principled reasons not to lend them out. The nature of the tools employed within the field are such that they restrict the natural and desirable criticism that a field must support. The current tension within Cognitive Science is, I think, partly a result of the dogma engendered by the formalist thrust one finds at its most stark in semantics. Secondly, there is quite a deep rift between the notions of evidence and explanation that different camps within Cognitive Science employ. Progress in a field is difficult when there is no general agreement about the sources and types of evidence used to test theories.

2. Origins of the Crisis

Primarily, the grip the computational program has on the field as a whole is a result of the misappropriation of logical formalism. In the workshops and everyday seminars on semantics, one hears the words "Fregean" and "Tarskian" mentioned with regularity. If one is to consult the works of the aforementioned, one finds a striking overall repellence to the regimentation of their formal creations as tools in the analysis of natural languages: something the computationalist program has adopted on a large scale. There was a general conception espoused by these progenitors of modern formal theory that natural languages were not sufficiently well defined and exact to allow treatment by formalisms that presupposed a certain amount of regularity and structure in their subject matter. For example, Frege says:

> Language is not governed by logical laws in such a way that mere adherence to grammar would guarantee the formal correctness of thought processes.[1]

So, here we see a concern, not for natural language, but with the thoughts lying behind it. Frege was skeptical about the application of his formalism to natural languages. While Frege was sometimes less than clear on this point, particularly during his early work, he makes quite strident remarks about the applicability of his formalism to natural language in places: more markedly in later work. However, we find this concern explicit in Tarski:

> ... the concept of truth (as well as other semantical concepts) when applied to colloquial language in conjunction with the normal laws of logic leads inevitably to confusions and contradictions.[2]

Tarski was of the opinion that the application of the formal methods to natural languages would necessitate a reform of the language; hardly something that an explicative semantics would aspire to. After all, the reform of a natural language results in an artificial one, thus defeating the object an explanatory enterprise.

Now, formalists generally either ignored these warnings or sought to prove them groundless by devious formal innovations designed to appropriate, in the spirit of Davidson's famous paper on "Truth and Meaning," more and more features of natural language for formal description. Around this time a few were beginning to worry about the assumptions inherent in the formal methods advocated by this approach. For example, Hubert Dreyfus published his famous book on the shortcomings of traditional approaches in AI in 1972. This pointed to the lack of progress overall and suggested certain underlying assumptions were to blame. It seems that the current crisis in Cognitive Science and in particular in semantics is of the same sort: an underlying inadequacy of assumptions inherited from formalisms unsuited to the task; but today it is even worse.

What makes things worse is that we have learnt some new tricks to prevent overt crisis; some new tricks that make everything seem alright. If one looks at the sorts of stock examples that formalist semanticists deal with today, it is quite disturbing to note how simple they still are given the supposed applicability of the formalisms employed. Disturbing too to note how similar they are to examples used ten or twenty years ago. The reason is that we have mastered the art of getting fatter instead of getting further. By that, I mean that a problem is something that results, not in a reevaluation of the foundations of a theory, but almost exclusively in revision of technical minutiae. If we cannot deal with a particular example, we tweak the formalism until we can. More dramatically, we invent another formalism specifically designed to deal with the problem. Progress is seen to be the accommodation of errant data with little respect for the implications for the assumptions of a theory. The crisis in Cognitive Science amounts to exactly this. Problems have come to be dealt with entirely within the scope of the dominant formalist research program. If your formalism starts to give you problems with the representation of the meaning of a certain sentence, make a new formalism. I have lost track in recent years of the number of different semantic formalisms. We now have Dynamic Predicate Logic, Dynamic Montague Grammar, Discourse Representation Theory, Situation Theory, Property Theory, Channel Theory, Linear Logic and many more. Many of these were explicitly motivated by problems with a particular natural language construction. I recall attending a workshop dedicated to formal semantics research just a few years ago where a prominent linguistic semanticist was challenged "But isn't this new theory just tackling the same examples as previous theories have been tackling for over ten years?" The answer was quite typical… "Ah but it's the *way* we do it that's important." To an extent this is a reasonable reply but it strikes me that it is not a reasonable reply when it is the only one ever made to this sort of objection, no matter how many times you meet a problem by generating a new formalism, within the same computational paradigm, specifically to deal with it.

Formalism has very few limits on its possible coverage because its constraints are things like consistency and completeness.[3] We are left with a large amount of choice about what to modify when we come up against a problem. Given that our job is to simply design a formalism that covers a certain construction, we are almost guaranteed to be able to do it, and in many different ways. For example, if you want to have a very compositional approach but your representation of determiners is not well formed, adopt lambda abstraction; if you want to be able to represent the meaning of multi-sentence discourse but the variables in the different sentences are unrelated, simply invent some formalism for variable threading. Only technical problems prevent this and they are not restrictive enough to prevent you from augmenting the formalism in arbitrary ways guaranteed to cover the data. Intractability of the computational approach is met with a new tactic today. We spread out into more and more formalisms that each end up facing the same problems again and again. We are, in effect, reinventing a square wheel. The question of substance here is: "what is it about the computationalist program that allows this?"

The crisis manifests itself because of the nature of the traditional logical approach. The whole allure of formal systems in performing their original normative role of reforming and clarifying language is that they are very flexible. They are our conscious and carefully designed creations and we allow that we may augment and improve them as we see fit. This sort of arbitrarily adaptable tool is not the sort of thing very conducive to an open debate on the foundations of a subject. The reason being that you will never feel compelled to deny your basic principles when you have a tool that can always be modified in some way to cope with recalcitrant experience. The traditional approach is supported by the logical tools it has adopted and these tools have turned out to be fantastically methodologically elastic.

I think this is an instance of a more general observation. A normative or prescriptive system is one not totally constrained by the evidence, but one that seeks to constrain it. Thus such a system is designed to do violence to the way things are. Such systems are by nature reformative. As a result, they are designed to be very flexible and accommodating of the desiderata of a good prescriptive theory. The trouble is that the desiderata for a good prescriptive theory are not dependent on features of the evidence they intend to prescribe; that is the whole point of them. This makes them quite naturally unsuitable for a Cognitive Science having an *explicative* ingredient that marks it out from the purely descriptive engineering practices of Computer Science and to some extent, contemporary AI. The root of the formal, computational approach so apparent in semantics is exactly a prescriptive system. For example, if you have a theory *X* that encounters a problem piece of evidence *Y* and your theory is based on a prescriptively designed formalism, you will never have difficulty in bending the theory to fit the problem because the underlying formalism was designed with the independence of the features of the evidence and finished theory in mind. So the theory is not really constrained to features of the evidence. If you have a problem with the logical independence of the terms in supposedly analytic sentences like "All bachelors are unmarried," you have the tools to construct complex expressions to serve as meaning postulates or perhaps you might construct lexical decompositions to square the data with theory. All that is necessary to sanction this move is the formalism: the reason you perform an arbitrary formal operation to solve a problem is because you are able to do so.

Now, it may be objected that it is rather strange to suggest that a tool that can always account for the data is a *bad* thing. Well, in the face of such a tool, you can make roughly two responses. Firstly, admit that it is a positively *good* thing to be able to always account for the data. It means you are doing the right thing. This is of course might be a legitimate tactic although one we might worry about this along methodological lines akin to Popper's famous concerns regarding Freudian and Marxian theory for example. Unfalsifiable theories are methodologically suspicious. Secondly and more accurately in my opinion, if the tool was designed as prescriptive, reformative and idealized, being able to account for all the data is no longer a virtue. Its "success" follows from the nature of the tool rather from the connection a theory employing the tool has with empirical reality.

Once might be tempted to suggest that surely some evidence can legislate between differing camps in Cognitive Science? After all, most people hold that it is *empirical* after all. This consideration leads me to the second main source of divide within the Cognitive Science community: the nature of evidence.

3. The Evidential and Explanational Schisms

Formalist semanticists still take to be evidence, in the explanation of the semantic aspect of human cognition, the intuitions of native speakers or understanders. Thus, in this Chomskian vein, we take note of intuitions about quantification scope, anaphoric resolution, relative clause nesting etc. The more psychological and neuropsychological camps do not take this as evidence as such. Evidence there is reaction times, discrimination task performance and the like. This is quite serious as the differing camps can barely agree on evidence to disagree about. Part of the reason that the computational program constrains the field to such a degree is that is has a monopoly on what is to count as evidence. For example, I have heard it said many times that the neuroscientific data is all very interesting but is entirely the wrong sort of level of explanation we should be concerned with in Cognitive Science. It is "too low" a level of explanation to be scrutable. In particular, it is too low level to be input to traditional logical formalism. The evidential scruples of the computational program are, again, a product of its appropriation of the logical formalisms. Logical systems were designed to function as tools to standardize and disambiguate in the service of science. The perfect logical language would be clear, concise and paradox free. Logic undertook to reform the propositional expression of information within the sciences therefore concepts such as "proposition" and "deduction" are appropriate elements. Also, one of the central features of the natural sciences is that they are *written* disciplines. The primary mode of communication of scientific ideas is literary. As a result, there is a heavy emphasis on clarity and portability of expression. Such material should not be particularly contextually and indexically dependent. This contrasts starkly with natural spoken language which actively *depends* upon contextual and indexical information to an enormous degree. A formal ramification is that the "canonical language" is not well suited to the task of accounting for context. Workers in the field have typically attempted to account for context using the same formal tools, resulting in famous systems that have floundered due to computational explosion when attempts have been made to extend them beyond their toy domains. Commonly, the list of predicate/argument expressions that embody the formalist approach to "context" simply get too big too quickly. So, it seems clear that here is a case where the design constraints of the chosen formalism have become a burden for the computationalist paradigm. However, once you have adopted the formalism, you get its inbuilt desire for a particular type of explanation for free. This type of explanation is far from inclusive of all relevant levels one might like to explore and thus the space of possible research is forcibly and questionably restricted.

The standard account of the veracity of computationalist style explanations has to do with the necessity of certain types of constraint on patterns of behavior. Pylyshyn's famous argument is that if we have a level of explanation E whose explanatory elements are insufficient to constrain any resultant model of reality to reality, then we require another level of explanation E' to explain such constraints.[4] Thus, the computationalist program depends on the idea that the formal level of explanation is necessary to account for regularities in our semantic life. Pylyshyn passes over rather quickly the response that the constraints on our behavioral patterns are merely contingent and accidental, arguing that the symbolic level's exclusion of certain types of explanation is not exclusion *by definition*. But I simply cannot see how it could exclude other explanations in any other way since the formal tools are, in their inception, *prescriptive*: they are not designed with features of the world in mind so much as features of a formally attractive model of the world. Furthermore, there seems to me to be a principled way of achieving some of the more reasonable constraints that computationalists desire without necessitating the generation of more "levels of description."

It is well known but still often underrated feature of evolutionary theory that nature is economical. The reuse of existent structures is accepted as an enormously fruitful way of regarding the genesis of features of body and behavior. The functions of many biological structures are forced onto them given changes of environment and genetics. In keeping with this, one would expect and desire an explanation of the intricacies of language and semantics expressed in terms of preexistent structures and concepts. Contrastingly, the modern trend has been towards a *sui generis* treatment of this aspect of human capacities.[5] We are reasonably well informed about the kind of structures that give rise to motor behavior and have, over the past fifteen years, proposed models empirically supported by work in neuroscience.[6] Given the, hardly implausible, assumption that motor behavior precedes linguistic in the evolution of the human species, we might predict a reuse of structures and strategies found in the temporally prior behavior. This is exactly what neuroscience began to suggest over ten years ago.[7] However, the kind of representations posited as underlying the reuse of motor coordination constructs are emphatically not of the sort formalists in Cognitive Science are used to dealing with. Inter-methodological incommensurability results: a point addressed above. This is not the place to undertake a lengthy exposition of the technicalities of the more geometrical approach suggested by these considerations but I should point out two substantial elements in its favor.

Firstly, the formalisms employed are explicitly descriptive. Geometrical mathematics is designed to best fit data within well known constraints of consistency with other branches of mathematics. It is designed to model how things are and not how we might like them to be. Thus its basic concepts are nowhere near as pregnant with explanatory biases as those of formal logic; "point," "line" and "space" hardly press one in the direction of a particular view of mind and language as forcibly as "proposition," "predicate" and "object/meta language." Secondly, the computational and explanatory advantages of this method have historical precedent. Einstein's rendering of gravitation as a *structural* feature of spacetime as opposed to a force within a structure is the modern archetype of good explanatory methodology. Not only does this cohere with the post Duhemian concern with parsimony, it has very important pragmatic implications for Cognitive Science. Dreyfus and Winograd have long held that the computational explosion one encounters in attempting to model inference using symbolic representation is the result of fundamental ignorance of 19th century phenomenology. What is not currently appreciated is how far the geometrical approach goes towards dealing with this. The computational explosion is a result of the syntactic intractability of *semantic* relation. For example, the truistic character of "No bachelor is married" is not initially syntactically explicable as it depends upon the meaning of "bachelor." The standard way of dealing with this is to either allow a lexical decomposition of "bachelor" into "unmarried man" or to propose a meaning postulate meta-theoretically linking models containing "bachelor" and "unmarried man." The trouble is that this problem is ubiquitous in language and explicit delineations of such relations along either line are just not plausible if we are to consider the amount of processing that the brain must perform under such models. It is exactly the problem that Husserl and even Carnap faced in attempting to provide formal models of phenomena and fails for exactly the same reasons. However, in a geometrical model we have the opportunity of following Einstein in rendering semantic relationships as *structural* features of the space in which an representation might occur, thus obviating any computational penalties associated with explicit representation. There is not space to detail this approach here but a small example should suffice. If we have some eggs in a square box with the lid closed, we can say little about the relations between their positions: the space they inhabit is fairly orthogonal. However, put them in an egg-box and the situation is very different. We know automatically *from the structure of the space they now inhabit* certain things about them. For example, we know that none of them lie on their sides. We know that the distance relations between eggs obey transitivity as egg-boxes are made to be all alike (so they stack well) and this defines a metric on the "egg-box space." None of this has been derived by any explicit rules or inference: the "conclusions" are facilitated by the structure of the space the eggs now inhabit. Computationally speaking, we have relations for free in the same way that gravitational effects come for free in General Relativity. This is a way of caching out Dreyfus' insistence on our ability to "just see" certain semantic relations without the need for any actual *inference*.

This conception of Cognitive Science also addresses the linchpin of the formalist approach: the productivity of language and thought. Actually, productivity has never been a particularly strong argument for formalism as proponents of the view have admitted.[8] Infinite and even truly massive finite productivity is an idealization never actually realized. If this is the case, then we hardly need a recursive formalism: an iterative one will do as long as there are enough iterations to cover the life of any given human. Also, given that we link the need for a productive formalism to the productivity aspect of language and thought, we spend a lot of time having to invent restrictions on our formalisms to prevent them being *too* productive. I am thinking here of non-monotonic and strongly typed formalisms etc. No, a much stronger argument for formalism is the well known argument from the systematicity of language and thought. This is one of those constraints that Pylyshyn thinks necessitate a formal level of explanation as it is a constraint not reflected in, say, sub-symbolic explanation. I think this is no longer the case. We are now beginning to see how the complexities of the structure of spaces employed in so-called "geometrical" models may well be able to provide a non-symbolic circumscription that fits in with the largely justified systematicity desiderata of the symbolist.[9]

Not only are there differences in what counts as evidence in the field: there is also a rather stark difference in what is to count as an explanation. This is brought out clearly in the disagreements between symbolists and connectionists. The latter are often happy to allow parts of their networks to have no interpretation under a particular theory.[10] The symbolists, however demand that their evidence be systematically interpretable and even go to great lengths, as in the case of providing "meanings" for determiners by using lambda abstraction, to ensure that it is. This desire for explanations and accounts that have something to say about every stage and aspect of a process seems to me to be a clear consequence of one of the central theses of semantics; that of compositionality. Drawn from Frege (indeed sometimes called "Frege's Principle") this requires that the meaning of a complex expression is a function solely of the meanings of the more basic expressions that comprise it. As a consequence, those accepting this principle prefer explanations of *X* that are parasitic upon explanations of parts of *X*. Thus every aspect of an account must be made clear in terms of the theory. Again, this seems to be a consequence of features of the adopted formalisms. The desire for compositionality is originally a formal one. We shall force this on our data; indeed Frege explicitly gave referents to non-referring expressions to accomplish this. This is not because non-referring expressions really do have referents and our formalism is telling us this, but because we would like them to have, in the name of formal consistency. This is a clear case of reformative formalism. As a result, I do not think it a coincidence that formalists came to believe in the compositional nature of mental representation. Their formal tools had this built in when they were adopted.

A general but underappreciated feature of formal sciences is the way in which their formalisms generate paradigms of explanation that mirror formal features. Formalisms have the reputation of being tools by which one implements a theory. As a result, their features as naturally seen as posterior to theory: a result of theory. Often, this inchoate view is mistaken. Adoption of a formalism promotes a two way process in which assimilation of theoretical distinctions into the formalism is only a proper half. The other half is the assimilation of formal distinctions into the theory. The traditional Fodorian modularity theses regarding the elements of inference are,[11] I think, as much a result of the formal distinction between object and meta-language as any putatively empirical evidence. Problematically, the formalism is not guaranteed to be a provider of good cues. Indeed, in Cognitive Science, given the attitudes of the formal progenitors, quite the opposite is seen to be the case. The maxim is: when using your formalism, be careful it does not use you.

Now, it is not always a bad thing to have your formalism suggest novel and unlooked for features of your data: indeed it is often a striking methodological bonus when this happens. However, the case before us is unduly troublesome in at least the following two respects. As mentioned above, the dictates of a formalism decidedly hostile in its inception are somewhat more suspect than a formalism designed to do what you are using it for. A formalism that suggests one finds theoretical significance in, for example, an object/meta language distinction had better be sympathetic to your overall purpose.[12] Otherwise, you are at the very least straining the application of the formalism to its natural limits. Secondly, the data we are considering here is of a quite different sort to that which we find in the natural sciences. The problem for the formalist in general, is that what is taken to be evidence shares a perniciously symbiotic relationship with the theories it is meant to inform. This is not to suggest that one should hope for a reinstatement of the long forsaken observation/theory language distinction. Rather that because native intuitions are such a strange sort of evidence, they are particularly prone to self-fulfilling prophecy effects. When you have tried to decide a few times whether or not a sentence has ten or twelve scopal readings, your intuitions become so very confused that they simply do not have the basic feature required of evidence: they are not particularly stable. If you want to test the current water level the last thing you want to stick your yardstick in is a raging sea. The battle between a reformative formalism and unstable data is a foregone conclusion. You always succeed in covering the data. By giving oneself by definition, in the manner of Chomsky, a theoretically stable but empirically inaccessible level of "competence," one does nothing to allay fears that the formalism is excessively driving the theory. Rather, the fears are confirmed.

So, the problem is compounded: not only do we have a formal assumption that allows us to continue along the same road in the face of every adversity; we also have a notion of evidence that is unstable enough to support serious critique of any particular theory. This adds up to a seemingly principled façade resulting in a difficulty in entertaining fundamentally different approaches: the position that Cognitive Science and particularly linguistic semantics finds itself in today.

Notes

1. Frege (1882).
2. Tarksi (1931), in Tarski, (1956).
3. In higher order systems, we often have to do without even these.
4. Pylyshyn (1984: 35–38).
5. A recent example of this is McDowell (1994).
6. Churchland & Sejnowski (1992) provides a good overview of recent work.
7. See, for example, Pellionisz & Llinás (1982).
8. See, for example, Fodor (1987).
9. See Gärdefors (1990) for the beginning of such an approach.
10. There is a current trend towards providing a type of compositional treatment for connectionist models nowadays. See, for example, Gärdenfors (1993). This strikes me as a consequence of the monopoly that the symbolic paradigm has on the concept of explanation.
11. See, for example, Fodor *et al.* (1980).
12. As is the case in Fodor *et al.* (1975); Fodor *et al.* (1980).

References

Churchland, Patricia S. and Sejnowski, Terrence J. 1992. *The Computational Brain.* Cambridge, MA: MIT Press.

Dreyfus, Hubert L. 1992. *What Computers Still Can't Do.* Revised edition of *What Computers Can't Do,* 1972. Cambridge, MA: MIT Press.

Fodor, Jerry A. 1987. Why there still has to be a language of thought. In *Psychosemantics*, 135–167. Cambridge, MA: MIT Press.

Fodor, J.D., Fodor, J.A. and Garrett, M.F. 1975. The psychological unreality of semantic representations. *Linguistic Enquiry* VI(4), 515–531.

Fodor, J.A., Garrett, M.F., Walker, E.C.T. and Parkes, C.H. 1980. Against definitions. *Cognition* 8, 263–367.

Frege, G. 1882. Uber den wissenschaftliche Berechtigung einer Begriffsschrift (on the scientific justification of a conceptual notation). *Zeitschrift für Philosophie und philosophische Kritik* 81. English translation by Bartlett in *Mind* 73, 1964.

Gärdenfors, Peter. 1990. Induction, conceptual spaces and ai. *Philosophy of Science* 57, 78–95.

Gärdenfors, Peter. 1993. How logic emerges from the dynamics of information. Lund internal paper.

McDowell, John. 1994. *Mind and World.* Harvard University Press.

Pellionisz, Andras and Llinás, Rodolfo. 1982. Space-time representation in the brain. The cerebellum as a predictive space-time metric tensor. *Neuroscience* 7, 2949–2970.

Pylyshyn, Zenon W. 1984. *Computation and Cognition.* Cambridge, MA: MIT Press.

Tarski, Alfred. 1931. The concept of truth in formalised languages. In Tarski, 1956.

Tarski, Alfred. 1956. *Logic, Semantics and Metamathematics.* Translation by Woodger. Oxford University Press.

Biomolecular Cognitive Science

Ajit Narayanan
Department of Computer Science
University of Exeter

1. Background

Cognitive Science, quite simply, attempts to provide solutions to the question of how mind and brain are related or, more generally, what constitutes mind/brain. *Classical* cognitive science (CCS), together with its subdiscipline of artificial intelligence (AI), is based on Newell and Simon's (1976) explicit commitment to the "Physical Symbol System Hypothesis" — the idea that all intelligent action and behavior can be necessarily and sufficiently described and explained by symbols and rules operating on those symbols, where the rules themselves can have symbolic form. Furthermore, these rules and symbols must be realized in any system for which claims of intelligent action and behavior are made.[1] AI's concern has been with *computational representations* of physical symbol systems.

CCS and AI have been attacked on the following three grounds: (a) that rule-following by itself is not sufficient (and may not even be necessary) for intelligence, awareness and consciousness; (b) that because CCS and AI are anti-materialist and perhaps anti-reductionist in nature they cannot explain how brain gives rise to mind and therefore cannot provide adequate accounts of mind/brain; and (c) that CCS and AI, because they succumb to the same formal limits that apply to computation and algorithms, cannot account for certain types of mental processes which fall outside the class of what can be computed.[2]

With regard to (a), the strongest expression of this objection to CCS and AI has come from Searle and his Chinese Room Argument (Searle 1980). The best reply to the actual Chinese Room scenario is the Korean Professor Argument (Rapaport 1988) which identifies a weakness in the Chinese Room scenario (the person in the room understands the instructions to be followed) before re-describing the scenario in a form acceptable to CCS and AI.[3] More generally, though, AI addresses this problem through a form of "Systems reply": intelligence, awareness and consciousness arise from computational processes and interactions between these processes. Whether these processes are mental, physical or behavioral is irrelevant, in that the system as a whole moves through various states, where the next state of the system is determined by the current state of the system and any input it receives. The stress here is on functionality and cognitive architecture (Putnam 1967; Fodor and Pylyshyn 1988): The states a system goes through are *representational* states, and a cognitive architecture is an architecture of representational states which involves the precise nature of the representations and the operations performed over them.

With regard to (b), neuroscientists claim that an understanding of the brain is required for any account of mind, where the claim is supported by evidence that so far it has not been possible to find an entity with a mind which/who does not also have a brain. In response to connectionist attacks (Rumelhart and McClelland 1985; Rumelhart, McClelland *et al.* 1986) Fodor and Pylyshyn (1988) further refined CCS to identify three characteristics which representations in any proposed cognitive architecture have to satisfy: systematicity (the ability of a system to produce/understand some expressions is intrinsically connected to the ability to produce/understand certain others), productivity (the ability to produce/understand expressions not previously encountered), and compositionality (the ability of an item to make the same semantic contribution to each expression in which it occurs).

There have been a variety of attempts to provide connectionist representational architectures which satisfy these characteristics (e.g. van Gelder 1990; Bodén and Narayanan 1993; Niklasson and Sharkey 1994; Christiansen and Chater 1994; Niklasson and van Gelder 1994).[4] But there are two types of neuroscience. On the one hand, *reductionists* in general accept that, even after reduction to a neuroscientific basis, mental processes do exist and can be described in their own terms.[5] This is to be contrasted with eliminative neuroscientists, who believe that the sort of reductionism canvassed by reductionists does not go far enough: "Eliminative materialism is the thesis that our common sense conception of psychological phenomena constitutes a radically false theory, a theory so fundamentally defective that both the principles and the ontology of that theory will eventually be displaced, rather than smoothly reduced, by completed neuroscience" (Churchland 1981: 206). This leaves neuroscientists with the problem of how to account for mind. The concept of emergentism is often appealed to at this point: a collection of relatively simple neuronal units, communicating with neuronal units at neighboring levels, together perform a global (holistic) computation that none of the individual units, or linear combinations of them, could do alone. Emergentism is the idea that what are called higher level cognitive processes can be accounted for by their emergence from the neurocomputational substrate. However, there is as yet no clear neuroscientific account of emergentism, except for references by way of analogy to the way that the microstructure of a physical object (in terms of atoms, molecules and lattice structure) can give rise to macro level physical properties (e.g. hardness).

The stance of CCS and AI on these issues has been that it makes as much sense to ask for details of the way the brain works when trying to understand the mind as it does to ask for details of hardware when trying to understand how a program works. That doesn't mean that an *implementation* of an algorithm or mind does not require hardware or a brain, respectively; rather, what is claimed is that details of the hardware/brain do not add anything to our algorithmic/mental accounts. The core question for neuroscientists, "How can the brain as material object evoke consciousness/mind?" can only be answered by appealing to representational states, and CCS and AI are best placed to offer an account of representational states, goes the argument.

With regard to (c), the Mathematical Objection (MO) is that machines will never be able to do everything human minds can do (Lucas 1961). This is because Gödel showed that any formal system of a sufficiently powerful kind cannot be both consistent and complete at the same time. This means that there will always be one statement which, if true, cannot be proved, and if proved, cannot be true. Since a computer and its program are an instantiation of a formal system, it follows that for any AI computer there will always be one statement (called the Gödel Formula) which the computer cannot see as true (or provable) but which we humans can see is true (or provable). Proposers of the MO claim that this argument prove that machines can never do everything that humans can do, that machines will always be one step behind human reasoning.

AI has traditionally replied to the MO in a variety of ways. For instance, the criticism that a computer cannot "jump out of the system" assumes that systems are logically separated onto separate levels, with simple systems at the bottom and increasingly complex systems at higher levels. But in CCS and AI the brain is at the bottom level, and the brain, if it is describable mathematically at all, will have a complex mathematical description. The only way to understand the brain is to "chunk" it on higher and higher levels, thereby *losing precision* until perhaps at the higher levels we have "informal" systems (Hofstadter 1979). That is, levels in a mathematical proof and levels in AI are not the same. Therefore, what a mathematician and an AI researcher jump out of and into are different also.[6]

2. Biomolecular Foundations of Mind/Brain

However,a fourth objection is now surfacing which presents serious difficulties for both CCS and neuroscience. The fourth objection — that consciousness is *biomolecular* and that any account of the mind/brain which does not take into account the behavior of biomolecules is doomed to failure — undercuts neuroscience which stresses *neurons* as the primitive computational element as well as classical cognitive science which stresses a cognitive system passing through various representational states. As Penrose (1994: 357) says:

> If we are to believe that neurons are the only things which control the sophisticated actions of animals, then the humble paramecium [a single-cell, eukaryotic organism belonging to the kingdom *protista*] presents us with a profound problem. For she swims about her pond with her numerous tiny legs ... darting in the direction of bacterial food which she senses using a variety of mechanisms, or retreating at the prospect of danger, ready to swim off in another direction ... Moreover, she can apparently even *learn* from her past experiences ... How is all this achieved by an animal without a single neuron or synapse?

The implication, quite simply, is that neuroscientists have got it wrong if they claim that networks of neurons (single cells) adequately account for mind/brain: while a cell is the basic unit of living systems, this does not mean that the cell is primitive.

Molecular computing, which stands in the same relationship to biomolecular science as AI does to CCS and connectionism to neuroscience, is the computational paradigm derived from and/or inspired by biomolecular processes within cells (Carter 1984; Conrad and Liberman 1982; Hameroff 1987).[7] So, what is molecular computing, and can it and its parent science provide an adequate account of mind/brain? A brief description of cell structure and function is required at this point.

A cell — typically 10–30 millionths of a meter across for humans — contains many specialized structures called organelles. The relevant ones here are the cell membrane (controls passage of substances into and out of the cell and encloses cell organelles as well as cell substances), cytoplasm (serves as a

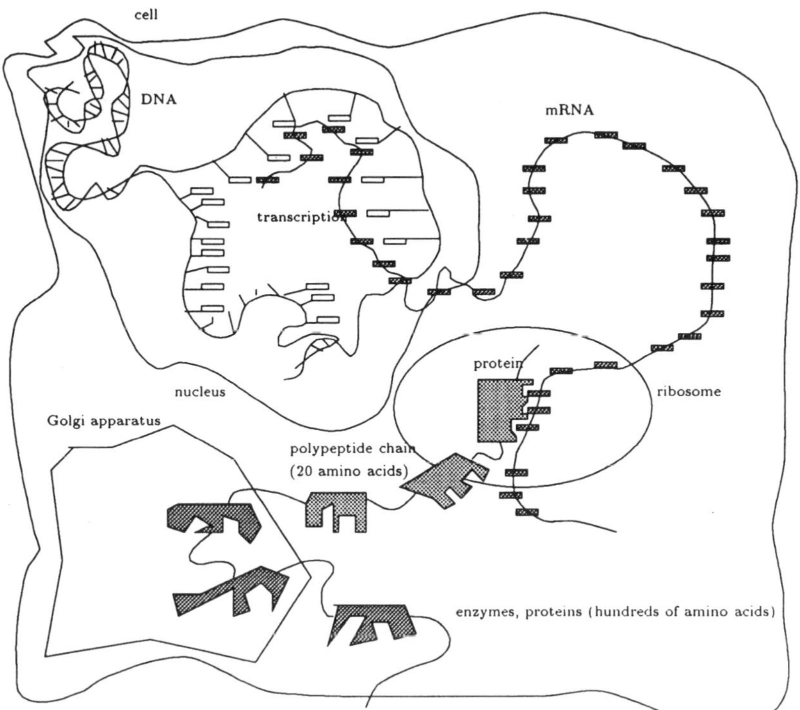

Figure 1. Only the relevant parts of the cell are shown here: the nucleus which contains DNA and RNA, ribosomes where protein construction (translation) takes place using the 20 basic amino acids, and the Golgi apparatus where individual amino acids are modified slightly to produce the variety of amino acids essential for life. A body cell is typically 10–30 millionths of a meter long and wide, and it is estimated that we have several trillion of such cells (for skin, muscles, liver, blood, heart, brain (a neuron is a brain cell), etc.). Each such body cell contains the full set of 46 chromosomes (discrete molecular structures of DNA) inherited from our mother and father (23 in each case, via sex cells). The "straight-line" length of the DNA in one cell is estimated to be 2 meters, which demonstrates the tightly packed nature of the chromosomes and their thinness. It is also estimated that the 46 chromosomes code for between 75,000 to 100,000 genes for humans, using about 8 billion bases (nucleotides). On average, about 100,000 bases are required for coding a gene, although this figure varies greatly from a few hundred to a few hundred thousand.

fluid container for cell organelles and other cell substances as well as assists in the transport of substances within the cell), nucleus (directs all cell activity and carries hereditary information), endoplasmic reticulum (serves as a transport network and storage area for substances within the cell), ribosome (manufactures different kinds of cell protein), Golgi apparatus (packages protein for storage or transport out of the cell), lysosome (digests or breaks down food materials into simpler parts and removes waste materials from the cell), mitochondria (serve as the power supply of the cell by producing ATP — adenosine triphosphate — which is the source of energy for all cell activities), microtubules (serve as the support system or skeleton of the cell) and microfilaments (assist in cell motility). Each organelle performs one or more special tasks to keep the cell alive. All the information directing every cell function is stored in large DNA molecules found in the nucleus.

A cell cannot function without DNA. The information it contains must be made available somehow to the rest of the cell as well as be passed on to all new cells. Although each cell contains the full complement of DNA, through some process which is not yet clearly understood certain parts of the DNA are switched on or off within cells, resulting in different types of cell producing different proteins for normal growth and functioning of the organism as a whole. The process by which the information in the DNA is carried out to the rest of the cell is through messenger RNA strands which leave the nucleus and attach themselves to the ribosomes, which then produce the protein for export from the cell (Figure 1). What is remarkable is that the DNA are large molecules made up of combinations of only four types of nucleotides — adenine, guanine, thymine and cytosine (called A, G, T, and C, respectively). It is estimated that the DNA in each one of our cells contains about 8 billion nucleotides, spread across 46 chromosomes (discrete molecular structures of DNA), each one of which takes the shape of a double helix. If all the DNA in one cell were stretched end to end, the length is estimated to be about two meters. Messenger RNA bang into these chromosomes and unzip part of the molecule, make a *complementary* copy of a certain length of the molecule, before leaving the nucleus for the ribosomes and protein manufacture. The process of DNA being mapped into mRNA is called *transcription*, whereas the process of duplicating all chromosomes is called *replication*.[8] Ribosomes produce the appropriate amino acids from the mRNA. For instance, the mRNA triplet *GCU* (guanine — cytosine — uracil), which is an mRNA transcription of the DNA triplet *CGT* (cytosine — guanine — thymine), is mapped onto the amino acid *alanine* by ribosomes.

It may appear from the above that the transfer of information from the nucleus to the rest of the cell is a highly organized affair. This is not correct. Random collisions millions of times a second between RNA polymerase (an enzyme, which is a large protein which helps make and break bonds) and the

DNA eventually lead to the RNA polymerase running into certain sequences of bases and latching onto them. These sequences of DNA bases are recognized by the RNA polymerase as start positions for transcriptions. The RNA polymerase then unravels the appropriate part of the DNA double helix. Free-floating bases in the nucleus attach themselves to the revealed DNA bases, forming a sequence which becomes the messenger RNA. The double helix is re-formed as transcription continues along the unravelled DNA molecule. When a terminating sequence of bases is found in the DNA, the resulting messenger RNA is dispatched to the ribosomes, where combinations of three bases at a time in the messenger RNA are used to produce one of 20 different amino acids. Sequences of these amino acids (varying in length from a few hundred to a few thousand) are called polypeptide chains, which are packaged in the Golgi apparatus and then secreted from the cell for use by other cells in the organism. These polypeptide chains therefore "represent" the sequence of bases unravelled in the DNA molecule (Figure 1). Again, it may appear that the production of polypeptide chains out of individual amino acids is a highly organized affair. This again is not true: there are so many millions of molecular collisions each second within ribosomes during polypeptide production that some of these must be the correct ones for the proper production of the polypeptides.[9] For instance, appropriate polypeptides (proteins/enzymes) for continually producing hair of a certain color for an individual are transferred from the individual's DNA in certain specialized hair-production cells.

3. Implications for Cognition

So, what happens to the enzymes/proteins produced by ribosomes and the Golgi apparatus? Proteins (enzymes) carry out many vital functions in living organisms. As structural molecules, they provide much of the cytoskeletal framework of cells. As enzymes they act as biological catalysts that speed up the rate of cellular reactions. The chemical composition of one of our cells could be placed in a test-tube and observed, We may, after some time, notice some chemical reactions naturally occurring in the test-tube. There will be a long delay because the activation energy required to start a chemical reaction acts like an energy barrier over which the molecules must be raised for a reaction to take place. An enzyme effectively lowers the activation energy required for a reaction to proceed. An enzyme locks onto a molecule, starts a reaction, and then is released unchanged. The rate of enzyme combination and release is called the *turnover rate* and is about 1000 times a second for most enzymes, with variation between 100 per second and 10 million per second. The increase in reaction rate achieved by enzymes ranges from a minimum of about a million to as much as a trillion times faster than an uncatalysed reaction at equivalent concentrations and temperatures. From this it can be seen that the process of enzyme/protein production, as determined by our DNA, is absolutely critical to our continued well-being, otherwise we as chemical beings would not produce chemical reactions fast enough to keep us alive (e.g. respiration, digestion).

What inheritance now means, according to biomolecular science, is the set of genes (DNA) which code for the production of appropriate enzymes which increase the rate of chemical reactions in our cells, where the nature and rate of reactions is determined by the nature of the enzymes. We are all essentially the same chemically: what differs is the enzymes produced by the DNA inherited by our parents and other factors (e.g. mutation of individual bases and genes by random means), and these enzymes control cellular processes differently for different people, thereby leading to different physical characteristics.

The applications of molecular computing are quite clear in the area of biomedical research. For instance, cloning is the process in which a diploid cell divides and produces a whole new organism through complete DNA replication (rather than the nuclei of two haploid cells merging to produce off-spring).[10] Also, various inborn errors of metabolism and chromosome errors which give rise to genetic diseases can be explained as errors in transcription or replication and through DNA mutation, such as sickle cell anemia (reduction in the solubility of hemoglobin in the blood), Tay-Sachs disease (absence of specific enzymes that hydrolyse specific lipid molecules), diabetes mellitus (insulin deficiency), hemophilia (improper clotting of blood), phenylketonuria (associated with mental retardation) and albinism (the production of skin pigment *melanin* is blocked). Viral infections (colds, flu, measles, chickenpox and mumps, for example) can be explained at a deep level and resulting computational models can generate hypotheses concerning their evolution and treatment.[11] Cancers of various sorts exhibit a wild, uncontrolled growth, dividing and piling over each other in a disorderly arrangement and pushing aside the normal cells in a tissue. Similarly, computational models can provide a useful service here in prediction and treatment. But what are the implications of molecular computing and its parent, biomolecular science, for mind/brain? Although a cell has a functional architecture and performs many functions, each of which is determined by the DNA information within a nucleus, the idea of a cell moving through various representational states as it processes information, where there representational states (as required by CCS) involve the manipulation of symbols, does not sit easily with the facts. Molecular computing is essentially a copying (translation and replication) process. However, if molecular computing and biomolecular science are to offer alternative accounts to CCS and neuroscience, there must be some method by which elements of molecular computing are tied up with representations, information processing and consciousness.

There are a variety of proposals in the molecular computing literature concerning the way that cells could give rise to information processing and consciousness. They can be split roughly into two types: the first type deals with the way that any cell can be regarded as an information processing device, and the second deals specifically with brain cells and attempts to use properties of neurons for accounting for consciousness. Among the proposals of the first type are claims that cells represent information through (i) reaction diffusion systems (chemical reaction waves within a cell propagate at uniform speed and interact with other waves within the cell to produce complex patterns (Conrad and Liberman 1982; Winfree and Strogatz 1984)) (ii) cellular automata (a large number of identical cells connected in a uniform pattern and communicating only with other cells in their neighborhood operate collectively to produce complex behavior (von Neumann 1966)) and (iii) the protoplasm (dynamic activities of cytoskeletal structures including cytoplasmic microtubules within a cell produce rudimentary consciousness (Hameroff 1987; Penrose 1994)). Among proposals of the second type are (i) holograms (the brain perceives sensory information by analysis of the interference of neural firing frequencies, resulting in a domain in which space and time are enfolded (Pribram 1986)), and (ii) cytoskeletal activity, but this time within neurons and at a quantum mechanics level (Hameroff 1987; Penrose 1994).[12] However, it must be said that all these proposals are highly speculative, leaving CCS with its stress on computation and neuroscience with its use of mathematically rigorous connectionist networks in the lead as far as clear proposals are concerned.

4. The "Mind Gene"

The question now is: Is there a mind gene? That is, is there a part of our chromosomes which produces enzymes/proteins which, when released in, say, neurons, give rise to consciousness and mind? The current approach to this question consists of appealing to an *evolutionary* account. As Crick (1994: 12) says about language:

> ... the understanding of the evolution of language will not come only from what linguists are doing, but from finding how language develops in the

> brain … and then finding the genes for it and trying to work out when those genes came in evolution.

Two approaches to this question can be predicted. The first depends on a contextual approach where, for instance, to account for, say, a type of sensation is to identify which part of one's DNA (hereditary information) is responsible for producing the polypeptide chains associated with that sensation and then to derive an evolutionary account based on neighboring DNA code. An account of desire may be based on identifying which part of one's chromosomes is responsible for producing the chemical proteins/enzymes associated with desire and then determining what is on either side of the DNA for desire. It may be that on one side in the chromosome is the code for producing polypeptides associated with goal-motivated behavior, and on the other the code for producing polypeptides associated with plan-producing behavior. An evolutionary account of desire would then be based on some story which related goals to plans by means of desire: At some stage in the evolution of consciousness, it was found beneficial for organisms to have desires as a way of bridging the gap between goals and plans for achieving those goals, for example. Such contextual answers are subject to the criticism that whole genes may be moved from one part of a chromosome to another through random displacement or peculiarities of DNA folding.

The second type of answer depends on identifying homologous (common ancestor) DNA. The gene for desire may be found to contain significant amounts of DNA associated with, say, goal-motivated behavior as well as its own specialized DNA. Desire can then be explained as having evolved from (inherited) goal-motivated behavior but also to have specialized with respect to goal-motivated behavior.[13] Such homologous answers are subject to the criticism that specialized genes may contain significant exceptions to what their common ancestor gene contains and may also inherit from more than one common ancestor.[14]

Irrespective of the approach adopted, a biomolecular approach to cognitive science implies a different way of looking at mind. A mind genotype is the genes a person has for mind, whereas a mind phenotype is the expression of these genes in actual thought. One possibility here is to predict actual thought processes of a person from a knowledge of their mind genotype, should such a genotype be discovered in our genes. However, mind phenotype may not be easy to predict, even with knowledge of mind genotype. While some phenotypes are discrete and can be predicted from their genotypes (e.g. blood type), other phenotypes are continuous (e.g. height) and may be dependent on environmental factors (e.g. nutrition). Similarly, it may be argued that mind phenotype is not like blood type but more like height.

Also, just as genotypes are inherited from parents who in turn inherited from their parents, and so on, mind genotypes, if they exist, must be inherited from parents and their parents ... The genetic make-up of one's mind consists of bits and pieces of mind genes of many ancestors, in the same way that color of hair or eyes consists of bits and pieces of inherited ancestor genes.

This raises the question of how many different types of mind genotype there are, and how they are manifest in mind phenotype. It is possible that a certain thought I have is of the same phenotype as a thought one of my ancestors (e.g. my grandmother) had, from whom I have inherited some bits of mind gene. But what distinguishes my thoughts from my grandmother's is my genotype, which is inherited from a wider pool of genes than just my grandmother. However, from a biomolecular point of view, the way my mind genotype is expressed in actual thoughts may be similar to one of my ancestors. Taking this to its logical conclusion, it could be argued that every one of my thoughts is an expression of certain aspects of my mind gene which in turn has been inherited from ancestors. I really do think like my grandmother, for some of the time anyway. Then I think like my mother, for some of the time, and so on. Why only some of the time? That's because during those times that part of my mind gene which I have inherited from my grandmother or mother is being transcribed to produce enzymes which, when active in my neurons, produces certain thoughts which are different in type from thoughts produced by enzymes transcribed from parts of my mind gene which I've inherited from, say, my father.

It's important to stress the difference between genotype and phenotype, and the continuous range of phenotype expressions possible for discrete genotype values. While mind is determined by genes, the expression of those genes in actual thought will be dependent on other factors also, such as current and previous experience, nutrition, and so on. The issue here is therefore not full genetic determinism. Rather, what is at issue is the classification of mind into distinct types, as given by the mind gene(s).

Of the estimated 75,000 to 100,000 genes in the human genome, many thousands have been identified and located on specific chromosomes. So far, there has been no identification of a mind gene, i.e. a specific location or set of locations in and across chromosomes which code for mind, but then it is not clear that molecular biologists are looking for a mind gene as such or would recognize it. A vast proportion of our DNA does not code for specific enzymes and therefore can be regarded as non-genetic. Their purpose is regarded as non-functional (except that such redundant DNA can be used for identifying individuals via DNA fingerprinting: the pattern of repeating "redundant" DNA in you is different from in me).

5. Conclusion

What the above has shown is that Rumelhart and McClelland (1985) were surely right when they state: "[T]here's more twixt the computational and the implementational than is dreamt of, even in Marr's philosophy ..." (1985: 196). The problem is that each level, to lend credence to its claim for accounting for mind/brain, posits a form of computation and representation not just appropriate for that level but also necessary and perhaps sufficient for explaining cognitive phenomena at that level. CCS proposes symbolic computation, algorithm and representation, neuroscience proposes mathematically based extraction of information and knowledge contained in connectionist networks (by means of hyperplane analysis, for example), and biomolecular science proposes molecular computing which is based on biomolecular processes within the cell nucleus. The recent interest in mind/brain issues shown by leading figures in the physical and biomolecular sciences indicates that cognitive science and neuroscience no longer have the field to themselves. There is already competition between the biomolecular and physical sciences as to which is going to prove to be the more suited for accounting for mind/brain.

The most immediate implication for CCS is that the notion of computation, tied as it is to the concepts of rule-following, algorithm, effective procedure and TM-computability (areas attacked by objections [a] and [c] described earlier), goes back into the melting pot. What may emerge is a concept of computation which is tied more closely to biomolecular principles (transcription, translation, replication, mutation, and so on) than to a formally specifiable and repeatable sequence of steps to reliably achieve a task. With regard to neuroscience (objection [b]) the most immediate implication is that neurons are at too high a level, and so any neuroscientific account of mind/brain in terms of layered networks will not be accurate. What is needed is a clearer understanding of the internal workings of neurons in biomolecular terms. Physical scientists may argue that biomolecular computing is still at too high a level and that its own computational arm, quantum computing (e.g. Deutsch 1992; Menneer and Narayanan 1995) provides a more appropriate computational level. A radical physical scientist may also claim that consciousness/thought is a feature of the brain's physical actions where these physical actions cannot even be adequately expressed in any computational terms (Penrose 1994) — the physical theory has no computational arm. The scientific foundations of mind/brain are currently up for grabs.

But the anti-materialist and perhaps anti-reductionist nature of CCS (objection [b]) may lead to it becoming isolated because of its unwillingness to accept computational paradigms and representations of levels lower than the algorithmic as real alternatives for an account of mind/brain. What is being proposed in this paper is that CCS should adopt, for the purposes of scientific hypothesizing, biomolecular paradigms. Even if this level is proved ultimately to be wrong, at least CCS will be contributing solutions to a fresh range of problems, some of which (e.g. explaining the biomolecular basis of diseases) have profound implications for humanity. More interestingly, though, if CCS adopts biomolecular paradigms, they can cut the ground away from under connectionists' feet by pointing out that, while networks of neurons may well perform certain tasks non-symbolically, within each neuron there are processes which can be described symbolically. Such processes are described using the symbol structures and processes of biochemistry (e.g. nucleotides, transcription, replication, enzyme production) and physical chemistry (e.g. molecule construction out of atoms, molecule folding), even if it is not currently clear whether these symbol structures and processes are computational. Nevertheless, it can pointed out that nonsymbolic behavior at the level of networks rests fundamentally on biochemical symbol structures and processes within each neuron making up the network. Nonsymbolic processes at the neural network level could be emergent properties of symbolic processes at the individual neuron level — an interesting twist.

Notes

1. Rules can be implicit rather than explicit, but symbols must be explicit (Fodor and Pylyshyn 1988).
2. Turing (1950) identified early versions of these objections as "The Argument from Informality of Behavior" and "Lady Lovelace's Objection" (for [a]), "Argument from Continuity of the Nervous System" (for [b]), and "The Mathematical Objection"(for [c]).
3. Imagine a professor in Korea who understands no English and who relies on the best available translations of Shakespeare's work in order to write, in Korean, deep, penetrating analyses of Shakespeare's plays. The Korean professor's writings are translated into English by translators and published in the best Shakespearian journals, leading to world recognition of the quality of the Korean professor's writings. If you

can imagine this, then, goes the argument, this is what CCS and AI are all about: the ability to reason and be creative, within a language that can be understood, even if that language needs translating for other to understand.

4. Searle (1987) proposed a form of biological materialism, where variable rates of neuron firing relative to different neuronal circuits produce all the different types of mental life we humans experience: "... mental phenomena, whether conscious or unconscious, whether visual or auditory, pains, tickles, itches, thoughts, and the rest of our mental life, are caused by processes going on in the brain" (p. 220 — stress removed), where these processes are "real biological phenomena" (p. 217).
5. This is analogous to a high-level program, even after being compiled into machine code, still existing as an entity in its own right as a textual entity about which certain judgements can be made (e.g. its complexity, structure, design).
6. Recently, another dimension has been added to the debate by Penrose (1989, 1994). CCS and AI have got it wrong, he argues, when associating thought, reasoning, consciousness and awareness with algorithms. Rather, algorithms, if they have any part to play at all in human intelligence, play a part at the "unconscious" or "subconscious" level, perhaps as a result of thousands of years of evolution. It is the role of consciousness to be non-algorithmic, i.e. to apply common sense, judgement, understanding and artistic appraisal to the results of our (unconscious) algorithms.
7. Hofstadter (1979) seems to have predicted this growing interest in biomolecular processes when he presented a simplified molecular computing system called Typogenetics.
8. Transcription differs from replication in that transcription involves the use of a fifth base — uracil — which is mapped onto by thymine. That is, during transcription, instead of DNA thymine being mapped onto mRNA adenine, DNA thymine is mapped onto mRNA uracil.
9. What has been described applies only to diploid cells, which are all body cells apart from the haploid, or sex, cells. In diploid cells, the nucleus of each body cell contains the full set of hereditary information. When a new diploid cell is formed, complete copies of all the chromosomes must be made through a process known as DNA replication.
10. Both translation of replication involve the unraveling of the DNA strands and the use of one strand as a template for joining together nucleic acid units in the proper sequence. After translation, the DNA recoils to its original form, but after replication the two original strands have separated, each with a new complementary partner strand.
11. A virus is a set of genes packaged inside a protein coat. Inside the coat of most human viruses is a single strand of DNA, coded to reproduce itself. A virus borrows the ribosomes of cells in a host organism to make proteins. New virus particles are formed which break out of the cell to infect other cells. A cell infected with a latent virus (unusual for humans) shows no sign of infection for a long time and may even have made many copies of itself (with the latent virus implanted in its DNA) before something triggers the virus into activity. AIDS is a highly unusual (for humans), latent retrovirus which injects RNA (and not DNA) into T4 white blood cells (an important

cell in a human's immune system). Reverse transcriptase then converts the RNA into DNA which is subsequently inserted into the T4 cell's normal DNA. White blood cells divide normally, carrying their infected DNA into new copies of themselves. When the latent infection is triggered, the emission of new copies of the AIDS virus causes the T4 cells to die, leading to a severe breakdown of the immunity system.

12. Interestingly, Hameroff and Penrose have opposite views on the adequacy of quantum computing (Deutsch 1992) for accounting for quantum mechanical effects, with Hameroff seeing no reason in principle why quantum computing should not provide adequate computational accounts of quantum behavior and coherence within microtubules, and Penrose believing that quantum computing, because it is still essentially computational, cannot in principle adequately account for how microtubules give rise to consciousness.
13. Object-oriented computational paradigms (e.g. the use of C++, CLOS, ONTOS), where inheritance is the basic mechanism for sharing information, can be predicted to be useful for modelling evolutionary accounts.
14. See Al-Asady and Narayanan (1993) for an overview of the problems associated with multiple inheritance with exceptions.

References

Al-Asady, R. and Narayanan, A. 1993. More notes on 'A clash of intuitions.' *Proceedings of the 13th International Joint Conference on Artificial Intelligence* (IJCAI–93), 682–687.

Bodén, M. and Narayanan, A. 1993. A representational architecture for nonmonotonic inheritance structures. In *Proceedings of the International Conference on Artificial Neural Networks* (ICANN–93), S. Gielen and B. Kappen (eds), 343–349. Amsterdam: Springer Verlag.

Brown, J. 1994. A quantum revolution for computing. *New Scientist* 1994, 1–24.

Carter, F.L. 1984. The molecular device computer: Point of departure for large scale cellular automata. *Physica* 10D, 175–194.

Christiansen, M.H. and Chater, N. 1994. Generalization and connectionist language learning. *Mind and Language* 9(3), 273–287.

Churchland, P.M. 1981. Eliminative materialism and propositional attitude. *The Journal of Philosophy* 78, 67–90. Reprinted in W.G. Lycan (ed), *Mind and Cognition: A Reader*. Oxford: Blackwell. The page reference is to the reprinted version.

Conrad, M. and Liberman, E.A. 1982. Molecular computing as a link between biological and physical theory. *Journal of Theoretical Biology* 98, 239–252.

Crick, F. 1994. Interview with Jane Clark. *Journal of Consciousness Studies* 1(1), 10–24.

Deutsch, D. 1992. Quantum computation. *Physics World* 5, 57–61.

Fodor, J.A. and Pylyshyn, Z.W. 1988. Connectionism and cognitive architecture: A critical analysis. *Cognition* 28, 1–71.

Hameroff, S.R. 1987. *Ultimate Computing: Biomolecular Consciousness and Nanotechnology*. North Holland.

Hofstadter, D.R. 1979. *Gödel, Escher, Bach: An Eternal Golden Braid.* Harvester Press.

Lucas, J.R. 1961. Minds, machines and Gödel. *Philosophy* 36, 120–124. Reprinted in *Minds and Machines,* A.R. Anderson (ed), 1964. Prentice Hall.

Menneer, T. and Narayanan, A. 1995. Quantum-inspired neural networks. Research Report 329, Department of Compute Science, University of Exeter.

Newell, A. and Simon, S.A. 1976. Computer science as empirical enquiry. *Communications of the ACM* 19, 113–126. Also in *Mind Design,* J. Haugeland (ed), 1981, 35–66. Cambridge, MA: MIT Press.

Niklasson, L.F. and Sharkey, N. 1994. Connectionism — the miracle mind model. In *Connectionism in a Broad Perspective*, L.F. Niklasson and M.B. Bodén (eds), 13–25. Ellis Horwood.

Niklasson, L.F. and van Gelder, T. 1994. On being systematically connectionist. *Mind and Language* 9(3), 288–302.

Penrose, R. 1989. *The Emperor's New Mind.* Oxford University Press.

Penrose, R. 1994. *Shadows of the Mind.* Oxford University Press.

Pribram, K.H. 1986. The cognitive revolution and mind/brain issues. *American Psychologist* May, 507–520.

Putnam, H. 1967. The nature of mental states. In *Mind and Cognition*, W.G. Lycan (ed), 1990, 47–56. Blackwell.

Rapaport, W.J. 1988. Syntactic semantics: Foundations of computational natural language understanding. In *Aspects of Artificial Intelligence*, J.F. Fetzer (ed). Kluwer Academic Press.

Rumelhart, D.E. and McClelland, J.L. 1985. Levels indeed! A response to Broadbent. *Journal of Experimental Psychology: General* 114(2), 193–197.

Rumelhart, D.E., McClelland, J.L. and the PDP Research Group. 1986. *Parallel Distributed Processing: Explorations in the Microstructure of Cognition — Volume 1: Foundations*. Cambridge, MA: MIT Press.

Searle, J.R. 1980. Minds, brains and programs. *The Behavioral and Brain Science* 3, 63–73.

Searle, J.R. 1987. Minds and brains without programs. In *Mindwaves*, C. Blakemore and S. Greenfield (eds). Blackwell.

Turing, A.M. 1950. Computing machinery and intelligence. *Mind* LIX 433–460.

van Gelder, T. 1990. Compositionality: A connectionist variation on a classical theme. *Cognitive Science* 14, 355–364.

von Neumann, J. 1966. *Theory of self-reproducing automata.* Edited and completed by A.W. Banks. University of Illinois Press.

Winfree, A.T. and Strogatz, S.H. 1984. Organizing centres for three dimensional chemical waves. *Nature* 311, 611–615.

The Search for Mind

A New Foundation for Cognitive Science

Seán Ó Nualláin
Dublin City University and NRC, Canada

1. Cognitive Science in Crisis

Paradoxically, it is a compliment to a young science such as Cognitive Science to state that it is in crisis. Kuhn's (1962) justly celebrated account of scientific discovery, using Heidegger's earlier terminology, distinguishes between periods of "normal" science and periods of crisis when fault-lines in the foundations of the discipline offer an opportunity for a paradigm-change, i.e., a change in its basic concepts which is irreversible. A great deal of progress in the discipline is necessary before its fundamental tenets become clear enough to be seen as being out of kilter with the field it claims to study. The concern of this section is to establish that, at least on an "as if" basis (Kuhn, we shall see, has his critics), such is the case for Cognitive Science, considered as the Science of Mind, at this moment.

What seemed at one point a naturally-occurring (super-) discipline with its subject a natural domain (i.e. "informavores"; see Pylyshyn 1984: xi) is now under attack on both theoretical and empirical grounds, with objections both to its methodological and substantive aspects (it would perhaps be relevant to point also to flaws in its pretheoretic specification of the domain, which we discuss below; however, that is more a philosophy of science issue). Let us very briefly examine the picture of Cognitive Science which is under attack, lest the assault be an attention-seeking fracas with a straw man. The domain of Cognitive Science is agreed as perception and knowledge, broadly defined (*ibid.* and Gardner 1985: 6); its tenets are the acceptance of a level of representation, the use of computers, de-emphasis (now nuanced) on social factors and affect, and an interdisciplinary ethos which is the only currently uncontroversial point. Usually, the only reason that materialism is not mentioned as a tenet is that it is regarded as *a priori* valid. Of these tenets, only the interdisciplinary ethos is likely to survive the decade.

Pylyshyn (*op. cit.*), with considerable ingenuity, attempted truly to found a discipline on these tenets. It might ignore consciousness and affect; it might take as its paradigmatic example of mind-world interaction the rarefied universe suggested by the interpretation of states in registers as binary numbers; however, apart from Von Eckardt (1993) which we discuss below, it remains the best worked-out foundation for computationalist Cognitive Science. It is in the epistemological assumptions inherent in Pylyshyn's model that we come across the most damaging attack on it. Edelman (1992: 230–234) argues persuasively that it contains many unproven assumptions about the structure of the world *and* the way we categorize it. Pylyshyn's Cognitive Science requires that a domain can be represented in a fashion which yields itself to syntactic analysis followed by semantic interpretation like the example involving states in registers he chooses. Edelman argues, correctly in my view, that this requires that the world be structured in classical categories, and that this structure informs our perception of it. (Searle 1992, echoing Kripke on Wittgenstein, is unprepared to admit that syntax is intrinsic to the physics; a human agent is necessary for the syntactic analysis as for the semantic interpretation). Rosch (1973) indicates that the notion that our perception of the world is "classical" is not true; the other point is perhaps better stated as "there exists a neat semantic description s for each cognitive domain c" which is also untrue. Wittgenstein's (1922, 1967) path from the neat to the scruffy camp is perhaps the most instructive demonstration of its untruth.

Edelman (*ibid.*) continues his epistemological attack on computationalist Cognitive Science with the argument that it requires that the world be in some non-trivial sense an infinite Turing tape, which of course it isn't. Moreover, Edelman argues, only with continual reference to evolutionary history and consequent neurological structure will any cognitive process be elucidated. However, his knowledge of many of the other constituent disciplines of Cognitive Science is very incomplete, showing the need for the administrative structures of the discipline whose conceptual foundations he has been so keen to criticize; Edelman would greatly benefit from an unbiased reading of Gardner.

Edelman's position on meaning, i.e. the insistence that it requires consciousness and embodiment, is a distant echo of the work of Michael Polanyi (1958) and all the more secure for this resemblance. The accent on consciousness is expanded on by Searle (1992), who finds it scandalous that a science allegedly concerned with Mind should ignore its conscious aspect. Moreover, Searle insisted that neither materialism or dualism is a tenable, or indeed a coherent position. However, Searle (1992: 228) is after bigger game; the limitation of inquiry in the sciences of mind to two levels i.e. the neurophysiological and the phenomenological. This is in explicit contrast to the central tenet of Cognitive Science, as proposed by Dennett (1993: 195):

> There is a level of analysis, the information-processing level, intermediate between the phenomenological level and the neurophysiological level.

Several good reasons exist for preferring Dennett's account of this particular event to Searle's. The first is that, having abandoned one level of analysis, there does not *a priori* seem to be any good principled reason for not abandoning others. For example, the neurophysiological level can successively be reduced to levels in which the explanatory frameworks of chemistry and sub-atomic physics are the most relevant. Secondly, Dennett's central tenet does not strongly constrain Cognitive Science; it can be interpreted as signalling that such a level is interesting and important without requiring one to buy into the notion that it is intentional and cashing out consciousness in this way. Thus interpreted, this tenet sits easily with the viewpoint on the study of mind outlined below.

The cognitive role of emotion is treated by de Sousa (1987), who insists that the fundamental role of emotion is to compensate for the insufficiency of reasoning, manifest *in excelsis* in the Frame problem (*op. cit.*, 195). Of the original pillars of AI, that leaves only representation standing. The work of Gibson (1979) seemed to have safely been dismissed (e.g. Pylyshyn 1984) until the successful use of its findings in computer vision, *inter alia*; Brook's (1991) provocative attacks on centralized symbolic representation forced the most diehard "Establishment" members at least to re-examine certain basic concepts.

The most recent thorough attempt to found Cognitive Science on classical lines is due to Von Eckardt (1993). Her argument on Cognitive Science is in itself powerful enough to merit attention; however, she explicitly denies that Cognitive Science is in Kuhnian crisis, but rather is an immature Science with an implicit set of commitments (30–31) which her book succeeds in making explicit (13).

Immature science is exemplified for Von Eckardt (353) by notions like the fluid theory of electricity. A paradigm change, on the other hand, could be provoked by a phenomenon like black-body radiation. Let's imagine that Physics had continued to ignore or explain away black-body radiation, and had continued in its long-accustomed path; such was the case for the Ptolemaic Universe (see Aizawa's paper in the next section). Is this self-blinkering not precisely analogous to the current attempts of Cognitive Science to establish a science of Mind without consciousness, emotion and social factors? What we have is an almost wilful ignoring of *explananda* in the manner of the Geocentrists. The ignoring of these factors has a history corresponding in its complexity and controversy to the cosmological issue, and by coincidence converging in the latter respect on the same individual. It was in fact Galileo's distinction between "primary" and "secondary" qualities which first exiled much of mental life from the scientific framework. We shall return to this point; it behooves us for the moment to consider Von Eckardt's exemplary account of Cognitive Science.

The central issue is this; if Cognitive Science really is to become the Science of Mind, it must include affect, consciousness and social factors. Yet the Foundations laid by Von Eckardt do not admit these factors any more willingly than do those laid by Pylyshyn; they can perhaps later be confronted (*ibid.*, 341). Von Eckardt does not relax this tension, nor is it her intention to do so. On the contrary; she is concerned with making explicit the assumptions with which Cognitive Science researchers have to date implicitly being working. In fact, her book can be read in this light as a rather more thorough demolition job on conventional Cognitive Science than that of which Searle, Edelman and their cohorts would have been capable.

Let us first consider her characterization of Cognitive Science as an immature science, rather than one in crisis. She is unwilling to accept that Cognitive Science is capable of crisis for two reasons; firstly, it is too immature as a science; secondly, Kuhn's notion of a paradigm is poorly formulated. She is willing only to accept his notions of a "disciplinary matrix" and "exemplar." In fact, Cognitive Science is to be viewed as a research framework, perhaps the precursor to a new science.

Let's consider the latter point first. The notion of a "paradigm" at least has the virtue of being generally recognizable; more importantly, we lack any more appropriate words to capture scientific revolutions like that involved in the transition from classical to quantum physics than "crisis" and "paradigm change." In the present article, the terms are being used on this "as if" basis; the alternative formulations to Kuhn's and Heidegger's like Laudan's need a similar concept. Secondly, though Cognitive Science may be immature in the sense Von Eckardt describes, the attempt to construct a science of mind is not. It has recently gone through stages where the major foci of study were the philosophy of mind, introspectionist and then behavioral psychology, and cognitive psychology. It is in this science of mind tradition that many of us, including the present author, are working; for the moment, we are content to call ourselves "Cognitive Scientists." (It is to be hoped that we can widen the terms of reference of the field to the point that Searle and Edelman can enter. Indeed, Searle suggests that his new neurophysiological/phenomenological field should be called "Cognitive Science").

For Von Eckardt (15) is proposing that Cognitive Science be considered "an approach to the study of Mind" with no expressed limitations. These enter only when she is outlining the metaphysical and methodological premises of her discipline. Suddenly, Mind becomes "the human cognitive mind/brain" (50) which consists of a set of cognitive faculties, for methodological purposes best considered as absolutely distinct from each other; the same purposes will require the exclusion of Consciousness, affect and social factors. It need not be emphasized that we have now left the study of mind; the field being described so thoroughly could perhaps be described as computational psychology. All the more so, since the two central assumptions are computationalism and representationalism. The functionalist ethos of modern mind science with its insistence of multiple realizability of mental process also permeates her discussion. The discussion of representationalism is superb (143 ff) and the introduction of different types of representation via Peirce's distinction between index, sign and symbol may save the concept from attacks like those of Stich. However, the tension between Cognitive Science as the science of mind and as something rather less has not been relaxed, nor will it be by Von Eckardt.

Whether deliberately or not, Von Eckardt does us a service by pressing the attack on this point. She refuses to allow its domain be simply propositional attitudes, as Fodor would prefer (65); Pylyshyn's "*ne plus ultra*" for the discipline is identified as unencapsulated faculties, before being rejected as too confining. However, her own analysis would suggest that the current *de facto* limitations of Cognitive Science will not allow it to handle many phenomena causative in human cognition; she has little more than aspirations to offer on that score. Finally, her analysis suffers from its complete ignoring of the Anthropological component of Cognitive Science, and the lack of detail on several of the other disciplines constituent of Cognitive Science. The constraints she sees Neuroscience imposing on Cognitive Science are strong (330); yet she refrains from referring to Edelman's conclusions, as noted above.

Von Eckardt, then, has out-Searled Searle in her criticism of Cognitive Science. An alternative path to constructing a foundation for Cognitive Science, and the one this author has taken, is first to review those disciplines which claim any province of the science of mind. Common themes, when they emerge, allow the construction of a common language as Halley (1992: 1) advocates, rather than its imposition, *de haut en bas*.

I have discussed the epistemological fault-lines exposed by Brooks *et al.* in (Ó Nualláin 1992). The task here is quite different. First of all, what remains of

Cognitive Science? Well, given the bewildering diversity of sources of attack, the interdisciplinary ethos seems alive and well. Two consensuses seem to have been reached by the attackers: a new emphasis on Consciousness (Ó Nualláin) and on something called "situated cognition" which works fine for tasks like those attempted by Brooks but is difficult to conceive of in symbolic activity. Let us call the tasks attempted by Brooks, where sufficient information is elicited by observing the organism's own movements with respect to certain environmental invariants, "egocentric cognition" (Kirsh 1991): the latter, symbolic case where one is operating in a more elaborate shared world, we can term "intersubjective" cognition.

The next task of this paper is to outline the fundamental substantive tenets of a new Cognitive Science; we shall use the terms just defined.

2. Foundational Considerations

The task of this section is to give details of valid tenets for Cognitive Science and briefly to indicate the corroborating evidence for each substantive tenet in the disciplines which comprise Cognitive Science. Following Aristotle, we take the domain of Cognitive Science as comprising those acts of mind of which we can predicate "true" or "false." Aristotle goes on to distinguish two types of cognition, roughly corresponding to perception and judgement. To predicate "true" or "false," it is necessary that we should be able to capture the content of the cognitive act in some informational manner. It can be argued that the Shannon formulation of information (i.e. its definition with respect to redundancy in a signal) is observer-relative; we need something that won't fall in the face of a Searleian argument. Thermodynamics provides us with an out; the informational state of a system can be defined as the number of different ways its constituents can be re-arranged while maintaining the same entropy. In other words, the vocabulary of bits and bytes we would like to use has a referent as objective, as anything in physics. Concepts on information across the physical sciences converge on the "bit."

Our focus, then, is any act of mind which can be characterized in informational terms. As de Lancey's paper in the next section abundantly demonstrates, this must include emotion. Likewise, attention has been demonstrated to be informative; we must include at least this, the focus of consciousness, in our Cognitive Science. These two facts, alone are enough to undermine the type of "minimal commitments" argument Scott makes later in this section. To study cognition properly, it is necessary to encompass all acts of mind of informational consequence.

Thus, Cognitive Science has no option other than to be a science of mind in order to fulfil its calling as a science of cognition. This is an arduous vocation; Cognitive Science must at some point address all phenomena of "mind" in order to determine which are informationally relevant. To do this is to run the risk of academic imperialism as well as to set the neophyte Cognitive Science researcher an impossible learning task. As Cognitive Science approaches each of its "constituent" disciplines, it must distinguish the subject-matter of the autonomous discipline proper, which is a matter for its specialists, from those findings of the discipline which are cognitively salient. To take linguistics as an example, the search for formalizations of language (GPSG, FUG, GB etc), the province of the linguist, should be distinguished from the findings indicating how language is used as a vehicle for content (informationally describable), a Cognitive Science task.

Similarly, the theory and practice of psychoanalysis should be set apart from the indubitable consequences arising from subconscious effects on cognition. The myriad conceptions and approaches to the study of consciousness, ranging from the anthropological to quantum mechanics, are likewise not focally our business; the fact that attention facilitates making of informational distinctions is our business. (The introduction to the third part of this book delineates the science of consciousness in more detail). This section ends with further marking out of the boundaries of Cognitive Science.

Consequently, the crisis in Cognitive Science has suggested a new conception of the project of Cognitive Science. It is an attempt to synthesize all findings from the sciences of mind which are of informational consequence; in so doing, it should suggest a new interlingua (common vocabulary) in which the scientists of the separate disciplines can couch their findings. My recent book suggests these syntheses and vocabulary:

a. Mind (and cognition) is manifest in the co-adaptation of species and environment over time.
b. At the individual level, we can usefully discuss cognition only in terms of the organism's being enmeshed in a Life-world. The adaptation at an individual level is best treated in Cognitive Psychology via a Principle of Rationality and in Neuroscience through a competitive principle of some sort like Edelman's (1992) Neural Darwinism. The paradoxical dynamic which compels the organism to seek stability in and yet increased mastery over the environment can be termed "equilibration" (Piaget 1972).

c. The focus of study of Cognitive Science is the combination of organism plus environment over time.
d. In the case of human cognition, a separate set of categories must be introduced to cater for symbolic behavior. In particular, we need our framework to have the capacity to "ground" symbols; moreover, a theory of situated cognition requires that the causal role of context must be both defined and explained.
e. Two sets of distinctions emerge in the analysis of human cognition. In the first place, we can distinguish between egocentric and intersubjective cognition, the latter of which also admits of an autistic mode. (It might be conjectured that the autistic mode is non-conscious egocentric mentation in an intersubjective domain). Secondly, we can distinguish between symbolic, operational (non-strictly symbolic e.g. conceptual in the case of language use Ó Nualláin *et al.,* 1994) and ontological (relating to one's role in the world e.g. speech-act knowledge in language use) dimensions in human use of any symbol-system (Ó Nualláin 1992, 1993a).
f. Context now can be explicated. In the first place, all cognition is contextual (Slezak 1993). In symbolic, intersubjective behavior, context relates to the interaction of the symbol system with other types of knowledge. (Context is handled by specialized neural hardware in egocentric mentation). With restriction of context, the interactions between the layers of the symbol system become altered; in effect, the layers compress (Ó Nualláin 1993b).
g. Consciousness is best treated in terms of projections of informational distinctions. As we see in Part 3, it is also related to at least one issue outside the normal domain of science, i.e. subjectivity. However, properly to address this issue may require the creation of an autonomous science of consciousness. (See introduction to Part 3 of this book).
h. Emotion has a causal role in cognition. It must therefore be included in Cognitive Science (de Sousa, *op. cit.*). However, affect, the subjective correlate of emotion, is outside our domain, as are "qualia" and all other phenomena which require consensual validation rather than informational characterization; these can perhaps be handled by a science of consciousness and (where relevant) depth psychology.

Cognitive Science is not yet a developed enough science to admit "proof" of its fundamental tenets by mapping onto ineluctable axioms. However, if a particular tenet accumulates credibility with analysis of the disciplines comprising Cognitive Science, it is greatly strengthened. Let's briefly look at each of the claims made above.

In the first place, we need evidence for the distinction between egocentric and intersubjective cognition. In philosophy, we find such in the tension between the work of phenomenologists like Merleau-Ponty on perceptual experience and that on symbolic cognition, where they have great difficulty. We also have the example of Berkeley who pointed out, in an early statement of the Frame Problem, that the attempt to treat perceptual and symbolic experience of the world in the same way leads to deep paradox. In Psychology, we find a similar tension between the Gibsonian and Establishment views of perception. From Neuroscience, we find evidence for the notion of specialized hardware in Egocentric mentation in such phenomena as the oculomotor reflex and the existence of neurons in the hippocampus with exquisitely directed roles (e.g. navigating in corners; see Berthoz *et al.* 1992). In AI, we find a tension akin to that in Psychology between Brooks and other subsymbolic researchers and the GOFAI establishment.

Next, we want some neuroscientific evidence for the existence of distinct linguistic and non-linguistic types of knowledge for full use of language; the task of generalizing this distinction to other symbol-systems as yet is beyond us. Damasio *et al.* (1992) provide us with the tools we need here. They distinguish between syntactic and lexical competence localized in the sylvian fissure of the left hemisphere and non-localized operational knowledge. Cognitive neuropsychology indicates pathologies specific to each of these components as well as to their interaction.

It is with points f–h that we must move from substantive to methodological prescription. With Pylyshyn, we can decide to neglect these matters; alternatively, in the option chosen here, we can decide to focus on at least the informationally-tractable part of conscious mental life.

But Cognitive Science is about computation! It is; all cognitive and perceptual processes can be phrased in the vocabulary of computation, given the emptiness of the original concept. (It is possible that the work of researchers such as Goel and Smith may in the future afford a richer concept.) The problems arise when, however ingeniously, an attempt is made to construct a more elaborate language with this vocabulary and to express the whole domain of cognition in this (Pylyshyn 1984). Goel (1992: 648) has established that Computer Science "cannot deliver an account of cognitive information processing" without intentional predicates. To extend this argument, the computational metaphor cannot support the burden of a science which includes conscious experience. (Searle 1992 argues that it cannot even support syntax).

But a theory of Cognition must have solid biological foundations, and the standard Cognitive Science theory makes erroneous assumptions both about the relation of mind and world, and the hardware-independence of mental processes! True; it must have biological foundations, and Edelman (1992) may win that particular argument. However, we lack sufficient neuroscientific data for explanation of higher-order cognition and certainly don't know anything substantial about how massively-complicated symbol-systems are neurally implemented.

Therefore, for language, that most essential of human faculties, the formal linguistic description (in some version or other) must take precedence! Or; the activity of Cognitive Science itself is embedded in a specific culture and we need a well-worked out anthropological theory to understand it! Alternatively; the personal motivations of researchers must be thoroughly researched in terms of their early experience to understand the form of their theories! There is a great deal of truth in every one of these positions.

Let's return to the basics. No assumption about Cognitive Science need be made other than it is the science that deals with cognition and thus, to this extent, Mind. In this version of Cognitive Science, we can encompass Gibson, Edelman and Fodor with the same sweep that netted Pylyshyn and Johnson-Laird. There is absolutely no doubt that the study of all of the separate disciplines comprising Cognitive Science yields a whole greater than the sum of its parts. Nor is there any doubt that specialists in each of the different disciplines should develop at least a passing acquaintance with each of the others. Cognitive Science is one of the most thrilling intellectual adventures in a century full of such. As it stands, it has the best claim on the mantle of "The Science of Mind." In order to maintain this status, it may have to jettison what is currently often perceived as its dominant paradigm; the gain is the possibility of including affect and Consciousness.

Historically, this may be the most valid path because, in a sense, Cognitive Science always has existed. In the last century, we can trace an experimental epistemology defined by Psychology before the domination by information theory, broadly defined, computationalism and now, at a guess, neurobiology. In other words, the subject was, is and always will be; a change in the dominant paradigm of the area is *not* its death-knell. So let us continue, with open minds and a willingness to admit at any stage that our pet area must take a back seat for a while, to study this bewilderingly complicated and fascinating super-discipline.

Finally, Cognitive Science academic departments may develop like new species looking for ecological niches, in which case it must have imperialist designs (over, *inter alia*, Psychology and philosophy of mind departments) or, like biochemistry and its gene, it may find a precise focus of study. In either case, it is likely to survive. One possibility is that, in the manner advocated here, it should see itself as synthesis of and overall framework for the sciences of mind. It might also decide to acknowledge social factors while focusing only on how they are processed, characterizing subject-environment relations in information-theoretic terms. This makes cognitive anthropology a cognitive science subject, while anthropology retains its autonomy. Likewise, emotion could be considered only as it biases information pickup, while the unravelling of the tangles in our psychic lives can be left to a separate science of psychopathology. The combination of solipsism as a research strategy and information as a theme may well afford a coherent science and in that sense Pylyshyn was correct. However, there must be a counterbalance in the ceding of power to other areas of inquiry as social factors enter the cognitive domain. Likewise, we can coherently discuss consciousness as projections of informational distinctions; however, that is only one aspect of a vast subject and we must concede this territory also, again waiting for the results of others' inquiry to become clarified enough for us to take it on board.

3. Cognitive Science is Dead: Long Live Cognitive Science!

We have now looked at the basic tenets of Cognitive Science, followed by a critique thereof. We then reviewed a proposal for a new set of fundamental tenets, followed by a brief indication as to their justification. It is being suggested that the nature of Cognitive Science is best thought of as an interdisciplinary search for mind. Let us conclude by examining the ramifications of this statement while summarizing the argument of this paper.

1. We explored the current debate about the domain and methodology of Cognitive Science and found a tension between the received foundations and current Science of Mind ambitions of the discipline. The tension is exacerbated by critics who fault the subject for its reluctance to address consciousness, affect and social factors and in so doing elevate itself in accordance with its current ambitions.
2. The viewpoint taken here is that addressing these issues still allows a coherent discipline to exist. Social factors can be handled by informational characterization of subject-environment relations (as done in, for example, situation semantics), emotion by studying its informational role, and consciousness by examining projection of informational distinctions.

3. These moves extend the discipline as required, while also allowing it a neat demarcation line from other disciplines. It can act as a reservoir of interdisciplinary knowledge while accepting that, for example, the analysis of social trends (as distinct from how they are processed by the individual) is not within its scope; likewise, those aspects of applied experientialism dealt with in consciousness studies are not informationally salient and so not within its own remit. The manufacture of those new tools for thought and action called "cognitive artefacts" is an engineering problem; the issue of their fit with human needs and abilities is an applied cognitive science issue.
4. At an individual level, several distinctions (e.g. egocentric versus intersubjective), arise naturally in the analysis of cognition, as does the notion of the importance of the Lifeworld, etc.
5. At a guess, the core subjects of cognitive science will remain cognitive psychology and those aspects of philosophy, anthropology, neuroscience, ethology, AI and linguistics which have specifically cognitive reference. They will remain in competition with each other in the sense that at any time only one of them will be perceived as having most promise.

References

Berthoz, A., Israel, I., and Wiener, S. 1992. Motion perception and spatial representation. In *Sciences Cognitives.* Le Courier du CNRS, No. 79, Halley (ed). Paris: CNRS-Editions.

Brooks, R. 1991. Intelligence without representation. *Artificial Intelligence* 47, 139–161.

Damasio, A.R and Damasio, H. 1992. Brain and language. *Scientific American* 267(3), 88–110.

Dennett, D. 1993. Review of John Searle's "The Rediscovery of The Mind". *The Journal of Philosophy*, 193–205.

de Sousa, R. 1987. *The Rationality of Emotion.* Cambridge, MA: MIT Press.

Edelman, Gerald. 1992. *Bright Air, Brilliant Fire.* New York: Basic Books.

Gardner, H. 1985. *The Minds' New Science.* New York: Basic Books.

Gibson, J.J. 1979. *The Ecological Approach to Visual Perception.* Boston: Houghton-Mifflin.

Goel, V. 1992. Are computational explanations vacuous? *Proceedings of the 14th. annual conference of the Cognitive Science Society.* Hillsdale, NJ: Erlbaum.

Halley, A. (ed) 1992. *Sciences Cognitives.* Le Courier du CNRS, No. 79. Paris: CNRS-Editions.

Kirsh, D.A. 1991. Foundations of AI: The big issues. *Artificial Intelligence* 47, 3–30.

Kuhn, T. 1962. *The Structure of Scientific Revolutions*. Chicago: The University of Chicago Press.
Ó Nualláin, S. 1992. On the problem of objectivity in Cognitive Science. *Proceedings of fifth annual conference of the French Cognitive Science Society*. Nancy: CNRS Editions.
Ó Nualláin, S. 1993a. Toward a new foundation for Cognitive Science. In *AI/CS '91* Sorenson (ed). London: Springer.
Ó Nualláin, S. 1993b. Language-Games and language-engineering: Theory and practice in computational linguistics. *Dublin City University: Second Workshop on the Cognitive Science of NLP*.
Ó Nualláin, S. 1995. *The Search for Mind: A New Foundation for Cognitive Science*. Norwood: Ablex
Ó Nualláin, S. and Smith, Arnold. 1994. An investigation into the common semantics of language and vision. AAAI Spring Symposium. Stanford, CA.
Piaget, J. 1972. *Principles of Genetic Epistemology*. London: Routledge.
Polanyi, M. 1958. *Personal Knowledge*. London: Routledge.
Pylyshyn, Z. 1984. *Computation and Cognition*. Cambridge, MA: MIT Press.
Rosch, E. 1973. Natural Categories. *Cognitive Psychology* 4, 328–50.
Searle, J.R. 1992. *The Rediscovery of the Mind*. Cambridge, MA: MIT Press.
Slezak, P. 1993. Situated Cognition. IJCAI workshop on using knowledge in its context.
Von Eckardt, B. 1993. *What is Cognitive Science?* Cambridge, MA: MIT Press
Wittgenstein, L. 1922. *Tractatus Logico-Philosophicus*. Oxford: Basil Blackwell.
Wittgenstein, L. 1967. *Philosophical Investigations*. Oxford: Basil Blackwell.

The Lion, the Bat, and the Wardrobe

Myths and Metaphors in Cognitive Science

Stuart Watt
Department of Psychology
The Open University, UK

1. Introduction

Cognitive science is about studying the mind, constructing explanations of human mental phenomena; and to provide these explanations, it uses a set of assumptions which provide a common ground for the participants in this study. But cognitive science is an open discipline, and there are substantial and public disagreements about some of these assumptions, for example in the schism between symbolic and subsymbolic approaches to this study (Papert 1988; Searle 1992). In this paper I will argue that the source of some of these disagreements lies in differences in the intuitive appeal of the metaphors which form the "substantive assumptions" (von Eckardt 1993) of these approaches, as well as in any qualities the accompanying scientific description might have.

Sciences of the mind are special because we, the people who study them, have minds, and our perceptions are indelibly tinted by this. One aspect of our having minds means that we have a tendency to empathize and identify with other people — and things — which appear to have minds. Unfortunately, one of the things we tend to identify with is the behavior of a model of or metaphor for a mental system (especially when it involves consciousness) and these are exactly what sciences of the mind are built on. The way we see things, even science, is affected by the nature of our human mental phenomena.

Pervasive phenomena like this natural empathy are hard to study. I would draw an analogy with gravity — although we can counterbalance the effects of gravity to some extent, we can't (with our current knowledge of physics) remove it, so theories of gravity can be difficult to falsify. Having a mind is similar, we can't just switch it off and see what cognitive science looks like without it.

In effect, there are two kinds of connection between the mind and a scientific model. On the one hand, there is the role of the model as a system description providing the substantive assumptions that underpin cognitive science. On the other, there is an intuitive psychological appeal — a reflection of how well people can see the model as something that could be psychological. These are intertwined themes, but they are fundamentally different, and the intuitive appeal of the second element can significantly affect the apparently objective scientific quality of the system description provided by the first. Moreover, this intuitive appeal is a true psychological phenomenon, a property of the human mind, so the models and metaphors that we use are, to an extent at least, subject to the nature of this human natural psychology.

Having a mind — and a human mind at that — is something pervasive. It colors the whole of cognitive science, acting as a continual pressure on the metaphors that we use. This is a pressure that we need to recognize, and as far as we can, counterbalance. But before we can try to counterbalance anything so insidious, we need to develop a better understanding of the kinds of pressures that are against us. That is the role of this paper.

2. Anthropomorphic Excursions

Wittgenstein (1953) wrote "if a lion could talk, we could not understand him." Wittgenstein's point was that using language is part of a "language-game" — part of an active "form of life" — outside which that language cannot be understood. He draws an analogy with a country with a foreign language and traditions: "even given a mastery of the country's language, we do not *understand* the people" (Wittgenstein 1953, original emphasis).

But somehow we intuitively feel that we should be able to interpret, to some very small degree, what a lion would say to us, even though we don't speak 'Lion.' What about paralinguistic communication in the foreign country analogy? Wittgenstein also wrote: "if I see someone writhing in pain with evident cause I do not think: all the same, his feelings are hidden from me" (Wittgenstein 1953), but can't we also make this connection with the people of this foreign country, and perhaps even with animals? We don't usually see lion pain with the intensity we see human pain, but we can and do recognize it, and we consider it something close enough to our own experience for it to be morally wrong not to prevent or alleviate it.

So here Wittgenstein and I might part company: I think it entirely possible for forms of life to overlap to a small extent, and, to this extent, we can have something like a common language with cats, even if only a language of the most simple form. After all, we share a common biological inheritance and a similar physical environment, and these shape our forms of life. Darwin (1872) said:

> in the case of lower animals involuntary bristling of the hair serves, together with certain voluntary movements, to make them appear terrible to their enemies; and as the same involuntary and voluntary movements are performed by animals nearly related to man, we are left to believe that man has retained through inheritance a relic of them.

And with the action we have retained an interpretation of the action.

Nagel's (1974) Bat also presents us with a case which also touches on human-animal communication, but from a rather different perspective. Nagel is principally interested in conscious experience. His conclusion is described by Akins (1993) as "the only possible access one could have to the phenomenal experience of another organism is by means of a kind of empathetic projection — by extrapolation from one's own case." Again, it is this very projection that makes us intuitively feel that we could indeed understand Wittgenstein's lion, should it ever speak. It is the empathetic nature of the projection that shows how important the psychology of the observer is in this ascription.

In the cases of both the lion and the bat we can see the same kind of projection clouding our analyses of the details of thought experiments that they embody. Anthropomorphism as a psychological phenomenon has not yet been seriously studied, although some have seen it as something worth looking at in its own right rather than as a hindrance to proper study (e.g. Caporael 1986; Eddy, Gallup and Povinelli 1993). One of the few pieces of research is that of Eddy *et al.* (1993) who suggest that there are two primary mechanisms involved:

> people are likely to attribute similar experiences and cognitive abilities to other animals based on (1) the degree of physical similarity between themselves and the species in question (e.g. primates,) and (2) the degree to which they have formed an attachment bond with a particular animal (e.g. dogs and cats).

(Eddy, Gallup and Povinelli 1993). Anthropomorphism probably has other factors and aspects, also depending on the social context and the behavior of the animal, for example, but for the purposes of this paper, the most important point is the strong connection between our ability to empathize with something and its degree of physical similarity to us.

Wittgenstein's and Nagel's experiments differ in one important respect: we ascribe human qualities to the lion but we try to share the bat's experience — it is more like identification than anthropomorphism. The first is projective and the second introjective, but Eddy *et al.*'s (1993) results show a strong correlation between the two, and the anthropomorphic principles underlying the two experiments are similar. In both cases they are clouded by anthropomorphism: while the arguments remain subject to philosophical analysis, they both open the same trap. We see the animals as other than they are and open, through the overlap in our forms of life, a narrow band of human animal communication which can trick us about the intended point.

Using this projection is seen as bad scientific practice, particularly in biology and ethology, but also in cognitive science. Searle (1992), for instance, comments:

> prior to Darwin, it was common to anthropomorphise plant behavior and say such things as that the plan turns its leaves towards the sun to aid in its survival. The plant 'wants' to survive and flourish and 'to do so' it follows the sun.

Searle then tries to "rescue" cognitive science by inverting its explanations from their original teleological forms to functional ones like "plants that turn their leaves towards the sun are more likely to survive than plants that do not." Surprising, he then goes on to say: "it is easy to understand why we make the mistake of anthropomorphising the brain — after all, the brain is the home of anthropos" (Searle 1992).

But while we may sympathize with Searle's criticism of anthropomorphism in cognitive science, it could just be endemic. Eddy *et al.* (1993) note that it is "almost irresistible." Krementsov and Todes (1991) comment that "the long history of anthropomorphic metaphors, however, may testify to their inevitability." And if anthropomorphism is at the heart of our point of view, banishing it won't help cognitive science. It is perhaps because we are totally unable to step outside our humanity that it is so tempting to regress to the Skinnerian vantage point (with apologies to Dennett) and deny the mentalistic terminology that connects the behavior of others to our own experience.

Many thought experiments are liable to the same kind of misdirection, even those not involving furry animals. Take Searle's (Searle 1980) "Chinese Room" argument, for instance. Hofstadter and Dennett (1981) make a variant on the "Systems Reply" and see the reader as "invited to identify with Searle as he hand-simulates an existing AI program." The word 'identify' is telling: we can clearly see Searle as human, but the argument asks us to see the *room* as human, and no matter how convincing the argument we rebel at the thought as it is against the grain of all the regularities. But when presented with an isomorphic framework in an existing brain or even in the head of a robot we suddenly find it easier — and all without thinking about the actual argument at all! Searle's thought experiment apparently invokes that very anthropomorphism that he has criticized so strongly. The intuitive power of Searle's experiment parallels, but is separate from, his logical argument. Using metaphors like this creates a tension between logic and intuition, and it isn't always the logic that wins. That is one of the problems with invoking intuitions. They are rather like loaded guns: in the hands of non-philosophers they can go off in almost any direction.

3. Myths and Metaphors

In Searle's thought experiment, the argument has been to an extent subverted by the apparently endemic phenomenon of anthropomorphism. While this may hint that thought experiments as a rule are dangerous, I want to show other, and perhaps deeper, implications.

Metaphors of an altogether grander scale come into play in the study of mind. There isn't a single science of the mind, but a variety of sciences, each with its different dominant metaphors. For one, there is the information processing metaphor with its roots in logic and symbol manipulation, but there are others, such as the connectionist metaphor with its neural analogies. As metaphors, these are ways of seeing cognitive science, but they are also ways of seeing minds.

In their strongest forms, these metaphors become what Turkle calls (1988) "sustaining myths" and "provide sciences of the mind with a kind of theoretical legitimation." Sustaining myths are often metaphors which work particularly well: the computer metaphor legitimated the use of words like 'memory' well enough to demolish behaviorism and successfully advance study of the mind (Turkle 1988), but the effects of a sustaining myth are not always positive.

Take the information processing metaphor: Searle's "Chinese Room" thought experiment is one variant on this metaphor — Searle in the room is clearly playing the role of a machine reading instructions and acting on them. It is the information processing metaphor that draws most from computer science: virtual machines, serial processing, and so on. This information processing metaphor has come in for a lot of criticism over the years. After all, how can it be that the mind is like a computer? Computers are made of silicon and grey plastic, and the similarity distance between a computer and me, made of protein and grey neurons, is perhaps just too large to accept.

Perhaps this is so, but if it is, it is partly our fault. Computers are our construction, and they were constructed in our image. "Von Neumann and Goldstine were *not* inventing the computer. They already knew what a computer was, what a computer did, and how long it took a computer to do it" (Bailey 1992, original emphasis). Our information processing metaphor is derived from this, the *human* computer that was the model for the electrical one. We constructed the computer in our image — or, rather, in the image of one kind of human mathematical and logical reasoning — before we started to consider our minds in its terms.

The information processing architecture was designed around a human model, and one where the roles were already clearly determined. It was possible to imagine *being* the central processing unit of a computer, acting as Searle did in the room, reading the next instruction and acting on it. Unfortunately, the strength of the metaphor is such that it is *only* possible to imagine being the central processing unit, because, originally, this central processing unit was modelled after a person acting sequentially. Although throughout the system there is a finely orchestrated and synchronized system of parts all collaborating, we see the system as serial while underneath the system is really parallel. The serial virtual machine (virtual machines are another kind of metaphor) supervenes on a parallel virtual machine, but psychologically it is more attractive. The central processing unit acts like a psychological magnet: the strength of the metaphor draws us to that one way of seeing the system at the expense of all others.

The connectionist metaphor has a similar pervasiveness, but in a radically different way. Connectionism's sustaining myth, according to Papert (1988), is "that a deeper understanding would reveal the naïveté of such everyday analogies." In particular, the theory of eliminative materialism claims that the illusory nature of folk psychology will eventually be revealed by a deeper understanding of human neurophysiology.

This difference in levels seems to redirect the identification. Woolgar (1985) discusses this with an example, a video recorder that responds to a signal to avoid recording advertisements. "Although on one level we could be perfectly happy with its 'intelligent operation,' we could also argue that the device was 'not really intelligent.' Importantly, the latter view redefines and thus reserves the attribute of 'intelligence' for some future assessment of performance." The need for mental consistency prevents us identifying with both levels, so if we are pushed away from one we will inevitably be drawn to the other. Eliminative materialism subverts its logical argument with this anthropomorphic effect: we are studying at the level of neurophysiology but rejecting identification at that level, leaving our intuitions free to associate with other levels.

There is a single theme running through all these cases. Parallel to the logical structure of each argument, there is an implicit test of intuitive plausibility. Ashmore (1993) describes a three player behavior modification scenario involving a cat, a catflap, and the cat's owner. In the first version the owner plays the actor. modifying the behavior of the cat as the behaver, with respect to the catflap as the environment. Ashmore then rotates the players through the roles of actor, behaver, and environment to generate different tales of varying plausibility, for example, in one variant the catflap modifies the owner's behavior through the environment in the form of the cat. Can a catflap be an actor? As Ashmore says:

> The only relevant question is: does the story work? Is it plausible? By which I mean how 'comfortably' does this story distribute its particular roles and statuses (Ashmore 1993).

As readers, the tales depend on our ability to see the players in their assigned roles.

Myths and metaphors compete for mind room in a truly Darwinian fashion. Those which are relatively successful, such as information processing and connectionism, can survive for many years. But this success isn't purely a matter of the corresponding theory's explanatory power, it is also a matter of intuitive plausibility. Translating metaphors may keep explanatory power unchanged but affect intuitive plausibility. Although we may hope that our sciences of the mind use functional rather than teleological explanations, we must always remember that any dominant metaphors within these sciences — like other sciences — are being evaluated through intuition as much as through critique.

4. Consciousness and Metaphors

So what of consciousness? Consciousness and anthropomorphism seem to be closely connected. When we remember Nagel's and Searle's thought experiments, the importance of the first person perspective highlights the connection. Flanagan (1993), for one, points out that the Chinese Room can be seen as a problem of absent consciousness, as well as absent intentionality.

In thought experiments like these, which ask us to take the first person stance, we are asked to try to identify with the system as well as taking the point of the philosophical argument, but this identification can change the argument subversively. This effect is particularly strong with models of consciousness, which frequently ask for this identification. I have a tendency — irrespective of any logical argument — to believe that you are conscious which gets stronger the more I perceive you as being similar to me. The same applies to models and thought experiments: the less we are able to see similarity in the behavior of a model — and this depends on us as much as the model — the more it resists identification. Of course, the model which least resists identification is that of another human; all other models resist identification more than this, but to differing degrees depending on our perceptions. The point is that when we can't identify with the behavior of a model, there is an especially strong psychological drift towards normalization — to restructuring or reinterpreting the model so we *can* identify with it.

There are several different ways of normalizing cognitive science with respect to consciousness, but in general there are two common strategies.

Strategy One: The Consciousness Box

The first strategy is simple: take consciousness out of the model by putting it in a separate box. This can be responsible for the 'self,' or for 'consciousness,' or it may show up in weaker forms, such as 'attention,' or 'supervisory system.' The trick here is to put everything somewhere else in the model, and we end up with something very like Searle's Chinese Room, a division between the computational model and a homunculus that we can identify with. It is as if we take Searle in the room, and put him in a Chinese Wardrobe in one corner of it: all this strategy does is to move the problem somewhere else — and so it is absolutely no help in resolving the problems of consciousness. It does, however, enable us to identify quite fully with a *part* of the model. This kind of normalization is especially common with the information processing metaphor.

Strategy Two: Radical Emergence

The second strategy also moves consciousness out of the model, but in a very different way. Here there is nothing like consciousness anywhere in the model at all: instead, somehow the action of the parts of the model causes consciousness to emerge at a higher level. Extreme versions of the strategy never explain how a model can cause consciousness, and often claim that it is impossible to explain this even in principle. Here the identification is through the change in levels: if an implementation layer is chosen which is radically dissimilar — the less similar the better — we are left with a kind of implicit encouragement to identify with the whole at a human level. This kind of normalization is especially strong in connectionism and emergent artificial intelligence, but it also appears the new approaches to quantum consciousness, and in religion.

These are typical of the normalizing pressures that exist today, but I don't believe that this list is comprehensive. We all have our dominant metaphor, and the normalizing pressure will act on us each individually.

I am not claiming that there aren't big issues of consciousness in a science of the mind: there are (e.g. Flanagan 1993; Searle 1992). My point is a methodological one: that we must be aware of the platform we are standing on when we look at these issues. Akins (1993) points out that we can "mistake our intuitive grasp of the visual perception of external events for an accurate description of internal attentional processes." It is especially important that when we study consciousness we are aware this intuitive grasp doesn't respond to all models and metaphors equally — and therefore that we might be accepting or rejecting these models for intuitive reasons rather than scientific ones without even realizing it.

5. Conclusions

The differences in the metaphors and the sciences that they shape has left the field of cognitive science in a rather disjointed state. The dominant computer and neural metaphors are competing within cognitive science in this Darwinian fashion. But this doesn't mean that the discipline is flawed in any important or significant way. Metaphors are continually in flux in all sciences. To claim that quantum physics is "in crisis" today, just because there are several different and competing metaphors for understanding quantum phenomena — and there are — would be taken as hyperbole, and yet physics is often taken as the gold standard against which disciplines are compared. The difference between physics and cognitive science is, I think, more an effect of the sociological perception of the disciplines than it is due to any fundamental difference between them.

Perhaps this difference in perception originates in subversion. Turkle (1988) points out that both artificial intelligence and psychoanalysis contain a subversive component, and that "a normalizing response is common to all subversive sciences of the mind." As cognitive science pushes harder at consciousness and affect, and we begin to see it becoming a little less of a mystery, there is a backlash which softens our models. The origin of the normalizing response is,

I believe, in the very anthropomorphism that started us on these issues: normalizing is an attempt to restructure the system so that its dominant metaphor is easier for us to identify with. Introducing an executive, or attention, into a cognitive model puts the remaining mysteries into a new box, with which again we can identify. The normalized model merely postpones the problem. The solution to the apparent impasse is not in trying to be less of a science of the mind, but in recognizing and resisting the normalizing response.

I think it extremely implausible that we could ever form a science of the mind — or any study of anything for that matter — without metaphors, but we must always try to be aware of the effects of the metaphors that we use, and take into account the effects of these metaphors on the models they accompany. Some metaphors do seem to go some way to providing the counterbalance which I suggest should be the correct response to these normalizing pressures. There is a trend to a new group of metaphors, those which, as Turkle (1988) describes it, "have a biological aesthetic — they are the kinds of things that could be going on in a brain." Note that this isn't always the sustaining myth of connectionists, that of levels of explanation and eliminative materialism.

Although I believe in the biological metaphor, and in evolutionary psychology, and that following Darwin, we really can learn about human mental life from animal mental life, we still need to tread with extreme caution, especially in the use of thought experiments. The new metaphor, whose slogan might be 'people are animals,' has new opportunities but also new dangers. A true science of the mind emerges as the competition between these myths and metaphors drives each into refinement, and into the creation of new metaphors. I don't want to be seen as arguing for a single metaphor: following the Darwinian meta-metaphor, that would be a kind of scientific eugenics. I believe that the best opportunity for cognitive science lies in the natural plurality that already exists. The error is to confuse plurality with crisis. The sociological pressures for normalization will not go away, and nor will our anthropomorphic tendency to see metaphors from inappropriate points of view, but if we recognize these issues we have overcome the worst part of the problem. The role of intuitions and metaphors in cognitive science is an important one, but we must remember it isn't always under our control.

Acknowledgments

I am deeply grateful to Arthur Stutt for coffee, corrections, and lessons on Wittgenstein; and especially to the anonymous reviewers for showing me where I needed them. Also to Seán Ó Nualláin and Paul McKevitt for organization and to Nelly Bourguignon for editing. All errors that remain are mine alone. John Domingue thought of the title.

References

Akins, Kathleen A. 1993. A bat without qualities? In *Consciousness,* Martin Davies and Glyn W. Humphreys (eds). Oxford: Blackwell.

Ashmore, Malcolm. 1993. Behavior modification of a catflap. Presented to 'non-human agency: A contradiction in terms' conference. September 1993, University of Surrey, UK.

Bailey, James. 1992. First we reshape our computers, then our computers reshape us: The broader intellectual impact of parallelism. *Daedalus* 121(1), 67–86.

Caporael, Linda R. 1986. Anthropomorphism and mechanomorphism: Two faces of the human machine. *Computers in Human Behavior* 2(3), 215–234.

Darwin, Charles. 1872. *The Expression of the Emotions in Man and Animals.* London: John Murray, London.

Eddy, Timothy J., Gallup, Gordon G. and Povinelli, Daniel J. 1993. Attribution of cognitive states to animals: Anthropomorphism in comparative perspective. *Journal of social Issues* 49(1), 87–101.

Flanagan, Owen. 1993. *Consciousness Reconsidered.* Cambridge, MA: MIT Press.

Hofstadter, Douglas R. and Dennett, Daniel C. 1981. *The Mind's I: Fantasies and Reflections on Self and Soul.* New York: Basic Books.

Nikolai L. Krementsov and Daniel P. Todes. 1991. On metaphors, animals, and us. *Journal of social Issues* 47(3), 67–81.

Nagel, Thomas. 1974. What is it like to be a bat? *Philosophical Review* LXXXIII, October, 435–450.

Papert, Seymour. 1988. One AI or many? *Daedalus* 117.

Searle, John R. 1980. Minds, brains, and programs. *Behavioral and Brain Sciences* 3, 417–424.

Searle, John R. 1992. *The Rediscovery of the Mind.* Cambridge, MA: MIT Press.

Turkle, Sherry. 1988. Artificial intelligence and psychoanalysis: A new alliance. *Daedalus* 117(1), 241–268.

Von Eckardt, Barbara. 1993. *What is Cognitive Science?* Cambridge, MA: MIT Press.

Wittgenstein, Ludwig. 1953. *Philosophical Investigations.* London: Basil Blackwell.

Woolgar, Steve. 1985. Why not a sociology of machines? The case of sociology and artificial intelligence. *Sociology* 19, 557–572.

Crisis? What Crisis?

Church's Thesis and the Scope of Cognitive Science

P.D. Scott
Dept of Computer Science
University of Essex

1. Introduction

The premise of this workshop is that there is a crisis in Cognitive Science caused by a tension arising from the constraints imposed by the computational metaphor which it has historically adopted. In this paper I shall argue that this premise is mistaken because no such crisis could have such a cause. In particular I shall argue that Church's thesis implies that the set of models of cognitive phenomena that may be constructed using the computational metaphor includes all models possible using any alternative scientific approach. Consequently no constraints arise from the use of the computational metaphor and we must seek elsewhere for an explanation of any conflicts that characterize the current state of the field.

In Section 2 I briefly review Church's Thesis and establish a familiar and uncontroversial result: that the set of models possible using the computational metaphor comprises the Turing computable functions. I then proceed in Section 3 to demonstrate that any genuinely scientific theory must be the embodiment of a Turing computable function and hence conclude that any scientific model of cognition would be compatible with the computational metaphor. Various possible objections to this conclusion are then examined. In section 5 I consider whether the fact that Church's Thesis concerns discrete sequential processes limits its applicability to continuous concurrent systems while in the following section I review the possibility that the computational metaphor implies something more than Turing computability. Section 7 contains a discussion of several different cognitive phenomena that previous authors have suggested are beyond the scope of the computational metaphor: in each case it is shown there is no reason to believe that they are legitimate objects of scientific study if they cannot modelled by a computer program. In the final section the principal conclusions of the paper are reviewed.

2. Church's Thesis

Church's Thesis, often called the Church-Turing hypothesis, emerged out of attempts to strengthen the logical foundations of mathematics by constructing a formal definition of the notion of an effective procedure: an intuitive concept that refers to a complete and unambiguous set of rules specifying what action should be carried out in every situation that may arise during the performance of some process. The main challenge to be overcome in formalizing the notion lies in providing an interpreter for the rules. During the 1930's, three very different systems were developed independently in an attempt to construct such a formalism: general recursive function theory (Kleene 1952), the lambda calculus (Church 1941), and Turing machines (Turing 1936). Somewhat remarkably it was found that all three were equivalent in the sense that any procedure which can be defined in and carried out by any one of them can also be defined in and carried out by both of the others. This led to the conjecture, known as Church's Thesis, that the set of procedures that can be expressed effectively in any language is just the set of those processes that can be specified in and performed by any one of these three systems. A common formulation is given by Minsky (1967):

> Any process which could naturally be called an effective procedure can be realized by a Turing machine.

Such processes are often termed the *computable functions*.

Because it is a conjecture concerning an informal notion, Church's Thesis is not amenable to mathematical proof. The general acceptance that it is correct derives from the numerous attempts to define formal systems that can realize processes that are not Turing computable. Such attempts have invariably failed and resulted only in yet another system that is computationally equivalent to a Turing machine.

Since it can readily be shown that Turing machines and digital computers are equivalent, it follows that Church's Thesis defines the scope of the computational metaphor: the set of models of cognitive function that can be expressed using the computational metaphor are the Turing computable functions. This much should be familiar to any student of computer science. We now turn our attention to more novel territory: the implications of Church's Thesis for scientific theories in general.

3. Church's Thesis and Scientific Theories

Would it be possible to construct a genuinely scientific theory that was not an effective procedure? Clearly the answer is likely to depend on what one means by a "genuinely scientific theory." There is of course a vast and controversy laden literature in the philosophy of science that attempts to formulate precisely what properties a scientific theory must have (e.g. Popper 1959) or alternatively tries to demonstrate that any such attempt is futile (e.g. Feyerabend 1975). Fortunately we shall not be concerned with the fortunes of this particular intellectual battleground since we need only consider a few attributes that are accepted as necessary (though not sufficient) properties of scientific theories by all protagonists.

Science is manifestly a social activity and consequently any genuinely scientific theory must be communicable to other scientists. The only medium scientists have for communicating with others who are remote in space or time is writing. Hence any scientific theory must be expressible in some written language. Furthermore, this written communication must be intelligible to a potentially large number of readers. Consequently only rather modest assumptions may be made about the interpreter of the written message. In this respect a scientific paper is the antithesis of a love letter, much of whose meaning may be inaccessible to all but the sole intended reader. It is also highly desirable that essentially the same meaning should be conveyed to all potential readers; unlike a poem, each of whose many readers may find a radically different interpretation.

Papers describing scientific theories also differ from love letters and poems in another very obvious way: the type of interpretation that the reader must be able to find. In order to be intelligible to its readership such a paper must communicate some assertion about relationships between entities in the world with sufficient precision for the reader to be able to determine what that assertion implies about any situation to which the theory is intended to apply. In other words the reader must be able to understand what consequences accepting the truth of the theory would entail. This is close to the widely accepted requirement that a scientific theory must generate testable predictions: it is weaker in that it does not stipulate that a test be possible.

Thus we have established that a scientific theory must be expressible in written language with sufficient precision and lack of ambiguity to enable many different readers to understand it sufficiently to determine what consequences the theory implies concerning any situation within its scope. This set of properties is essentially the same as the informal notion of an effective procedure provided at the start of Section 2. Of course, scientists usually couch their theories in declarative rather than procedural terms but this is of no significance: many Turing equivalent formal systems are also declarative in form. Hence any genuinely scientific theory must embody an effective procedure that will allow its intended audience to determine what it entails about the range of situations to which it applies.

Note that this does not exclude the possibility of scientific theories that are vague and imprecise. For example a law of gravity that said merely that all objects of finite mass are attracted towards each other would satisfy the criteria: it would simply be a much less useful theory than Newton's inverse square law.

Theories in Cognitive Science form a subclass of all scientific theories and hence must embody effective procedures. Thus we have proved the claim made in the introductory section: the computational metaphor includes all models possible using any alternative scientific approach. It is important to note that this does not imply that all cognitive phenomena can be satisfactorily modelled using the computational metaphor. There may indeed be aspects of the mind that cannot be fully described by some Turing computable function. All the argument concludes is that any such phenomena are also outside the scope of any other form scientific theory: there can be no "science of the mind" that goes beyond the computational metaphor.

4. Objections to the Argument

This is a rather startling conclusion: it is therefore prudent to examine the various types of objection that might be raised against it. There are at least three. First it might be asserted that Church's theses only applies to processes that consist of a sequence of discrete steps while the world manifestly contains phenomena that are both parallel and continuous. Alternatively a critic might accept the validity of the argument yet claim that it took an overly simplistic view of the computational metaphor: if the latter implies something more than Turing equivalence then the fact that all scientific theories must be Turing computable no longer implies they can necessarily be accommodated within the computational metaphor. Finally it could be argued that there are phenomena that appear to be legitimate objects of scientific study yet cannot be adequately described as computable functions. In the remainder of this paper we examine each of these objections in turn.

5. Continuity, Concurrency, and Non-determinism

The notion of effective procedure embodied in Turing machines and other equivalent formal systems is a process comprised of a sequence of discrete steps. However, there are clearly innumerable examples of processes in the natural world in which many entities interact simultaneously and continuously. It is virtually certain that important aspects of brain function are examples of such continuous concurrent processes. Does this invalidate the argument that all scientific theories of cognition must be effectively computable?

This objection cannot refute the argument because it does not address it. If one were to accept the suggestion that the discrete nature of Turing computations implies that they cannot provide an account of continuous concurrent processes one would in fact be forced to conclude that there could be no scientific theories about phenomena involving such processes. However this is clearly not the case: essentially all of classical physics is concerned with continuous concurrent processes.

Physical scientists have been able to develop theories that are effectively computable to describe such processes by creating languages for modelling continuous systems in terms of discrete symbols and operations. For example, calculus provides the means by which Newton's Laws can be shown to provide a good model of the motions of the planets. Of course, such models may have limitations: calculus does not provide exact solutions to the many body problem but only approximations whose precision depends on the amount of time devoted to the computation. It would take an eternity to calculate precisely where the moon will be one second from now, although the moon, with a little help from the rest of the universe, manages to derive the exact answer in just one second. Nevertheless it is generally accepted that the Newtonian theory provides a satisfactory model of planetary motion because the approximation involved in a finite calculation has no practical significance.

Our ability to predict the future state of the solar system depends upon the fact that it is not chaotic: that is, small differences in initial conditions do not lead to large differences in subsequent behavior. Other physical systems do not have this property. A well known example is the behavior of the earth's atmosphere: although the physics of weather is well understood, prediction is difficult because small events may have major consequences.

It is clearly possible, perhaps even likely, that brain function is significantly dependent on chaotic phenomena. Would this imply that it is beyond the scope of the computational metaphor? It does not place any constraints on the models one can formulate but it would limit their ability to generate precise predictions about behavior. What it would mean is that we could not hope to build a complete model that replicated the behavior any particular individual even if there were no practical difficulties in obtaining a complete description of that individual's mental state at a particular time. This is not a significant restriction: cognitive science has no pretensions to precisely modelling particular individuals. Just as a science of meteorology does not imply the ability to predict the weather at 3:15 pm a week next Tuesday, so a science of the mind does not imply the ability to determine what John Smith will be thinking about if it happens to be raining at that time.

6. Does the Computational Metaphor Imply More Than Turing Equivalence?

The argument presented in Sections 2 and 3 necessarily rests on an assumption about what is meant by "the computational metaphor." The particular interpretation chosen was that it refers to the approach in which theories of cognition are formulated in terms of processes that could be emulated by programs running on a digital computer. Since digital computers are Turing equivalent, the argument is valid given this assumption: it would not be if a narrower or radically different interpretation were made.

To take an extreme example, suppose someone were to hold that the computational metaphor actually meant assuming the mind had the same structure as the traditional Von Neumann computer architecture: a store holding programs and data plus a processor unit that executes instructions in the program. Clearly such an assumption would drastically constrain the range of possible models of cognitive processes in a way that would exclude many perfectly reasonable (and computable) theories.

No reasonable person could hold such a view but the possibility exists that someone might maintain that the computational metaphor implies more than mere Turing equivalence. Clearly such a restricted alternative would itself have to be Turing equivalent but this leaves plenty of scope: every programming language is a potential candidate. The difficulty with this view is that there is no obvious contender, or set of contenders for the role. The computational metaphor really does seem to mean different things to different people and the only thing the interpretations share in common is Turing equivalence. As Pylyshyn (1989) remarks:

> It may come as a surprise to the outsider then to discover that there is no unanimity within cognitive science on either the nature (and in some cases the desirability) of the influence or what computing is, or at least on its essential character, as it pertains to cognitive science.

The present author is not among those who find this lack of consensus a surprise. The theoretical finding that a Turing machine can compute any effective procedure has a thoroughly pragmatic counterpart which is at the heart of practical computer science. When a programmer designs and writes a program what he or she is doing can invariably be viewed as creating a new machine with the properties that the end user requires by contriving that the computer should behave as if it were the desired machine. Thus the programmer creates a *virtual machine*. In all probability this virtual machine will itself be implemented on another virtual machine. For example, a COBOL compiler creates the illusion that the computer is a COBOL machine, and a programmer may use it to provide a customer with a system that appears to be a pay roll generator. Every day, all over the world, tens of thousands of new virtual machines are being implemented. Even the majority of those cognitive scientists who believe that the connectionist approach is in some way radically different from alternative computational methodologies test their ideas by implementing a neural virtual machine on a conventional digital computer. It is the protean nature of computers that makes them so useful: it is thus hardly surprising that they serve as source for a wide variety of metaphors.

6.1. *The Symbol-Processor Interpretation of the Metaphor*

Pylyshyn (1984, 1989) himself presents what he terms the "classical view" of the relationship between computers and minds in which each is characterized by three distinct levels of organization: the physical, the symbolic and the semantic. It is difficult to see how any complex system could be understand without some such set of alternative levels of description. Indeed the notion of virtual machines discussed above demonstrates that in practice many more than three levels may be needed to provide an adequate account of even relatively simple computer programs. Thus if the analogy has any substance it must derive from the nature of the particular levels of description. Of the three levels specified, only that labelled "symbolic" is sufficiently contentious to be a potentially fruitful source of metaphor. Thus the core of this view is that the mind can usefully be regarded as some kind of symbol processing device. An extreme form of this position has been stated with admirable directness by Newell and Simon (1976). Having defined the essential characteristics of a physical-symbol system to imply a machine that is Turing equivalent, they then propose:

> *The Physical-Symbol System Hypothesis.* A physical-symbol system has the necessary and sufficient means for general intelligent action.

The claim for sufficiency is essentially a restatement of Turing's (Turing 1950) view that artificial intelligence is possible; an opinion that has widespread but far from universal support. It is very closely related to the interpretation of the computational metaphor adopted in other sections of the present paper.

The claim for necessity is more interesting, since it is this that would, if correct, make the computational metaphor stronger. Unfortunately it is ambiguous. It could be interpreted to mean that any generally intelligent agent must be Turing equivalent. Such a claim is easy to refute because it can readily be demonstrated that human behavior falls far short of Turing equivalence: for example, our ability to comprehend nested sentence structures is so limited that we clearly lack even the computational power of a stack automaton.

Alternatively the necessity claim may only imply that any intelligent agent must contain at least one component that is Turing equivalent. There is no contradiction in postulating a device that falls far short of Turing equivalence which nevertheless contains a Turing equivalent component: numerous domestic appliances whose behavioral repertoires are exceedingly limited are controlled by microprocessors. However, the microprocessor in a washing machine could be replaced by a computationally weaker device. The necessity claim implies that there are functions performed as part of cognition that can only be carried out by Turing equivalent devices.

This is a coherent, if rather extreme, conjecture. Newell himself was faithful to it: the Soar architecture (Laird, Newell and Rosenbloom 1987) is a magnificent instantiation of this approach to cognition. However, very few other cognitive scientists have attempted to develop such universal theories of behavior. Consequently most of the work in cognitive science that has drawn on the computational metaphor has not involved components capable of computing all the computable functions. Hence it is reasonable to conclude that the computational metaphor does not imply the very strong assumptions of universality made by Newell and his colleagues.

6.2. *Why has the Metaphor been so Fertile?*

An interesting variation of the objection that the computational metaphor implies more than Turing computability argues that the very fact that it has proved so fruitful in cognitive science demonstrates that it must be exerting some pressure on the type of models proposed. If it really does encompass all genuine scientific theories how and why has it generated so much worthwhile research activity? One answer is that it has served to force ill-formed and ambiguous conjectures into precisely stated forms whose consequences can be elucidated through simulation. In other words, the chief value of the metaphor is to constrain researchers to producing genuine scientific theories.

There are some parallels between the situation in contemporary cognitive science and the state of the physical sciences in the first half of the seventeenth century. The physics of the schoolsmen was essentially verbal. Some appreciation of the difficulty they faced in developing models of physical phenomena without the aid of mathematics can be gained from the fact that the Merton Mean Speed Rule was one the major accomplishments of later medieval physics (Crombie 1959). The discovery of the power of mathematics was as important as the shift to empiricism in producing the scientific revolution of early modern times. As a consequence many leading scientists of the time leaned to the view that the mathematics was more than a useful tool; that, as Galileo put it, "the book of nature is written in the language of mathematics." This Platonist viewpoint still has its adherents, but the majority of working scientists have taken the pragmatic view that mathematics provides a powerful system for expressing and investigating models of the world without making metaphysical assumptions about the underlying mathematical nature of reality. It is this practical application of mathematics that has proved so fertile.

Until recently behavioral scientists had to struggle with little more than verbal models of cognitive phenomena in much the same way as the schoolsmen physicists struggled several centuries ago. The advent of the digital computer finally provided a tool that could be used in the same way as physicists had begun to use mathematics four hundred years earlier. It is therefore not surprising that the last three decades have seen a continuing debate about whether the mind actually is a computer (whatever that might mean). As in the case of physics, ultimately it may not matter very much: the principal contribution of computer science to the understanding of cognitive phenomenon is likely to be in providing a language in which theories can be formulated and their consequences investigated.

7. Proposed Limitations to the Computational Metaphor

The third type of objection to the argument asserts there are cognitive phenomena that are legitimate objects of scientific study yet cannot be adequately described as computable functions.

7.1. *Creativity, Insight and Understanding*

A number of authors have argued strenuously that there are certain aspects of intellectual activity that cannot be realized as a program on a digital computer. The best known examples are the claim that Gödel's Theorem implies that human beings have an insight into what is true that cannot be formalized (Lucas 1961; Penrose 1989), and Searle's Chinese Room (Searle 1980) which, it is claimed, demonstrates that a rule based system might appear to understand language without any such understanding being present.

These claims have often been refuted in print (See for example Whitely 1962 on the Gödel argument and Boden 1988 on the Chinese Room) but their validity is not a central issue here. Let us suppose that these or similar objections to the computational metaphor were well founded. What would that imply, given the argument that all scientific theories must embody effective procedures? None of the authors of these objections propose alternative scientific theories of the cognitive phenomena concerned. Consequently, even if correct, these arguments do not contradict the conclusion of the present paper, that all scientific theories fall within the computational metaphor. Thus if one accepts these arguments one has also to accept that the cognitive phenomena concerned are outside the scope of scientific study.

7.2. *Emotion*

One of the difficulties to be overcome in addressing the question of whether the computational metaphor could provide an account of emotion is that the term is very ill-defined. A widely respected textbook (Strongman 1987) catalogues over thirty different theories of emotion, each of which is based on a different view of what emotion actually is. For present purposes it is useful to distinguish two aspects of emotion: behavior and affect. When a man loses his temper he will both act and feel differently than he would in a calmer frame of mind.

It is plausible that adequate models of the behavioral aspects of emotion involving complex and conflicting goal states could be implemented as computer programs. Several research programs with this objective are already under way (e.g. Beaudoin and Sloman 1993; Moffat, Frijda and Phaf 1993). Such systems would exhibit behavior that would be described as emotional in a human being and might well have internal states that corresponded to the states of affect that a person would experience in a similar situation. If the implementation were sufficiently anthropomorphic observers might be strongly drawn to believing the system actually had the corresponding feelings; although when the underlying mechanism was explained the same observers might retract this judgement. Certainly such a model would be a good theory of emotional behavior.

However, it might reasonably be argued that the attempt to build such a computational model of emotion is doomed to failure unless some means is found to incorporate affect. Proponents of this view would say that one cannot build a program that successfully emulates the behavior associated with complex emotional states unless the system was actually able to experience the feelings that constitute those states. This is certainly a reasonable position to hold: it is difficult to see how one might construct an android that could laugh at jokes without endowing it with a sense of humor. If this view is correct then it implies that a computational account of emotion is only possible if we can build systems that have awareness. I suggest below that consciousness falls outside the scope not only of the computational metaphor but also of any scientific study. How complete a scientific theory of emotion can be built remains an empirical question.

7.3. *Social Factors*

It is far from clear that there is any real problem associated with the use of the computational metaphor to model social behavior or the influence of social factors on other aspects of behavior. Relatively little work has been done in the area. The most obvious exceptions are the extensive body of work on natural language understanding, research on distributed artificial intelligence, and attempts to model entire societies (Gilbert and Doran 1994). Many of the classic results of social psychology — for example Asch's (1956) work on conformity and visual perception — could be modelled by a computer program, but it is not clear whether any useful purpose would at present be served by doing so.

The appearance of conflict with the computational metaphor probably has its origins in a much deeper conflict within the social sciences, notably sociology. These disciplines have both a scientific and a literary-humanistic tradition with conflicting objectives and epistemologies. Proponents of the literary-humanistic perspective are, by their own declaration, anti-scientific. Consequently it is not surprising that they find the computational metaphor uncongenial.

The core of the literary-humanistic objection to the scientific approach appears to be that the objects of study in the social sciences are conscious entities and consequently the methodologies developed for the study of inanimate matter are inappropriate. Thus we are led to a similar conclusion to that we reached when considering emotion. Because consciousness may play a critical role some socially related behavior may be beyond the scope not only of the computational metaphor but of any scientific study.

7.4. *Consciousness*

Consciousness is a philosophical problem of the first magnitude that has been the subject of a number of recent publications intended for readers outside the confines of academic philosophy (See, for example, Dennett 1991; Edelman 1992; Crick 1994). The difficulty has two roots. First we have no way of determining whether or not an entity has the capacity of being aware. Since we ourselves are aware, most of us attribute both a similar capacity to other people and a degree of consciousness to other entities more or less in direct proportion to how much they resemble us. We have no method of testing an entity to see if it has conscious awareness. All clinical tests of consciousness actually measure behavioral surrogates and would be better described as tests of arous-ability.

The second problem is that there appears to be no functional purpose for consciousness. For example no biological function appears to be served by the fact that painful stimuli hurt. Mechanisms can readily be built that withdraw from dangerous stimulation and learn to avoid situations where such stimulation might occur without any need to implement awareness. Nor is consciousness needed to juggle tasks of different priorities competing for cognitive resources (Scott 1979, 1994): all but the simplest operating systems do this continuously but I have never heard anyone claim Unix was conscious. One cannot evade the problem by asserting that consciousness is something that happens when a system contains a representation of itself: numerous programs ranging from simple self-replicating programs to, again, operating systems, include such a representation.

These two characteristics seem to me to present insuperable problems to developing any scientific account of consciousness, computational or otherwise.

8. Conclusion

The principal conclusion of this paper is that any scientific theory of cognition can be realized as a computer program and hence there can be no conflict between the computational metaphor and alternative scientific approaches to the phenomena of mind. The argument does of course rest on Church's Thesis which is an unprovable conjecture. Those who find the conclusion unacceptable therefore have the option of rejecting the thesis. To do so is to assert that there are procedures which can be unambiguously and completely described that cannot be realized by a Turing machine. Such a procedure if discovered would have wide reaching consequences since it would imply the existence of machines of greater (or at least different) computational power than digital computers. In view of the overwhelming evidence that no such procedure exists, it seems reasonable to place the onus on those rejecting Church's Thesis to produce a counterexample.

The argument does not imply that all aspects of the mind could be modelled using the computational metaphor. The possibility exists that there are cognitive phenomena that cannot be emulated computationally: several candidates were discussed in Section 7. However, the argument does imply that any such phenomena are outside the scope of any form of scientific theory. There can be no science of the mind that goes beyond the computational metaphor.

References

Asch, S.E. 1956. Studies of independence and conformity: A minority of one against a unanimous majority. *Psychol. Monogr.* 70, 80–90.

Beaudoin, L.P. and Sloman, A. 1993. A study of motive processing and attention. In *Prospects for Artificial Intelligence: Proceedings of AISB93,* Leeds, UK, A. Sloman, D. Hogg, G. Humphreys, A. Ramsay, and D. Partridge (eds), 229–238. Amsterdam: IOS Press.

Boden, M.A. 1988. Escaping from the Chinese Room. In *The Philosophy of Artificial Intelligence,* M.A. Boden (ed), 89–104. Oxford, UK: Oxford University Press.

Church, A. 1941. The calculi of lambda-conversion. *Annals of Mathematics Studies* 6, Princeton University Press.

Crick, F.H.C. 1994. *The Astonishing Hypothesis: The Scientific Search for the Soul.* London: Simon and Schuster.

Crombie, A.C. 1959. *Augustine to Galileo 2: Science in the Later Middle Ages and Early Modern Times 13th–17th Century.* Harmondsworth, UK: Penguin Books.

Dennett, D.C. 1991. *Consciousness Explained.* USA: Little, Brown and Company.

Edelman, G. 1992. *Bright Air, Brilliant Fire: On the matter of the Mind.* London: Penguin.

Feyerabend, P. 1975. *Against Method.* London: Verso.

Gilbert, N. and Doran, J. 1994. *Simulating Societies: The Computer Simulation of Social Phenomena.* London: UCL Press.

Kleene, S.C. 1952. *Introduction to Metamathematics.* Princeton, NJ: D. Van Nostrand.

Laird, J.E., Newell, A. and Rosenbloom, P.S. 1987. Soar: An Architecture for General Intelligence. *Artificial Intelligence* 33, 1–64.

Lucas, J.R. 1961. Minds, machines, and Gödel. *Philosophy* XXXVI, 112–127.

Minsky, M. 1967. *Computation: Finite and Infinite Machines.* London: Prentice-Hall International.

Moffat, D., Frijda, N.H. and Phaf, R.H. 1993. Analysis of a model of emotions. In *Prospects for Artificial Intelligence: Proceedings of AISB93,* Leeds, UK, A. Sloman, D. Hogg, G. Humphreys, A. Ramsay, and D. Partridge (eds), 219–228. Amsterdam: IOS Press.

Newell, A. and Simon, H.A. 1976. Computer science as empirical inquiry: Symbols and search. *Communications of the ACM* 19, 13–126.

Penrose, R. 1989. *The Emperor's New Mind.* Oxford, UK: Oxford University Press.

Popper, K. 1959. *The Logic of Scientific Discovery.* London: Hutchinson.

Pylyshyn, Z.W. 1984. *Computation and Cognition: Toward a Foundation for Cognitive Science.* Cambridge, MA: MIT Press.

Pylyshyn, Z.W. 1989. Computing in cognitive science. In *Foundations of Cognitive Science,* M.I. Posner (ed), 49–92. Cambridge, MA: MIT Press.

Scott, P.D. 1979. The brain as an operating system. In *Proceedings of EuroIFIP'79,* London, UK, P. Samet (ed), 281–286. Amsterdam: North-Holland Publ. Comp.

Scott, P.D. 1994. Cognitive resource scheduling: Attention mechanisms in intelligent agents. In *Proc. SSAISB Workshop on Computational Models of Cognition and Cognitive Function*, R. Cooper and S. Grant (eds), 3–17. Leeds, AISB.

Searle, J.R. 1980. Minds, Brains and Programs. *The Behavioral and Brain Sciences* 3, 417–424.

Strongman, K.T. 1987. *The Psychology of Emotion.* third edition. New York: John Wiley and Sons.

Turing, A.M. 1936. On computable numbers, with an application to the Entscheidungsproblem. *Proc. Lond. Math. Soc.* 42, 230–265.

Turing, A.M. 1950. Computing machinery and intelligence. *Mind* LIX No. 236, 433–460.

Whitely, C.H. 1962. Minds, Machines, and Gödel: A Reply to Mr. Lucas. *Philosophy,* XXXVII, 61–62.

What's Psychological and What's Not?

The Act/Content Confusion in Cognitive Science, Artificial Intelligence and Linguistic Theory

Terry Dartnall
Faculty of Science and Technology
Griffith University

> A proposition may be thought, and again it may be true; never confuse these things. (Frege, *The Basic Laws of Arithmetic*)

1. Introduction

There is something strange about saying that a belief is true and sincere, or that it is tautologous and four years old. Suppose that I sincerely believe that it is cold at the north pole. Then it would seem that my belief is both true and sincere. Or suppose that I have believed for four years that it is raining or not raining. Then it would seem that my belief is a tautology and four years old. After all, my belief is the proposition 'It is raining or it is not raining,' which is a tautology, and I have had this belief for four years.

But the statements are odd. They are odd because they treat different kinds of thing as if they were the same kind of thing. 'Belief' (like all mentalistic terms) is ambiguous between a psychological act, state or process, and the *content* of that act, state or process. My belief as a psychological state can be sincere and four years old, but it cannot be true or tautologous. On the other hand, my belief as the *content* of my psychological state is the proposition 'It is raining or it is not raining.' This can be true and tautologous (and shared by others), but it cannot be sincere or four years old.

This confusion between these two senses of 'belief' illustrates a more general confusion between what is psychological and what is not. Belief *states* are psychological objects and can be subjects of psychological enquiry. But the propositions that express the *content* of such states are not psychological and cannot be subjects of psychological enquiry. Psychologists do not (and should not) concern themselves with the propositions 'It is cold at the north pole' and 'It is raining or it is not raining.' These propositions belong to the domain of the geographer and the logician, respectively.

I shall try to show that this confusion between the psychological and the non-psychological is widespread in classical AI and cognitive science, and can be found in linguistic theory and elsewhere. It is a serious confusion because it concerns the conceptual foundations of disciplines. In the nineteenth century it gave rise to a position known as 'psychologism,' in particular to psychologism in logic and mathematics. This is the belief that we can find out about logic and mathematics by studying the mind. Logic and mathematics are seen as branches of psychology, and therefore as *empirical* disciplines. This clearly makes radical claims about the nature of logic and mathematics.

Now, Frege and Husserl successfully criticized psychologism. The thrust of their criticism is captured in Frege's cryptic remark that "A proposition may be thought, and again it may be true; never confuse these things." (1967, introduction.) That is, we should not confuse the psychological act of thinking a thought or a proposition with the thought or proposition itself. We should not confuse the psychological and the non-psychological.

I shall try to show that we have not learned this lesson. It is true that psychologism is usually seen as a position to be avoided, but there is little understanding of *what is wrong with it*. In the absence of such an understanding, a mirror image of psychologism has emerged that has its roots in the same psychological/non-psychological confusion. I call this mirror image 'reverse psychologism.' *Psychologism* says that we can find out about the content of thought by studying the mind. *Reverse psychologism* tries to find out about the mind by examining the public symbolisms (such as logic and language) that express the content of thought. Much of cognitive science, for instance, tries to model the mind by positing internalized models that manipulate our public, communicable, symbol structures. Some linguistic theory tries to find out about the mind by studying language. Knowledge representation tries to give us machines that *know* by internally representing symbol structures that express and embody the *content* of knowledge.

The paper is in two parts. Part One is about psychologism. It describes psychologism, provides examples, and spells out its ontological and methodological requirements. It then examines what is wrong with psychologism, how psychologism originates, and how the confusion is compounded. Part Two is about reverse psychologism — what it is, and how it comes about. It provides case studies from linguistic theory, cognitive science and AI. The paper concludes by showing why there is no cognition in the Chinese Room.

2. Psychologism

2.1. *What is Psychologism?*

'Psychologism' originally referred to the doctrine that philosophy is a study of the mind. Macnamara (1986) acknowledges this broad connotation but confines it to the doctrine that *logic* is a study of the mind. I shall use it to refer to the belief or attitude that all or part of any discipline other than psychology is a subfield of psychology — meaning, that all or some of the objects of study of that discipline fall within the domain of psychological enquiry.

Three things need to be explained here.

First, the standard characterization of psychologism treats it as an explicitly held belief. But someone may have a psychologistic attitude without explicitly subscribing to psychologism — they may employ a psychologistic methodology and ontology. This is why I characterize psychologism as a belief *or attitude*. I will explain what I mean by a psychologistic methodology and ontology later on.

Second, proponents of psychologism have usually maintained that *all* of the objects of study of the discipline in question fall within the domain of psychology. In fact for a position to be psychologistic, it only has to be claimed that *some* of the objects do. For instance, it might be claimed that logic is the study of valid inference, but that only psychology can study the axioms of logic. This would be psychologism as well.

Finally, since a psychologistic belief or attitude is normally about some particular discipline, it would be possible to talk about 'psychologism-in-logic,' 'psychologism-in-linguistics,' etc. I shall not always be so thorough. It will be clear which discipline I am talking about, and plain 'psychologism' makes it easier to talk about the things that psychologism-in-logic, psychologism-in-linguistics, etc, have in common.

2.1.1. *Some examples.*

Psychologism emerged in Germany in the first half of the nineteenth century as a reaction against the excesses of Hegelianism. Jakob Fries and Friedrich

Beneke maintained that introspection is the only way of establishing truth in philosophy, so that psychology is the fundamental philosophical discipline and logic, ethics, metaphysics etc are merely parts of it. Beneke (1833) said

> With all of the concepts of the philosophical disciplines, only what is formed in the human soul according to the laws of its development can be thought; if these laws are understood with certainty and clarity, then a certain and clear knowledge of these disciplines is likewise achieved.

Psychologism was later defended in the field in which it seems most alien — logic and mathematics. In his (1843) Mill maintained that introspection is the only basis of the principles of logic and the axioms of mathematics, and in his (1865) he classified logic under psychology. In a work called *An Investigation of the Laws of Thought on Which are Founded the Mathematical Theories of Logic and Probabilities*, George Boole construed the laws of logic as the laws of thought:

> The laws we have to examine are the laws of one of the most important of our mental faculties. The mathematics we have to construct are the mathematics of the human intellect. (1854, preface)[1]

A contemporary psychologistic claim is Chomsky's thesis that linguistics is a subfield of psychology:

> I would like to think of linguistics as that part of psychology that focusses its attention on one specific cognitive domain and one faculty of mind, the language faculty. (1980: 4)

2.1.2. *Ontology and methodology.*

To be more than an empty claim, psychologism must be accompanied by an ontology and a methodology that purport to show how the objects of study can *in fact* be studied as part of empirical psychology. The objects must be psychological (such as psychological states, acts or processes, or cognitive capabilities) and the methodology must be empirical.

Suppose that the claim is that logic is a subfield of psychology. Now we want to know how we can *in fact* study logic as part of psychology. Mill provided an account. Logic, he said, is the empirical study of the belief-states that can and cannot co-exist in the mind. Of the law of non-contradiction (that a proposition cannot be both true and false) he said:

> I consider it to be, like other axioms, one of the first and foremost generalizations from experience. The original foundation of it I take to be, that Belief and Disbelief are two different mental states, excluding one another. (1843, bk. 2, ch. 7, sec. 5.)

Thus according to Mill the law of non-contradiction is an empirical generalization about the way that people think. It is the empirically based claim that anyone who is in the belief-state characterized by believing A will not also be in the belief-state characterized by believing not-A.

Mill treated the law of the excluded middle (that a proposition is either true or false) in a similar way:

> The law on Excluded Middle, then, is simply a generalization of the universal experience that some mental states are destructive of other states. It formulates a certain absolutely constant law, that the appearance of any positive mode of consciousness cannot occur without excluding a correlative negative mode; and that the negative mode cannot occur without excluding the correlative positive mode… Hence it follows that if consciousness is not in one of the two modes it must be in the other. (*ibid.*)

Thus Mill provided both a methodology and an ontology for his psychologism. The objects of logical enquiry are psychological states, and the methodology is to conduct an empirical investigation into the consistency-relations holding between them.

2.2. *What's wrong with psychologism?*

The principal historical opponents of psychologism in logic and mathematics are Frege and Husserl. Husserl's first book, *The Philosophy of Arithmetic*, attempted to base arithmetic on psychological foundations. Frege reviewed it and criticized its psychologism. Husserl acknowledged this criticism and spent the following years studying psychologism in logic and mathematics and formulating arguments against it. I shall draw heavily on the work of Frege and Husserl.

In this section of the paper I look at four arguments against psychologism *in logic*. I abandon two of them, *the contingency argument* and *the contingently false argument*, not because they are unsound, but because they become complicated, and there are sound, uncomplicated arguments available. These are Husserl's *a posteriori* and existential arguments.

Next I turn to diagnosis and argue that Frege and Husserl correctly identified the origins of psychologism in general as a failure to distinguish between a psychological act and its content.

Then I provide an overview that ties the two sections together. The act/content confusion leads to the falsifying consequences that are so apparent in the case of logic. Consequently, psychologism in logic is a counterexample to psychologism in general.

In the remaining sections I consider three factors that contribute to the confusion: the ambiguity of mentalistic terms, the state/object confusion, and the use of what I call 'spurious idealization.'

2.2.1. *Four arguments against psychologism.*

The contingency argument. Frege was a realist about truth and believed that propositions are objectively true or false, depending on whether or not they correspond to an objective reality. He distinguished this from the claim made by "psychological logicians" that, as he put it,

> 'objective certainty' is merely a general acknowledgment on the part of the subjects who judge, which is not independent of them but susceptible to alterations with the constitution of their minds. (1967: 15)

Frege maintained that we cannot establish truth by examining people's beliefs and looking for general agreement. We cannot, for instance, study geography by studying what people believe to be true. Most people believe that it is cold at the north pole, but this is a fact about people, not about the north pole. Mass derangement might make it true that most people believe that it is hot at the north pole, but this will not make it hot at the north pole.

This "objective certainty" of truth was most obvious for Frege in the case of the laws of logic (such as the law of non-contradiction). He believed that such laws are necessarily true, whereas basing them on an empirical survey of beliefs would at best produce contingent truths. I call this 'the contingency argument.' A sharper reformulation is this. Saying that the laws of logic are contingently true amounts to saying that it is possible for a necessary truth to be false. This, however, is a contradiction. Take as an example the law of non-contradiction, $\neg$ (A & $\neg$ A). To say that this is only contingently true is to say that it is possible that it is false: possibly $\neg\neg$ (A & $\neg$ A). This is a contradiction in itself.

The psychological logician may now say that this begs the question. The argument I have just given amounts to:

P1. It is not possible for necessary truths to be false.
P2. The laws of logic are necessary truths.
Therefore, it is not possible for the laws of logic to be false.

The psychological logician may say that P2 is exactly what she is denying, that her claim is just that laws such as the law of non-contradiction are not necessarily true. This leaves us with a stand-off, with the psychological logician making a claim and her opponent denying it. However obvious it may seem to us that the laws of logic are necessarily true, we should try to do better than mere assertion.

I think that the correct response to the psychological logician is to say that she cannot provide an account of modal concepts *at all.* Since she denies that the laws of logic are necessarily true she must be able to say what she means by 'necessity' and 'possibility.' Presumably she wishes to replace logical possibility with *conceivability* and logical necessity with *inconceivability*, where these terms mean that states can or cannot co-exist in the mind. Thus, for instance, 'not possibly (A & ¬ A),' means 'not conceivably (A & ¬ A),' meaning that states characterized by believing A *cannot* co-exist with states characterized by believing ¬ A. But what does 'cannot' mean here? It cannot mean 'given the laws of human psychology it is *logically impossible* that …,' for this uses the notion of logical impossibility. And it cannot mean 'given the laws of human psychology it is *inconceivable* that …,' for this is circular. All that she can say is that as an empirical generalization certain states *do or do not co-exist.* But this, of course, loses the modal status of the claim. The upshot is that she cannot give currency to modal concepts, and thus cannot say what she means when she says that it is *possible* for the laws of logic to be false. She cannot say that the laws of logic are only contingently true, for to say they are contingently true is to say that it is *possible* for them to be false.

I shall not pursue the contingency argument, because less contentious arguments are available.

The contingently false argument. This argument states that if the laws of logic are empirical generalizations about consistency relations between belief states, then they are not only contingent, but contingently *false*, since some people are inconsistent some of the time.

Commenting on an earlier version of this paper, Jay Garfield suggests that the psychological logician can say that the laws of logic are empirical generalizations about consistency relationships holding between psychological states in the mind of an Ideal Thinker. The point of introducing ideality, as he says in (1988:315), is that "we know that there will be deviations from the ideality we posit, but they will be sufficiently minor so as not to vitiate the explanatory utility of the competence we ascribe".

I think it is a disanalogy to liken the concept of an Ideal Thinker to that of, say, an ideal gas (and in fact Garfield agrees (*ibid.*)), for there is more to be accounted for in the case of thinking and reasoning than minor error. For one thing, there are no grounds for believing that people's psychological states are consistent *at all* where these states are attitudes towards complex logical expressions. There are indefinitely many inconsistent pairs of expressions that the average person will not recognize as inconsistent, and indefinitely many valid forms of argument that the average person will not recognize as valid. In fact, people may actually *reject* complex logical or mathematical theorems. Garfield cites the work of Kahneman and Tversky, which suggests that most people reject Bayes' Theorem.[2]

It is also important to remember that the notion of an Ideal Thinker is an empirical and not a regulative idealization, and that empirical idealizations must be abandoned in the light of sufficiently strong adverse empirical evidence. We would abandon Charles's Law or Boyle's Law if there was a large enough gap between the behavior predicted by those laws about an ideal gas and the behavior of real gases that we actually observe. In the same way, it is possible for sufficiently many people to think inconsistently that we would have to abandon the laws of logic even as idealizations about Ideal States.

I shall not pursue this argument either. Again, this is not because it is unsound, but because, with the introduction of idealization, it becomes complicated, and there are uncomplicated arguments available.

We find these in Husserl. I call them 'the *a posteriori* argument' and 'the existential argument.'

The a posteriori and existential arguments

The a posteriori argument. Husserl observed that the laws of logic are *a priori*, whereas psychological laws are *a posteriori*. That is, the laws of logic can be determined independently of experience — we do not need to look in the world (that is, to do an empirical survey) to determine the truth of such laws. But we do have to do this in the case of psychological laws. Therefore logic cannot be based on psychology. Husserl said:

> No natural laws can be known *a priori* … The only way in which a natural law can be established and justified, is by induction from the singular facts of experience … Nothing, however, seems plainer than that the laws 'of pure logic' all have *a priori* validity. (1970: 99)

The existential argument. Husserl made the related point that the laws of logic are not about anything in the empirical world and are therefore not about mental states. If, he says, "the laws of logic have their epistemological source in psychological matters of fact [then] ... they must themselves be psychological in content, both being laws for mental states and also be presupposing or implying the existence of such states." (1970: 104). But, he continues, "No logical law implies a 'matter of fact'." In fact Husserl said that psychological laws not only imply matters of fact, but are laws for matters of fact and are therefore to be confirmed by reference to matters of fact.

I think that these two arguments are sound and straightforward. Logic proceeds *a priori*: we do not establish theorems by empirical survey, nor are the theorems of logic about anything in the world. Psychologism in logic is false both because it has the wrong ontology and the wrong methodology: logic is not about psychological states, and is not based on empirical enquiry.

2.2.2. *Diagnosis: The act/content distinction.*

So far we have seen what is wrong with psychologism *in logic*. In this section I ask where the proponent of psychologism goes wrong *in general*, and I argue that Frege and Husserl successfully identified the origins of psychologism as a failure to distinguish between a psychological act and its content.

Frege and Husserl both suggested that psychologism in logic arises from a failure to distinguish between two senses of 'the laws of thought.' Frege observed that in one sense "a law asserts what is; in the other it prescribes what ought to be. Only in the latter sense can the laws of logic be called 'laws of thought'" (1967: 13). That is, in one sense the laws of thought are empirically-based generalizations about how people actually think. In the other they are prescriptive laws telling us how we *ought* to think, and only in this sense are we talking about the laws of logic.

But Frege and Husserl provided another explanation that I think has greater explanatory power. This is the distinction between a psychological act and its content. Macnamara (1986) only attributes this distinction to Husserl, but Frege drew it as well — in fact it follows naturally from his realist notion of truth. Since propositions are true or false independently of our belief in them we must distinguish between 'belief' in the sense of *what is believed* (for Frege this is always a proposition) and 'belief' in the sense of a psychological attitude towards that proposition. As Frege put it:

> A proposition may be thought, and again it may be true; never confuse these things. We must remind ourselves, it seems, that a proposition no more ceases to be true when I cease to think of it than the sun ceases to exist when I shut my eyes. (1967, introduction).

What is believed or thought is a proposition. It can be true or false. It can be a tautology (if I believe that it is raining or not raining) or a contradiction (if I believe that it is raining and not raining). In this sense people can have the same belief, since they can subscribe to the truth of the same proposition. In the other sense a belief or a thought is a psychological state. It can be sincere and heartfelt (and four years old), but it cannot be true, false, tautologous or contradictory.

Husserl called this the "act/content distinction," though he also characterized it as the distinction between the objects of thought and our consciousness of them (e.g. 1962: 61). Macnamara puts it clearly:

> Husserl felt that Mill and the psychological logicians generally did not see the objectivity of logic because they failed to distinguish clearly the contents of mental acts, in the sense of what the acts were about, from the acts themselves. Had they done so, they might have seen that the content presented itself as objective, whereas the acts presented themselves as subjective. But they failed to discriminate between mind and what the mind knows. Mind, Husserl concluded, is the study of psychology: What the mind knows (except, of course, when it is studying mind itself) is the study of some other discipline, be it economics, physics, geology, or logic. (1986: 18)

In other words, most disciplines are about what the contents of thought are about, but logic is about the relationships holding *between* the contents of thought, i.e. between propositions. The contents of thought are not psychological entities, but the acts or states of thinking them are.

Locating the source of psychologism in the act/content distinction has greater explanatory power than locating it in the ambiguity of 'the laws of thought.' For one thing, the act/content distinction accounts for the ambiguity of 'the laws of thought,' since it is the laws applying to psychological *acts* that are empirical and imprecise, and it is the laws specifying consistency-relations between the *contents* of psychological acts (that is, between propositions) that are stipulative and regulatory. The act/content distinction also provides a more general explanation, since it accounts for all psychologisms, whereas the laws-of-thought explanation only holds in the case of logic.

2.2.3. *An overview.*

We can now tie the two previous sections together.

Frege criticized psychologism because he believed that it leads to a kind of 'consensus theory of truth.' We find out whether it is cold at the north pole by doing a survey of north-pole-belief-states. Frege felt that such a consensus lacked "objective certainty," and he believed this was most obviously true in the case of logic.

However, the case against psychologism is stronger than he made out. Psychologism confuses the contents of psychological acts with the acts themselves, and psychological acts (states etc) are not the sorts of things that can be true or false *at all*: 'believing it is cold at the north pole' cannot be true or false, though the *content* of the belief can be. In order to reintroduce truth and falsity the proponent of psychologism has to ascend a level and generalize *about* belief-states — 'everyone has the belief-state believing-it-is-cold-at-the-north-pole.'

This ascension of levels fails most obviously in the case of logic, where more is lost in the act/concept confusion than truth and falsity. To begin with, necessary truth and falsity are lost as well. Mill simply bit the bullet on this and accepted that the laws of logic are contingent. But *a priority* is also lost. The ascension of levels commits the psychological logician to saying that the laws of logic are *about* mental states, and these (of course) cannot be known *a priori*. *A priority*, unlike truth and falsity, cannot be reintroduced by an ascension of levels, because empirical generalizations about belief states cannot be known *a priori*.

Frege said that the loss of objective certainty is most obvious in the case of logic. What he should have said is that the consequence of *confusing act and content* is most obvious here. This confusion and the subsequent ascension of levels force the psychological logician into saying that logic is neither necessary nor *a priori*. The same act/content confusion and ascension of levels occur with psychologism in other disciplines, but here there is no change of modality. There is no change of modality, for instance, in trying to base geographical truths on generalizations about belief-states, for both geographical truths and generalizations about belief-states are contingent and *a posteriori*.

In sum — psychologism *in general* arises out of the act/content confusion. It leads us to believe that by studying the mind we can access the declarative structures that are the *contents* of mental acts and states. This is most obviously false in the case of logic, whose declarative structures do not quantify over objects, and which are knowable *a priori*. Psychologism in logic is therefore a counterexample to psychologism in general, even though other disciplines are about things in the world and proceed *a posteriori*.

2.2.4. *Compounding factors.*

Propositional attitude terms. A major contributing factor in the act/content confusion is the systematic ambiguity of propositional attitude terms. Verbs of propositional attitude appear in contexts of the form 'A <verbs> that p.'

Examples are 'to know,' 'to think' and 'to believe.' The corresponding propositional attitude terms are 'knowledge,' 'thought' and 'belief.' All of these are ambiguous between act and content. We have already noted the ambiguity of 'belief,' which is ambiguous between the act or state of believing, on the one hand (I do not distinguish between act and state) and the content of the belief, on the other.

Knowledge of abstract objects: the state/object confusion. The semantics of 'to know' and 'knowledge' is highly complex. 'To know' frequently functions as a verb of propositional attitude,[3] so that 'knowledge' is ambiguous between act and content. Even here, however, 'knowledge' differs from other propositional terms inasmuch as 'to know' is factive: if A knows p, then p is true. Consequently, knowledge as content is always true: we cannot have false knowledge. This content-knowledge is Popper's "objective knowledge," which he tells us is found in books and libraries.[4]

But 'knowledge' can be ambiguous in another way (other than between act and content). When we are talking about abstract objects our content-knowledge of the object can actually *be* the object, so that we have an ambiguity between an act or state of knowledge on the one hand and *the object of knowledge* on the other. By 'object of knowledge' I mean the object that is known about. It is surprising that content-knowledge can be the object of knowledge, because this violates the distinction between ontology and epistemology (that is, the distinction between how things are and what we can know about them), but I submit the following as evidence:

- John knows that Mae West said at time t 'Come up and see me some time.' John's knowledge of what Mae West said at time t is 'Come up and see me some time.' So what Mae West said and John's knowledge of it are one and the same — 'Come up and see me some time.'

- The history contained in a history book and the author's knowledge of the history expressed in that book are one and the same: they are the historical propositions found in the book.

- I know the rules of Scrabble and write them down. I have now written down my knowledge of the rules of Scrabble — and what I have written down are the rules themselves. My knowledge of the rules is not a set of propositions *about* the rules (such as 'there are 20 of them,' 'they are difficult'). It is the rules themselves.

A first approximation to what is going on here is that 'to know' is not functioning as a verb of propositional attitude, but as a fully relational verb such as 'to love' or 'to desire.' There is a strong temptation to resist this claim because we see 'to know' as a verb of propositional attitude, and we are influenced by the Frege/Quine tradition that says that terms in propositional attitude contexts do not refer. For instance, 'Santa Claus' in 'Johnny believes that Santa Claus brings presents' does not occur in referring position, so that it is not being asserted that Johnny *stands in a relationship* to Santa Claus, since he cannot stand in a relationship to something that does not exist. In contrast to this I am claiming that 'A knows 'o,'' where 'o' is an abstract object, is fully relational and functions like 'to love' and 'to desire.'

Why 'to love' and 'to desire'? Because these verbs exhibit the same behavior as the relational sense of 'to know' and are ambiguous in the same kind of way. Consider 'My love is unrequited and works in a bank.' Here 'my love' is equivocal between a psychological state and the object of a psychological state. It is my love as a psychological state that is unrequited, whereas it is my love as the *object* of that state (let us call her 'Maria') who works in a bank. If we confuse the two we may end up believing that Maria is unrequited, or that I have a psychological state that works in a bank.

I am suggesting that we speak of knowing the rules of Scrabble, or French history, or what someone said, as if we stand in a relationship of *knowing* to them in the same kind of way that I stand in a relationship of *loving* to Maria. And we get the same kind of ambiguity. My knowledge of abstract object 'o' is ambiguous between a psychological state and 'o' itself. If we switch from talking about an *object* of knowledge to a *state* of knowledge (that can be studied by psychology) we get psychologism. If we switch from talking about a state of knowledge to talking about an object of knowledge (that can be studied by some other discipline, such as linguistics) we get reverse psychologism. I will provide examples of this in Part Two.

Spurious idealization. We can compound the state/object confusion by introducing the concept of an Ideal Knower, such as Chomsky's Ideal Speaker or Macnamara's Ideal Thinker. Chomsky maintains that linguistics is concerned with the competence of an Ideal Speaker, and Macnamara uses Chomsky's framework in his competence theory of human reasoning, replacing the notion of an Ideal Speaker with that of an Ideal Thinker.

Katz (1981) observes that linguistic *knowledge* must be true and *perfect* knowledge must be complete, but he does not notice that the concept of an Ideal Knower cements into place the identification of knowledge-of-an-object with the object itself. To say, for instance, that linguistics should study the knowledge of the Ideal Speaker ('who knows its language perfectly') is to say no more and no less than 'linguistics should study the language itself.' This is the same kind of spurious idealization that we find in the claim that we should study the universe as God sees it, which is a redundant way of saying that we should study the universe as it really is. God drops out of the equation as a redundant explanatory device, as does the concept of *any* Ideal Knower. What we lose in such an analysis is the impression that we are doing psychology and studying Mind.[5]

2.2.5. *Terminology.*

We emerge from Part One with the fundamental distinction between psychological states, acts and processes on the one hand and their contents or objects on the other. Sometimes it is useful to refer to the former with '-ing' words such 'believing,' 'thinking' and 'knowing,' and to use words such as 'belief,' 'thought' and 'knowledge' to refer to the contents or objects of thought. For the sake of clarity, however, I shall often use neologisms such as 'state-knowledge' and 'content-knowledge.'

3. Reverse Psychologism

Reverse psychologism is the mirror image of psychologism. Both stem from state/content or state/object confusions and both are compounded by spurious idealization. But whereas psychologism tries to study the contents or objects of thought by studying mental states, reverse psychologism tries to study the mind by looking at the contents or objects of thought: it talks about in-the-head entities as if they were public and open to scrutiny by disciplines such as logic or linguistics.

The best way to understand reverse psychologism, and how it differs from psychologism, is to look at examples. I will provide case studies from linguistic theory, cognitive science and AI. I will begin by looking at Chomsky's mentalism during his Standard Classical period.[6]

3.1. *Chomsky's Mentalism*

Chomsky's mentalism during this period appears to be (and is seen to be) psychologistic. In characterizing mentalism he says:

> linguistic theory is mentalistic, since it is concerned with discovering a mental reality underlying actual behavior (1965: 4).

These mental realities are the "actual subject matter of linguistics." (*ibid.*, see also pp. 193, 194) Later he says:

> Linguistics ... is simply the subfield of psychology that deals with these aspects of mind. (1968: 24). I will concentrate here on some of the issues that arise when we try to develop the study of linguistic structure as a chapter of human psychology. (1968: 59)

Katz says:

> Chomsky's theory [of grammars and linguistic theories] undeniably occupies center-stage in discussions of psychologism in linguistics. It was Chomsky who refuted the nominalism of the structuralist tradition, who made conceptualism [effectively, psychologism — TD] the dominant position in linguistics, and who today stands as the chief proponent of the view that linguistics is psychology. (1981: 11)

Macnamara agrees with Katz "that Chomsky's position is psychologistic":

> We do not have to infer this from [Chomsky's] statements because he has been quite explicit: 'I would like to think of linguistics as that part of psychology that focusses its attention on one specific cognitive domain and one faculty of mind, the language faculty.' (Chomsky 1980: 4).
> (Macnamara 1986: 27)

But when we look more closely we find that Chomsky's approach is *reverse* psychologistic. It is easier to see this if we distinguish between two types of mentalism — weak and strong — and focus on the latter. Strong mentalism says that the object of linguistic enquiry is a mental entity and that this entity is the grammar of the language. Weak mentalism says that the mental entity is just a *representation* of the grammar. There is clear evidence of both positions in Chomsky's Standard Classical work, and clear evidence that he does not distinguish between them.[7] Both positions are reverse psychologistic, but the situation is clearer in the case of strong mentalism.

A *psychologistic* position would say:

> The native speaker knows the grammar of the language. Therefore the grammar is part of her knowledge. Psychology studies human knowledge. Therefore psychology can study the internalized grammar. Therefore linguistics is a sub-field of psychology.

In fact Chomsky says:

> The native speaker knows the grammar of the language. Therefore the grammar is part of her knowledge. Linguistics studies grammars and can therefore tell us about the speaker's knowledge. Therefore linguistics can tell us what is going on in the head.

The first passage goes from talking about a grammar as a content or object of knowledge to talking about a *state* of knowledge (that can be studied by psychology). The second goes from talking about a state of knowledge to an object or content of knowledge. This is reverse psychologism.

Here it is in more detail. Psychologism first:

> The native speaker knows the grammar of the language.
> *True — the speaker is in a state of knowledge about the grammar.*
> Therefore the grammar is part of her knowledge.
> *Another sense of knowledge is introduced: knowledge in the sense of 'what is known,' the content or object of knowledge.*
> Psychology studies human knowledge.
> *Switches back to knowledge as a psychological state.*
> Therefore psychology can study the internalized grammar, so that linguistics is a sub-field of psychology.
> *Trades on the ambiguity of the two senses: goes from saying that psychology studies state-knowledge (which is true) to saying that it studies content/object knowledge (which is false).*

Now look at what happens in the case of reverse psychologism:

> The native speaker knows the grammar of the language.
> *As for psychologism. True — the speaker is in a state of knowledge about the grammar*
> Therefore the grammar is part of her knowledge.
> *As for psychologism. Another sense of knowledge is introduced: knowledge in the sense of 'what is known,' the content or object of knowledge.*
> Linguistics studies grammars and can therefore tell us about the speaker's knowledge.
> *The argument now breaks with psychologism and stays with the newly introduced sense of knowledge — knowledge as content or object.*
> Therefore linguistics can tell us what is going on inside the head.
> *Trades on the ambiguity of the two senses: goes from saying that linguistics studies content/object knowledge (which is true) to saying it studies state-knowledge (which is false).*

Thus reverse psychologism really is the reverse of psychologism, but is based on exactly the same act/content confusion.

Nevertheless, it still has to be asked why Chomsky goes this way. The ambiguity of mentalistic terms makes it easy to do so, but *why* does he do so? He gives two main reasons for his mentalism.

The first is that he believes that only mentalism can account for the speaker's ability to produce and understand indefinitely many sentences. The child is exposed to a finite number of sentences, many of them degenerate, yet within a comparatively brief period acquires the ability to produce and understand indefinitely many new and well-formed ones. Chomsky says this can only be explained by saying that the child 'internalises' a body of rules that gives it this ability.

The second reason he gives for mentalism is that a speaker may not initially understand a sentence, or may not recognize an ambiguity,but may be coaxed into doing so without being given fresh information. Chomsky says that we can only explain this by saying that the speaker has an imperfect access to an internalized grammar that assigns these readings to the sentence. "Few hearers," he says "may be aware of the fact that their internalized grammar in fact provides at least three structural descriptions for ['I had a book stolen']." (1965: 21–22)

Now, these abilities may show that the child/speaker has implicitly grasped the grammar of the language, or at least that her implicit knowledge is not accurately reflected in her performance, but it does not follow that she has *internalized* the grammar, any more than the fact that I have grasped the rules of Two-handed Five Hundred means that I have internalized the rules of Two-handed Five Hundred. I regularly forget the rules of Two-handed Five Hundred and have to look them up. It is not that the rules sometimes exist in my head and sometimes do not. It is that sometimes I know them and sometimes I do not. To say that the user 'internalizes' the grammar is to say that the grammar is a mental entity. In the same way the rules of Two-handed Five Hundred would be mental entities, which clearly they are not. (Neither the grammar nor the rules of the game can be destroyed by destroying everyone who knows them, yet a strong mentalist is committed to saying that they can!)

Chomsky's mistake is to confuse object-knowledge with knowing, or state-knowledge. He first identifies the rules of a grammar with an (idealized) content-knowledge of them. Then he confuses this knowledge (now identified with the grammar) with knowing or state-knowledge, thereby locating the grammar in the head.

Here it is in more detail. Chomsky distinguishes between competence and performance. Performance is "the actual use of language in concrete situations" (1965: 4). This is marred by "memory limitations, distractions, shifts of attention and interest, and errors," which are "grammatically irrelevant." What matters is the underlying *competence*, which Chomsky takes to be "the speaker-hearer's knowledge of his language." (*ibid.*) Thus, he says, "a generative grammar attempts to specify what the speaker actually knows" (1965: 8), and the linguist is concerned with "the theoretical (that is, grammatical) investigation of the knowledge of the native speaker." (1965: 19)

Here there is no attempt to distinguish between the content-knowledge of an object and the object of knowledge. In this case, where we are dealing with an abstract object (a language or its grammar), this may be harmless. But Chomsky also identifies the content-knowledge with state-knowledge, thereby locating the object of knowledge in the head. 'Competence' now means both an inner state underlying behavior, and an inner *object* (the grammar) that is in the head: "we must regard linguistic competence — knowledge of a language — as an abstract system underlying behavior…" Such a system is a generative grammar (1968: 62).

This reverse psychologistic reading of Chomsky accords with a more general view of his work, which is linguistic rather than psychological. *Aspects of the Theory of Syntax*, the main work of this period, begins with the well known 'Methodological Preliminaries' chapter in which he says that linguistics is a subfield of psychology, but the rest of the book is about syntactic theory, deep structures and grammatical transformations. In other words, standard linguistic stuff.

It is interesting to contrast Chomsky with Mill. Mill at least talked about psychological states and was prepared to pay the price for making logic part of psychology — the price of saying that logical truths are contingent. In contrast, Chomsky's methodology for doing linguistics is not psychological at all, and his objects of study are not mental entities (cf. Mill's "mutually destructive psychological states"). He is caught between two worlds, on the one hand saying that linguistics is part of psychology, which commits him to psychologism, and on the other trying to find out about the mind by doing linguistics, which is reverse psychologism. (He would probably reject my analysis and deny the distinction between linguistics and psychology. See especially (1988: 6). But that would deny the distinction between what is psychological and what is not — and we have seen that there is a clear distinction between these things.)

If any doubt remains about Chomsky's approach, consider his use of idealization. He maintains that linguistics is concerned with the competence of an Ideal Speaker in a completely homogeneous speech community, and that it is only under this idealization that performance is a direct reflection of competence and that the subject-matter of linguistics (the internalized rule-set) is available to us.[8] But we have seen that the concept of an Ideal Knower is a redundant device that falls away under analysis. To say that we should study the rules internalized by an Ideal Knower is like saying that we should study the universe as God sees it. It amounts to saying that we should study the rules themselves, the rules as they really are. When we realize this, the illusion that we can do linguistics and psychology *at the same time* falls away. Saying that we should study the Ideal Speaker apparently enables us to:

- study what is in the head (in a manner that abstracts away from trivial deviations and that overcomes the problems that what is in the head is hard to access and varies from person to person), *and*
- study the properties of the language or grammar itself.

It apparently enables us to study mind by studying language, to do *a priori* psychology. Chomsky says that a speaker may *deny* things that she tacitly knows to be true: the internalized grammar may assign readings to sentences which the speaker *denies* can be assigned to them (see especially 1965: 8, 21–22). This bears on the disanalogy between an ideal gas and an Ideal Thinker. The concept of an ideal gas allows us to accommodate minor deviations and trivial errors, but it is quite another thing for a speaker to *positively deny* claims that are made about her logical or linguistic intuitions. Idealization gets its respectability from its use in the established sciences, but in fact there is no precedent for idealization such as this.[9]

Chomsky's mentalism — his belief in the psychological reality of grammars — established a rich tradition. In her Introduction to *The Mental Representation of Grammatical Relations* (1982) Joan Bresnan says:

> A longstanding hope of research in theoretical linguistics has been that linguistic characterizations of formal grammar would shed light on the speaker's mental representation of language. One of the best-known expressions of this hope is Chomsky's *competence hypothesis*: '... a reasonable model of language use will incorporate, as a basic component, the generative grammar that expresses the speaker-hearer's knowledge of the language ...' (Chomsky 1965: 9)

But, she says, despite many expressions of hope by linguists, and despite intensive efforts by psycholinguists, generative grammars have not yet been successfully incorporated in psychologically realistic models of language use.

Her response to this is to say that psycholinguists have been trying to incorporate the wrong sort of grammar, and she proposes a grammar that she believes can be incorporated: Lexical-Functional Grammar (LFG). As Marilyn Ford (1988) puts it, Bresnan, Kaplan and Ford developed "A psycholinguistic theory that incorporates LFG as the competence grammar used to construct representations in sentence perception…" Ford concludes "The psycholinguistic theory thus takes Noam Chomsky's (1965) *competence hypothesis* seriously."

I will not examine the obvious reverse psychologism of this approach. Instead I will move slightly further afield and look at an attempt to apply Chomsky's theory to the study of human reasoning.

3.2. *Macnamara's Theory of Logical Competence*

In *A Border Dispute: The place of logic in psychology* (1986) John Macnamara attempts to apply Chomsky's notion of competence to the study of human reasoning. He maintains that "every argument that Chomsky adduces for the relevance of linguistics to the psychology of language applies, *mutatis mutandis*, to the relevance of logic for psychology" (p. 30). His thesis is that "logical competence is the key element in the psychology of human reasoning" (*ibid.*). Here are his main claims:

> A logic that is true to intuition in a certain area constitutes a competence theory for the corresponding area of cognitive psychology… The mind in part of its functioning applies the principles of that logic. It is this that entitles us to say that to each ideal logic (true to intuition) there corresponds a mental logic. By mental logic I mean (1) linguistic resources in the mind sufficient to express propositions, (2) the ability to understand sentences formed with those linguistic resources, and (3) the ability to grasp inferences among such sentences. (p. 22)

Other salient features of his theory are that logical competence is an abstraction from error and an idealization concerned with the competence of an Ideal Reasoner. Individual differences are treated as "differences in the ability to work with a common, uniform logic or sets of logics." (p. 28).

Now, Macnamara agrees with Katz that Chomsky's position with respect to linguistics is "squarely psychologistic" (p. 27), and he is anxious to avoid psychologism himself. He characterizes psychologism in the same way that I have. It is the belief that some discipline (that is not in itself psychology) is a subfield of psychology, and he characterizes psychologism with respect to logic as "the doctrine that logic is a study of the mind" (p. 10). In order to avoid psychologism he examines the arguments against it and concludes, correctly, that it violates the act/content distinction. His account of Husserl's act/content distinction is so lucid that I quoted it in my account of the distinction earlier in this paper.

However, having drawn the distinction, Macnamara violates it by committing *reverse* psychologism. Rather than saying that we can do logic by studying the mind he says that we can find out about the mind by studying logic — rather than finding out what the mind knows by studying mental states we can find out about mental states by studying what the mind knows: "logical studies supply a characterization of psychological competence. For this reason, the work of logicians is essential to psychologists." (p. 33). These claims violate the act/content distinction just as much as psychologistic ones. Macnamara is so intent on avoiding psychologism that he does not realize that it is possible to avoid it and still violate the act/content distinction.[10] He says:

> we have agreed to accept the main conclusion of the psychologism debate, that is, that logic is not a branch of psychology receiving its basic principles from psychology ... we must discover a theory that brings logic and psychology into a relationship close enough to explain how we can know the laws of logic and employ them in our reasoning without our attaining them by generalizing over empirical data. (p. 22)

But the belief that logic is a branch of psychology and that the laws of logic are empirical generalizations is a *consequence* of the act/content confusion, and avoiding the belief will not automatically avoid the confusion underlying it. Macnamara falls into the trap by saying that the study of logic tells us about the mind, which is a claim that is the mirror image of the position that he is trying to avoid.

Diagnosing the confusion in more detail is largely a rerun of our account of Chomsky. As with Chomsky, the key concept is 'competence.' If by 'competence' Macnamara means 'what is known,' that is, the object of knowledge, then he is saying that the logic known by an Ideal Reasoner is an object of psychological enquiry. But 'the logic known by an Ideal Reasoner' is a redundant way of talking about the logic itself,[11] and there is no reason to believe that the study of logic tells us anything about human psychology — about states, acts, processes, abilities, etc.

If on the other hand 'competence' means 'a state of knowing' then Macnamara is saying that the knowledge-states of the Ideal Reasoner with respect to logic(s) are objects of psychological enquiry. Now *this* appears to be a psychological claim, but is not a claim about *mental logics*. It is a claim about perfect state-knowledge of logics, and state-knowledge of a logic is not itself a logic. It seems that when Macnamara says that "to each ideal logic (true to intuition) there corresponds a mental logic" he means that there is a corresponding *knowledge* of the logic. Like Chomsky, he appears to believe that to know a logic is to have it instantiated in the head. He says that because we can reason validly and have intuitions about logical validity "thus the foundations of the logic(s) at which the logicians aim, the ideal logic(s), *must be psychologically real in the sense of being instantiated in some form in the mind*." (p. 31, my emphasis.) We have already considered the case of Two-handed Five Hundred. A more graphic analogy is that I do not have an internalized Eiffel Tower when I know about the Eiffel Tower (or to use Descartes' example, I do not have an internalized chiliagon — a plane figure with a thousand angles — when I know about chiliagons).

The upshot is that Macnamara's thesis founders on the horns of the act/content dilemma. If by "the competence of the Ideal Reasoner" he means 'what the Ideal Reasoner knows about logic' then he is talking about logic but not about mind. If he means 'the psychological states of the Ideal Reasoner' then he is talking about mind, but not about logic.[12]

3.3. *AI and Cognitive Science*

The act/content confusion is endemic in classical, symbolic AI and in much of cognitive science. The Physical Symbol System Hypothesis that underlies and drives classical AI says that a necessary and sufficient condition for a system to be intelligent is that it is a physical symbol system (Newell and Simon 1976). The equally important Knowledge Representation Hypothesis (Smith 1985) says that a system knows that p if and only if it contains a symbol structure that means p to us and that causes the system to behave in appropriate ways. In keeping with these beliefs, knowledge engineers put knowledge structures and belief structures (frames, semantic networks, production systems, sentences, logic) into Belief Bins and Knowledge Bins, in the belief that this will give the systems knowledge and belief. It does give them knowledge and belief, of course, but only in the sense of contents or objects, not in the sense of *states*. Books contain contents or objects of knowledge and belief, but they do not know or believe anything. An analogy would be that we want to build a system that *desires*, and put objects of desire, such as Brenda, or a Porsche, or a bottle of Nuits St.-Georges (1990) into its Desire Bin.

There is a similar confusion in cognitive science. Fodor's Language of Thought Hypothesis was the only game in town in cognitive science for a long time, and it characterizes cognition as the manipulation of symbols in an innate, inner language that Fodor calls "mentalese" (see especially Fodor 1975).

I will look at reverse psychologism in cognitive science by looking at the diagnostic modelling of subtraction skills and at some recent work in qualitative physics. I will then look at the Knowledge Representation Hypothesis.

3.3.1. *The diagnostic modelling of subtraction skills*

Diagnostic modelling is a method used in the construction of Intelligent Tutoring Systems. 'Overlay' or 'differential' models represent the student's knowledge as a subset of the knowledge of a hypothetical Expert, so that the student is depicted as thinking in the same way as the Expert, but as knowing less. 'Diagnostic' models recognize that the student may think differently to the Expert and have misconceptions rather than a mere lack of knowledge. Diagnostic models of subtraction skills (e.g. Young and O'Shea 1981) construct a model of what a hypothetical Expert knows and then perturb it in the hope that this will generate characteristic human errors. The model of the Expert performs atomic tasks such as 'compare,' 'borrow,' 'pay back,' and 'add 10,' and the program provides a running report of the subskills it is performing. This is seen as 'looking in the mind of the Expert.' Characteristic errors can be generated by perturbing the program. For example, children often fail to borrow when the top digit is lower than the bottom one. Instead they subtract the lesser number from the greater. According to the theory, they are running a procedure from which 'borrow,' 'pay back' and 'add 10' have been omitted. The Expert's program can be modified to do the same thing.

Such modelling attempts to provide cognitive models by modelling our manipulation of a public, communicable symbolism that embodies the *content* of cognition. When we watch one of these programs running we see the manipulation of numbers according to rules: the units in the unit column are compared, 10 is borrowed from the 10s column and added to the top number in the units column, and so on. This is reverse psychologism. It models the content or object of a psychological process, not the process itself.

Of course, such models might genuinely model the way in which *numbers are manipulated by us*. Some people do subtraction by decrementing the top number in the 10s column after they have borrowed. Others 'pay back' by adding to the bottom number. Some people 'think in blocks' (to subtract 378 from 432, subtract 378 from 400, subtract 400 from 432, and add the results together). These are differences that a number-manipulating model can capture. But the models capture the content of cognitive processes, not the processes themselves.

Another way of looking at this is to say that the notion of a cognitive model is ambiguous between a *model of the student* and a *model used by the student*. A model used by the student embodies such things as her perception of the problem and her perception of how symbol-structures can be manipulated according to rules and procedures to solve the problem. The model outlined above is a model in this sense. We can say of it as it runs 'this is the way in which the student believes that symbols should be manipulated to get the solution.' Properly speaking it is our model of the model used by the student. It is not a *model of the student herself*, of her acts, states or processes.

The concept of the Expert plays its usual role in compounding the confusion. We are told that the Expert is an *Ideal Knower* (Miller 1982; Burton 1982) so we would expect it to be a device for importing 'out there' structures into the head. And this is exactly what we find. The knowledge of the Expert "merely provides a computational machine that performs the skill and is of no particular interest" (Burton and Brown 1978). It "is not meant to be a cognitive construct, but simply a framework for relevant pieces of information" (Burton 1982). Yet we are told that the misconceptions of the skill are represented in a network that is *psychologically real* (Burton and Brown, 1978). Young and O'Shea (1981) call the claims to psychological reality "strong claims." Thus *cognitive structures* (states, processes etc.) ostensibly emerge as perturbations of a perfect, 'out there,' body of rules! What is going on here, of course, is that the Expert's knowledge is just *the rules and procedures themselves*. There is nothing psychological about it. The student model is an impoverished or deviant version of these rules and procedures, and there is nothing psychological about it either.

3.3.2. *Qualitative physics.*

This example can be seen as a sequel to the previous section, since some of the work done in qualitative physics developed out of an interest in building mental models for intelligent tutoring systems. De Kleer describes qualitative physics as follows:

> Qualitative physics, like conventional physics, provides an account of behavior in the physical world. The vision of qualitative physics, in conjunction with conventional physics, is to provide a much broader formal account of behavior, an account rich enough to enable intelligent systems to reason effectively about it. However, unlike conventional physics, qualitative physics predicts and explains behavior in qualitative terms. (1987: 807)

He says that the reason for studying qualitative physics is that "one wants to identify the core knowledge that underlies physical intuition." He says that there are two approaches to qualitative physics, the physical and the psychological. "The physical approach seeks physical theories that accurately predict behaviors. Although observations of human behavior are used as hints, there is no concern for modelling human foibles. This approach seeks to understand the common sense of an ideal expert." In contrast, "the psychological approach seeks physical theories that can conform to observed human behavior. This approach seeks to model the behavior of experts as well as neophytes." (Naive physics is concerned with neophytes and non-experts.)

De Kleer's 1987 account plays down the psychological approach. His account of it makes no mention of psychological modelling, but says that the approach seeks models that conform to observed human *behavior*. He then concentrates on the physical approach. Now this is curious. His earlier work with Brown (1981, 1982, 1983) was concerned with the construction of mental models for intelligent tutoring systems. These were meant to be models of *mental processes*. As Wenger (1987) said, "The strength of de Kleer and Brown's approach … resides in its attempt at formalizing the representations and processes involved in constructing and using mental models …" (p. 75) But as he goes on to say "Recently the research has taken on a more general and ambitious character; de Kleer and Brown (1984; Brown and de Kleer 1984) have set out to devise a 'framework for a qualitative physics' as an alternative to the usual mathematically based physics."

The research did not just take on a 'more general and ambitious character.' It has totally changed its focus. What began as an attempt to model psychological states and processes become an attempt to model the world.

This change of focus is consistent with an ambivalence in the original research, between modelling states and processes on the one hand and modelling content on the other. Figures 1–3 provide an example. Figure 1 is supposed to represent phases in the process of constructing a causal mental model or 'envisionment' of a device and mentally running it. Figure 2 represents a device (a buzzer) and Figure 3 is supposed to represent the final model.[13] Now, Figure 1 refers to psychological processes, such as envisioning and inferring, but Figure 3 does not. It depicts a causal model of the device. So 'mental model' is again ambiguous between a *model of the student* and a *model used by the student*. Figure 1 is a (very incomplete) model of the student's psychological processes. Figure 3 depicts a model that may or may not be used by the student. It is not, however, a model *of the student*.

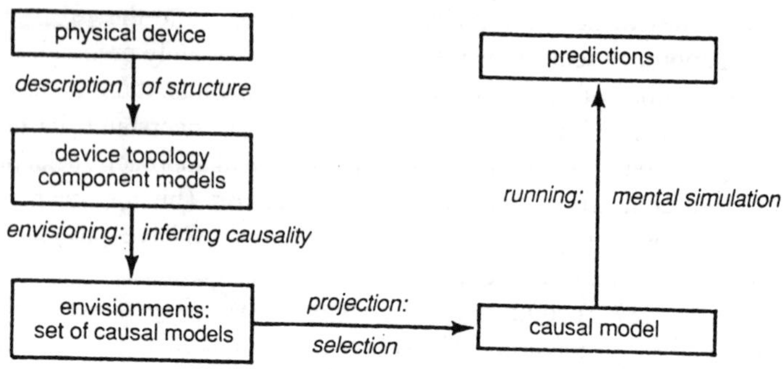

Figure 1.

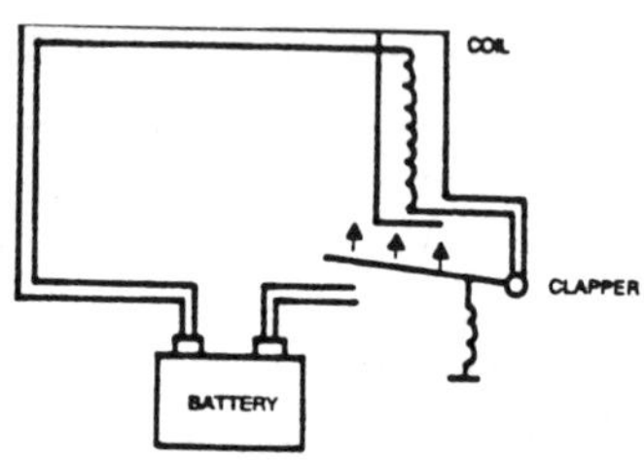

Figure 2.

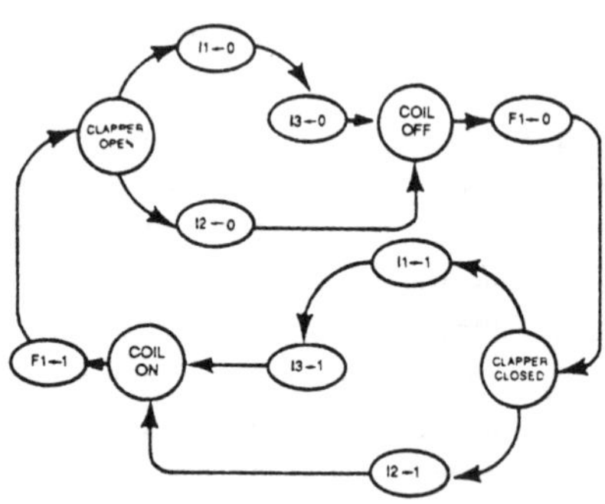

Figure 3.

Now recall how Chomsky and Macnamara trade on the ambiguity of 'knowledge' to locate content structures in the head (grammars and logics, respectively). Qualitative physics does the same thing. De Kleer says that it attempts "to identify the core knowledge that underlies physical intuition." It "concerns knowledge about the physical world." Such terminology is common in the field. In the first and second Naive Physics Manifestos Hayes says that naive physics involves a formalization of our knowledge of the everyday physical world (see especially 1985: 1; also 1978). Hobbs' opening sentence in his Introduction to *Formal Theories of the Commonsense World* (1985) is "We are capable of intelligent action, in part because we *know* a great deal. If intelligent programs, or robots, are to be constructed, they too must have a great deal of knowledge" (my emphasis). He asks what an intelligent agent *needs to know*, and indicates that knowing is "encoding knowledge," and that expert systems know.

Here we have the same equivocation on 'knowledge' that we find in Chomsky and Macnamara. *It is assumed that to be in a state of knowing is to have content-knowledge in the head.* Chomsky talks about an internalized grammar, Macnamara about an internalized logic, and qualitative physics about an internalized model of the world. This is reverse psychologism — at least if it is claimed that we can find out about the mind by examining such models.

The confusion is compounded by talking about the knowledge of the Ideal Expert. De Kleer says that the physical approach "seeks to understand the common sense of an ideal expert" (1987: 808). But to say that we should understand something as the expert sees it is to say that we should understand it as it really is. Talk about the knowledge of the Ideal Expert fosters the illusion that we are studying mind.

3.3.3. *The knowledge representation hypothesis*

Nowhere is the confusion between act and content, or state and content, clearer than in Knowledge Representation. Here again the assumption is that knowing is content-knowledge in the head. Brian Cantwell Smith formulates the "Knowledge Representation Hypothesis" (KRH) as follows:

> It is widely held in computational circles that any process capable of reasoning intelligently about the world must consist in part of a field of structures, of a roughly linguistic sort, which in some fashion represent whatever knowledge and beliefs the process may be said to possess. For example, according to this view, since I know that the sun sets each evening, my 'mind' must contain (among other things) a language-like or symbolic structure that represents this fact, inscribed in some kind of internal code. (1985: 33)

Additionally, the syntax or morphology (Smith calls it the 'spelling') of this internalized symbolic structure is presumed to play a causal role in the generation of intelligent behavior. This gives us the full statement of the KRH:

> Any mechanically embodied intelligent process will be comprised of structural ingredients that a) we as external observers naturally take to represent a propositional account of the knowledge that the overall process exhibits, and b) independent of such external semantical attribution, play a formal but causal and essential role in engendering the behavior that manifests that knowledge.

Thus a system knows that p if and only if it contains a symbol structure that means p to us and that causes the system to behave in appropriate ways. It knows, for instance, that tigers bite if and only if it contains a structure such as 'Tigers bite' that causes it to climb trees in the presence of tigers. Smith goes on to distinguish between a weak and a strong version of the KRH, and he is skeptical of both of them. Throughout his paper he emphasizes the need to analyze and clarify concepts and issues in the knowledge representation problem. What follows is an attempt to contribute to the task that he started.[14]

The story is a familiar one. The KRH does not distinguish between knowing (as a state) and knowledge (as the content or object of a state). To know something is to have an internalized data structure. To have mental states is to have "a set of formal representations" (p. 37). This is the now familiar move of treating knowing as knowledge in the head.

Now, Smith reports that the AI community is divided in its opinion of what these internalized structures stand for. He expresses surprise at the outcome of a survey conducted by Brachman and Smith (1980) which shows that most respondents believe that the structures represent *the world, or situations*, rather than facts or propositions about the world or about situations.[15]

I suggest that this confusion arises from characterizing knowing as internalized content-knowledge. Inasmuch as internalized knowledge is seen as a state or structure it can be a model and can represent a state of affairs or an entity. Like any model it would have no meaning in itself, but would be given one by accompanying declarative knowledge that mapped it onto a state of affairs or an object. If on the other hand it is seen as something like an inner sentence, that is, like something that already has a meaning, then it would not be a model, but it *would* express a proposition. If we run these readings together we will end up saying that the symbol system *represents a proposition*. I do not know what "representing a proposition" means, and I can only think that it is a confusion between an uninterpreted state or structure that can be used as a model to represent something, and something akin to a sentence, that expresses a proposition. There is, after all, no reason to believe that sentences *represent* anything. The early Wittgenstein worked such a notion hard with his picture theory of meaning, and ultimately it failed. Equally, there are no grounds for believing that *anything at all can represent* a proposition — though some things (such as sentences) can express them. This appears to be yet another case of trying to identify a state with the content or object of the state.

This becomes clearer when we look at the second clause of the KRH — that the internalized symbol structure plays a causal role in generating intelligent behavior. The KRH tries to have it both ways: the symbol structures associated with knowledge are at once meaningful to us and causally efficacious for the system. This is the standard state/content confusion: states, but not contents, are causally efficacious; contents, but not states, are meaningful. The state of knowing that tigers bite might cause me to exhibit intelligent behavior in the presence of tigers (such as climbing a tree), or to say intelligent things about them in their absence (such as 'When in their presence, get up a tree').[16] But the *content* of my knowledge is not causally efficacious: the proposition expressed by the sentence 'Tigers bite' cannot cause anything.

Now, this argument cheats a little, because the KRH does not say that knowledge structures cause intelligent behavior by themselves. It says that they play a *role* in the causal process: intelligent behavior is caused by a combination of knowledge structures *and the procedures that act upon them.*

We need a clear perspective here. AI has discovered that intelligence requires knowledge: for a system to be intelligent it must know a great deal. But AI does not distinguish between knowing and knowledge, and it assumes that a system knows if it contains a representation of the content of knowledge and if it has procedures that can act upon that representation, such that, together, these produce intelligent behavior.

We have had ways of *representing what is known* for a long time. First there was speech, then there was writing, then there were libraries, now there are databases. Books, libraries and databases have knowledge represented within them. But no-one believes that they know and are intelligent. The KRH proposes something like a fancy book that responds to input on the basis of the causal efficacy of internal structures that express the content of knowledge — structures that are meaningful to us but not to the machine. There is no reason to believe that such a fancy book would *know* anything.

The ambiguity of 'knowledge' carries over into the ambiguity of 'knowledge structure.' This is ambiguous between 'cognitive structure' and 'data structure.' Knowledge structures are commonly regarded as data structures, possibly accompanied by search algorithms. Feigenbaum says, "A representation of knowledge is a combination of data structures and interpretive procedures …" (1981: 143). Elaine Rich says, "we will discuss a variety of knowledge structures. Each of them is a data structure in which knowledge about particular problem domains can be stored." (1983: 203) Tore Amble says, "A picture of tomorrow's computer vocabulary can be imagined, if all the words containing 'data' or 'information' are replaced by the word 'knowledge.' " (1987: 11) Once we have replaced 'knowledge' by 'data' it is easy to regard cognitive structures as data structures in the head.

We need to distinguish between two questions. The first is 'How can we construct machines *that know*?' This is not a technological question. It is a philosophical question. Traditional epistemology asks 'Under what conditions does agent A know that p?' The standard answer, which philosophers have never been entirely happy with, is 'A knows that p if A believes that p, p is true, and A has grounds for believing that p': knowledge is justified true belief.[17] 'How can we construct machines that know?' amounts to 'Under what conditions does machine M know that p?' This is a question in what we might call 'machine epistemology.'

Knowledge Representation asks another question: 'How should we represent knowledge in machines?' This is a practical, engineering question that assumes an answer to the first question. I have argued that that answer is wrong.

4. Envoi: Illuminating the Chinese Room

I will finish this paper by revisiting John Searle's infamous Chinese Room Argument (Searle 1980, 1990). This argument aims to show that symbol manipulation cannot cause cognition. I will outline the argument and argue that, although its conclusion is correct, the argument is unilluminating, for it does not tell us *why* there is no cognition in the Chinese Room. I will argue that the state/content distinction shows us why there is no cognition, and I will also argue that it shows why Searle's parallel attack on connectionism in his Chinese Gym Argument fails.

Here is an outline of the argument (which everyone knows!). A monolingual English speaker is in a room which has input and output slots. Chinese characters are passed through the input slot. The speaker manipulates them by consulting a rulebook that is written in English, and passes the results through the output slot. The rulebook is so accurate that from the point of view of a Chinese speaker outside the room, the room appears to understand Chinese. But, says Searle, the room understands nothing, because the person in the room does not understand Chinese, nor does anything else in the room, nor do the room and its contents as a whole.

Now, if manipulating symbols according to formal rules cannot yield understanding, a digital computer cannot understand, for all that it does is to manipulate symbols according to formal rules.

That is the main result: digital computers cannot understand, or have any other cognitive states, because formal symbol manipulation is not sufficient to produce cognitive states.

Searle claims another result: *brains cause minds*. The brain 'causes mental events by virtue of specific neurobiological processes' (Searle 1990: 23).[18] Brains are made of the right stuff and can cause cognition, but digital computers are not, and cannot. Searle is therefore committed to saying that connectionist machines cannot cause cognition either, for they are made of the same stuff as digital computers.

Ten years after the publication of the Chinese Room he provided an argument to this effect — the Chinese Gym Argument (Searle 1990).

This is a variant of the Chinese Room. Multitudes of monolingual English speakers in a vast gymnasium carry out the same operations as the nodes of a connectionist system. The gymnasium appears to understand Chinese, but Searle says that no-one in the gym understands Chinese, nor does anything else in the gym, nor do the gym and its contents as a whole.

Searle's argument is accompanied by a Commentary by the Churchlands. They argue that it is irrelevant that no unit in the system understands Chinese, for the same is true of nervous systems: no neuron understands English, but *brains* do. They also say that it would require the population of over 10,000 earths to simulate the brain in such a way, but that if we were to simulate it in such a way, there are no *a priori* grounds for believing that the simulation would not think: the question is an empirical one, and the default assumption is that the system would think.

My sympathies are with the Churchlands on the Chinese Gym, but with Searle on the Chinese Room. The Chinese Room, however, needs to be unilluminated: some people are convinced by it, others are infuriated, but no-one has provided a satisfactory explanation for why there is no cognition in the Chinese Room. The debate has degenerated into an exchange of analogies, which is where it came in, for the Chinese Room is itself an argument from analogy.

The state/content distinction provides the explanation. There is no cognition in the Chinese Room because we cannot generate cognitive states by manipulating the symbols that express the content of those states. This explains how, in Searle's words, Classical AI 'got into this mess' in the first place. (Searle 1990: 25)

Commenting on a version of this paper, Brian Smith suggests that I focus on the content/cognition (act/content, state/content) distinction, but ignore the distinction between content and the symbols that express it. I plead guilty. I ignored the distinction to prevent the paper from becoming more complicated than it already was. Smith goes on to say: "AI and cognitive science in fact rest on an identification ... between symbols and cognition, *not* ... between content and cognition."

Here I disagree. *That* is what the Chinese Room *apparently* shows us, but after fifteen years of debate we have been unable to make this explanation stick: we have been unable to show why the symbol-handling in the Chinese Room does not generate cognition. The point is this. We are fascinated by symbols *because they have content*, and we confuse content with cognition. Reverse psychologism believes we can study cognition by studying symbolism, *because the symbols express the content that it confuses with cognition.* AI believes that we can generate cognition by internalizing symbols, for exactly the same reason.

That is the nub of it, though more can be said. For one thing, the psychological/non-psychological distinction is deeper than the symbols/content distinction, for sometimes the psychological/non-psychological distinction doesn't involve symbols at all. This is the case with the distinction between state and object, as we have seen: the distinction between my love as a state and my love as an object has nothing to do with symbols. And strictly speaking we don't need symbols even in the case of content. Adrian Cussins (1990) distinguishes between what he calls 'conceptual and non-conceptual content.' When we say that Jo believes that Fred is a bachelor, we attribute the concept 'bachelor' to Jo. But when we say that Fido thinks that the sound came from the south, we do not attribute the concept 'south' to Fido. We can talk about the content of Fido's thought without attributing the concept to him, let alone the symbols that express that concept. These, however, are difficult issues, and it is easier to talk about symbols, but to accept that we do so because they carry content. It is the content that matters. We think that content is cognitive and we *bring its vehicle along for the ride*.

What about the Chinese Gym Argument? This doesn't work, because connectionism does not address itself to computation on the symbolic level and consequently does not try to generate cognitive states by internalizing the symbols that express their content. As the Churchlands say, it is immaterial that individual units don't understand, because this is true of individual neurones, and the argument provides us with no *a priori* grounds for believing that a full-size gym wouldn't think.

The failure of strong AI is therefore a conceptual failure to do with the act/content distinction, and not the result of silicon being 'the wrong stuff.'

Notes

1. Charniak and McDermott cite this passage with approval in their *Introduction to Artificial Intelligence* (1982). They preface it with:

   "If we want more scientifically respectable ancestors [than Frankenstein! — TD], we may press into service the many logicians and philosophers who have labored to formalize, and ultimately to mechanize, the 'laws of thought.' One of the classic names here is that of George Boole.

2. There are numerous mathematical conundrums where the majority of people will reject the answer, even though it follows deductively from the premises. If, for instance, there are more trees in the forest than there are leaves on any particular tree then if there are no barren trees there are at least two trees in the forest that have the same number of leaves. My experience is that most people will reject this result, and are unhappy about it even when they are shown a proof. They are even less happy with the following. A piece of string is tied around a billiard ball. The string is cut, a meter is added to its length and the string stands out evenly around the billiard ball. Let the distance from the edge of the ball to the string be N. Now imagine the same thing with a very long piece of string tied around the earth's equator, which has been smoothed down with sandpaper. What is the value of N now? The answer is that it is exactly the same as it is in the case of the billiard ball. The size of the sphere makes no difference, since the answer is one meter divided by 2 π, which is about 6 inches.

3. It is not a verb of propositional attitude in cases such as 'A knows how to do such and such,' and 'A knows Lloyd George.'

4. See especially his (1972). He maintains that objective knowledge resides in a Third World akin to Plato's World of Forms (though the contents of the Third World are the products of human contrivance, whereas the contents of the World of Forms are not).

5. The last two sections (on the state/object confusion and spurious idealization) leave many questions unanswered. Most obviously, they make no attempt to explain how (perfect) content-knowledge of an object can be the object. Strictly speaking, I do not need to provide an explanation since I only claim to be reporting usage. Nevertheless it would be nice to see the broader picture. Anything like an adequate treatment would need to take account of what Evans (1973, 1980) calls "Russell's Principle" (that it is not possible to make a judgement about an object without knowing what object you are making a judgement about) and Millikan's criticisms of it (1984, 1986). Dennett (1987) observes that the attempt to preserve the principle led Russell to his doctrine of knowledge by acquaintance, which says that for the mind to make judgements about abstract objects it must be acquainted with them. Dennett considers this principle to be "pernicious." An adequate treatment would also need to cover recent work on the de re/de dicto ('knowledge of/knowledge that') distinction originating in the work of Quine. Perfect-knowing has curious logical properties. 'a' occurs transparently in 'A perfectly-knows that Fa' on any criterion of transparency. It passes the quantifying-in test as follows. Let Oedipus be an Ideal Knower. He wants to marry Jocasta, and therefore knows that he wants to marry his mother, since he knows everything about

the domain, including the fact that Jocasta is his mother. Since we can substitute 'his mother' for 'Jocasta' *salva veritate*, these terms occur transparently. We might take this as evidence that we should reformulate the use of quantifying-in as a criterion of transparency by adding the rider that we should assume that A does not know that the terms co-designate. Even on such a reformulation, however, objects of perfect-knowledge would behave like non-intentional, fully relational, objects. I am inclined to think that in the case of perfect-knowledge we witness a collapse of the de re/de dicto distinction and that (and this is a stronger claim) having perfect propositional knowledge of an abstract object 'o' is logically indistinguishable from knowing 'o' by acquaintance. Having perfect propositional knowledge of the rules of Scrabble, for instance, is to be fully acquainted with the rules of Scrabble.

6. Chomsky's Standard Classical period is roughly from *Aspects of the Theory of Syntax* (1965) to *Language and Mind (extended version, 1972).*
7. For strong mentalism see 1965, p. 8. The phrase 'internalized grammar' occurs constantly during this period. We are told that the innate Language Acquisiton Device *actually becomes* the grammar (1965: 33; 1968: 64, 67; 1970: 450, 467). For weak mentalism see 1965: 25; 1968: 81; 1969: 263. For evidence that he does not distinguish between them, see especially 1965: 25.
8. The most commonly cited reference is 1965: 3, but see also 1980, 1984.
9. It seems that with spurious idealization the distinction between weak and strong mentalism collapses, for the Ideal Speaker's representation of a grammar is a perfect copy of the grammar and therefore indistinguishable from it. This collapses the distinction between a grammar and its representation, and with it the distinction between weak and strong mentalism.
10. This would explain Garfield's observation that Macnamara does not seem to appreciate the force of some of the standard objections to psychologism (Garfield 1988).
11. I should make it clear that I am not denying that psychology can use the notion of an Ideal Thinker, if this is an empirical idealization about cognitive states. Such a notion would be an abstraction from human error and an idealization inasmuch as the psychologist would be prepared to accept minor deviations in her observations of actual behavior on the grounds that the notion provides explanatory power for the cases that do fit.
12. Garfield comes to a not dissimilar conclusion and says that there is "a real equivocation on 'logical competence' " (1988: 316). However, he sees this as an equivocation between an empirical and an ideal sense of the term:

    > "If by [logical competence] Macnamara intends an empirically discoverable basic set of mental capacities [then] ... there is no reason to believe that logic can tell us what these structures are ... On the other hand, if by 'logical competence' Macnamara intends a sound logical system meant to characterize an ideal reasoner, then there is no real reason to think that such a system will play any significant role in empirical psychology.

13. Figure 1 is from Wenger (1987), p. 71. Figures 2 and 3 are from Wenger, p. 72, adapted from de Kleer and Brown (1983).

14. See his excellent (1991) for an account of the reasons for his skepticism.
15. There is a striking similarity between this situation and the dilemma facing advocates of the Correspondence Theory of Truth. The latter say that the main sorts of things that are true or false are sentences or propositions, and that these are true if and only if they *correspond to the facts*. But what are 'the facts'? There are two accounts. One says that facts are 'what are expressed by sentences or propositions.' This is circular, since it is now being claimed that a sentence or proposition is true if and only if it corresponds to what it expresses. The other account avoids this circularity by saying that facts are not linguistic entities but are 'states of affairs' (sometimes called 'Sachverhalten,' after Wittgenstein's use of the term in the *Tractatus*). This leads to a bloated ontology, for now we have to talk about not only states of affairs, but negative states of affairs, states of affairs inside other states of affairs, hypothetical states of affairs, and so on. (For a brief but clear account of these problems see Richards (1975), pp. 142–143.)
    The KRH faces a similar dilemma: 'Do knowledge structures represent propositions or meanings, or do they represent states of affairs?' (cf. 'Do sentences correspond to propositions or meanings, or do they correspond to states of affairs?')
16. Do tigers climb trees?
17. See Gettier (1963) for the classic list of counterexamples to this claim.
18. 'Brains cause minds' is Searle's Axiom 4.

References

Amble, T. 1987. *Logic Programming and Knowledge Engineering*. Wokingham, England: Addison-Wesley.

Beneke, F.E. 1833. *Die Philosophie in ihrem Verhaltnis zur Erfahrung, zur Spekulation, und zum Leben*. Berlin: Verlag Dokumentation.

Boole, G. 1854. *An Investigation of the Laws of Thought on Which are Founded the Mathematical Theories of Logic and Probabilities*. London. Reprinted 1954, New York: Dover Publications.

Bresnan, J. 1982. *The Mental Representation of Grammatical Relations*. Cambridge, MA: MIT Press.

Brown, J. and de Kleer, J. 1984. A framework for qualitative physics. *Proceedings of the Sixth Cognitive Science Society Conference*, 11–17. Boulder, Colorado. Available from Lawrence Erlbaum, Hillsdale, NJ.

Burton, R. 1982. Diagnosing bugs in a simple procedural skill. In *Intelligent Tutoring Systems,* D. Sleeman and J. Brown (eds). London: Academic Press.

Burton, R, and Brown, J. 1978. Diagnostic models for procedural bugs in basic mathematical skills. *Cognitive Science* 2, 155–192.

Charniak, E. and McDermott, D. 1982. *Introduction to Artificial Intelligence*. Reading, MA: Addison-Wesley.

Chomsky, N. 1965. *Aspects of the Theory of Syntax*. Cambridge, MA: MIT Press.

Chomsky, N. 1968. *Language and Mind.* New York: Harcourt, Brace and World. (Extended edition, 1972.)

Chomsky, N. 1969. Some empirical assumptions in modern philosophy of language. In *Philosophy, Science, and Method: Essays in Honor of Ernest Nagel.* S. Morgenbesser, P. Suppes and M. White (eds). New York: St. Martin's Press.

Chomsky, N. 1970. Problems of explanation in linguistics. In *Explanation in the Behavioral Sciences,* R. Borger and F. Cioffi (eds). Cambridge: Cambridge University Press.

Chomsky, N. 1980. *Rules and Representations.* Oxford: Basil Blackwell.

Chomsky, N. 1984. Changing perspectives on knowledge and use of language. Paper presented at a Sloan Conference, MIT, May, 1984. Available from Lawrence Erlbaum, Hillsdale, NJ.

Chomsky, N. 1988. *Language and Problems of Knowledge: The Managua Lectures.* Cambridge, MA: MIT Press.

Churchland, P.M. and Churchland, P.S. 1990. Could a machine think? *Scientific American* 262(1), 26–31.

Cussins, A. 1990. The connectionist construction of concepts. In *The Philosophy of Artificial Intelligence,* M. Boden (ed). Oxford: Oxford University Press.

Dennett, D. 1987. *The Intentional Stance.* Cambridge, MA: MIT Bradford.

de Kleer, J. and Brown, J. 1981. Mental models for physical systems and their acquisition. In *Cognitive Skills and their Acquisition.* J. Anderson (ed). Hillsdale, NJ: Lawrence Erlbaum.

de Kleer, J. and Brown, J. 1982. Some issues on mechanistic mental models. *Proceedings of the Fourth Cognitive Science Society Conference.* Available from Lawrence Erlbaum, Hillsdale, NJ.

de Kleer, J. and Brown, J. 1983. Assumptions and ambiguities in mechanistic mental models. In *Mental Models,* D. Gentner and A.L. Stevens (eds). Hillsdale, NJ: Lawrence Erlbaum.

de Kleer, J. and Brown, J. 1984. A physics based on confluences. *Artificial Intelligence* 24, 7–83.

de Kleer, J. 1987. Qualitative physics. In *The Encyclopedia of Artificial Intelligence,* 1149–1158. New York: Wiley.

Evans, G. 1973. The causal theory of names. *Aristotelian Society Supplementary Volume* XLVII, 15–40.

Evans, G. 1980. Understanding demonstratives. In *Meaning and Understanding,* H. Parret and J. Bouveresse (eds), 280–303. New York/Berlin: Walter de Gruyter.

Frege, G. 1967. *The Basic Laws of Arithmetic.* Berkely: University of California Press.

Garfield, J. 1988. Review of Macnamara, J. A Border dispute: The place of logic in psychology. *Journal of Symbolic Logic* 53, 1, 314–317.

Gettier, E. 1963. Is justified true belief knowledge? *Analysis* 23, 121–123.

Hayes, P. 1978. The naive physics manifesto. In *Expert Systems in the Microelectronic Age,* D. Michie (ed), 242–270. Edinburgh: Edinburgh University Press.

Hayes, P. 1985. The second naive physics manifesto. In *Formal Theories of the Commonsense World,* J. Hobbs and R. Moore (eds), 1–36. Norwood, NJ: Ablex.

Husserl, E. 1962. *Ideas — General Introduction to Pure Phenomenology.* New York: Macmillan.

Husserl, E. 1970. *Logical Investigations, 1.* New York: Humanities Press.

Katz, J. 1981. *Language and Other Abstract Objects.* New Jersey: Rowman and Littlefield.

Macnamara, J. 1986. *A Border Dispute: The place of logic in psychology.* Cambridge, MA: MIT Bradford.

Mill, J. 1843. *A System of Logic.* London: Longmans, Green, Reader and Dyer.

Mill, J. 1865. *Examination of Sir William Hamilton's Philosophy.* Boston: William V. Spencer.

Miller, M. 1982. A structured planning and debugging environment for elementary programming. In *Intelligent Tutoring Systems,* D. Sleeman and J. Brown (eds). London: Academic Press.

Millikan, R. 1984. *Language, Thought and Other Biological Categories.* Cambridge, MA: MIT Bradford.

Millikan, R. 1986. Thoughts without laws: cognitive science without content. *Philosophical Review* XCV, 47–80.

Popper, K. 1972. *Objective Knowledge: An evolutionary approach.* Oxford: Clarendon Press.

Rich, E. 1983. *Artificial Intelligence.* Auckland: McGraw-Hill.

Richards, T. 1978. *The Language of Reason.* Rushcutters Bay, Australia: Pergamon.

Searle, J. 1980. Minds, Brains, and Programs. *The Behavioral and Brain Sciences* 3, 417–424.

Searle, J. 1990. Is the Brain's Mind a Computer Program? *Scientific American* 262, 1, 20–31.

Smith, B.C. 1985. Prologue to Reflection and Semantics in a Procedural Language. In R. Brachman and H. Levesque (eds), *Readings in Knowledge Representation,* 31–41. Los Altos, CA: Morgan Kaufmann.

Smith, B.C. 1991. The owl and the electric encyclopedia. *Artificial Intelligence* 47, 251–288.

Wenger, E. 1987. *Artificial Intelligence and Tutoring Systems.* California: Morgan Kaufmann Publishers.

Wittgenstein, L. 1961. *Tractatus Logico-Philosophicus.* London: Routledge and Kegan Paul. (Original German edition published 1921.)

Young, R. and O'Shea, T. 1981. Errors in children's subtraction. *Cognitive Science* 5, 153–177.

Is Cognition an Autonomous Subsystem?

Mark H. Bickhard
Department of Psychology
Lehigh University

In standard views, representation consists of elements in correspondence with what they represent. Only certain kinds of correspondences will do — not all correspondences are representational — but correspondence is the basic category within which representations are differentiated (B.C. Smith 1987). The representational elements are taken to encode that which they are in correspondence with. This assumption regarding the nature of representation is held in common among those who consider such encodings to be transduced (Fodor and Pylyshyn 1981) or innate (Fodor 1981) or designed (Newell 1980) or trained into a net (Churchland 1989) — these positions differ regarding the presumed *origin* of encoding correspondences, not regarding their basic representational nature.

In such a view of representation, cognition is taken to consist of various stages of the input and processing, and sometimes the output, of such encodings. The fundamental backbone of cognition is assumed to be the sequence from perception to cognitive processing to the re-encoding into utterances in some language.

This view lends itself to, if not forces, a strong modularization of models of the mind. This modularization is *in addition* to the potential modularizations within cognition itself (Fodor 1983). Cognition per se is seen as being organized around one or more storage banks for encodings. Various processes enter data into such banks, process the contents of such banks, and re-encode selected contents into utterances.

For any model of a real agent, with a real mind, additional modules are required. Some subsystem is required to control action. Another subsystem interprets the cognitive contents for their relevance to action. Another engages in the motivational selections of actions to be performed, again informed by the cognitive contents. Still further fragmentations of mentality are common, but I will focus in this paper on these three: representation, action, and motivation.

In particular, I will argue that the standard view of representation as some kind of correspondence, as an encoding, is wrong. I outline an alternative model of representation that emerges naturally in agents, biological or designed, that actually engage the world (Beer 1990, 1995, In press; Beer, Chiel and Stirling 1990; Bickhard 1980, 1993; Bickhard and Terveen 1995; Brooks 1991a, 1991b, 1991c; Cherian and Troxell, 1995b; Malcolm, Smithers and Hallam 1989; Smithers 1994). One primary consequence of this alternative model of representation — called interactivism — is that functions that are standardly taken to reside in separate modules, such as representation, action, and motivation, are inherently integrated as separate functional *aspects* of one single underlying ontology. They are not inherently distinct modules. If standard models that permit such modularization are in error, then so are such modularizations per se.

1. Encoding Models of Representation

Encoding models of the nature of representation are wrong. The deepest errors, however, are not perspicuous, and attempting to locate them can lead to explorations of endless mazes of blind alleys and fruitless pursuit of red herrings. The consequences are fatal to any aspirations of understanding mental processes, and can be devastating even to some strictly practical design goals in artificial intelligence (Bickhard and Terveen 1995).

I will not be arguing that encodings do not exist. They clearly do: Morse code is a paradigm example. What I do argue, however, is that encodings cannot be the basic nature of representation. Genuine encoding must be derivative from some other, more fundamental, form of representation. If the assumption is made that the fundamental nature of representation is encodings (Palmer 1978; cf. Bickhard and Terveen 1995) — whether as an explicit assumption or a deeply buried implicit assumption — an incoherence results.

There is a large family of corollary arguments showing the incoherence of encodingism — of the assumption or presupposition that representation is encoding. I will outline only a few.

External forms of representation, such as Morse code or a statue or a blueprint, require interpretation and understanding. The dots and dashes of Morse code must be understood and interpreted into characters; the statue must be understood as a statue, and relevant resemblances noted and interpreted. This is unexceptional so long as the requirement for such an interpreter is un-problematic.

When attempting to account for the inner representations of such an interpreter, however — the inner mental representations of any real mind — presupposition of still another interpreter is lethal. The assumption that internal representations are of the same nature as external representations yields infamous regresses of interpretations requiring interpretations which require further interpretations, and so on, with the corresponding regress of interpretive homunculi to perform these feats. Whatever internal representation is, it cannot be the same kind of thing as external representation (Bickhard and Terveen 1995; Clancey 1991).

Considering Morse code in a different respect, we note that the dot and dash patterns of the code are stand-ins for the characters that they encode. They stand-in for those characters in the sense that we define them and use them and interpret them that way. It is useful to do so because, for example, dots and dashes can be sent over telegraph wires while characters cannot.

Such stand-in relationships are what constitute the Morse encodings as representations at all. The dot and dash patterns borrow their representational content — the specification of what they represent — from what they stand-in for. "Dot dot dot" represents the same thing as does "s." Again, this is not problematic so long as there is something to borrow representational content from.

But if we assume that all representations are encodings, are stand-ins, then we encounter an incoherence. The stand-in relationship can be iterated. X can be defined in terms of Y, which can be defined in terms of Z, but this iteration must be finite, so long as we are considering finite systems. Therefore, there must be a bottom level — a level of representations that do not stand-in for some other level.

If we assume that this level is constituted as encodings, however, we encounter the incoherence. Consider some presumed element, say "X," of this presumed grounding level of encodings. It cannot be defined in terms of any other representation by assumption. The only alternative is to define it in terms of itself, but that yields something like " "X" represents X" or " "X" stands-in for "X"." This does not succeed in providing "X" with any representational content at all, and, therefore, does not succeed in constituting "X" as a representation at all. But there are, by the encodingism assumption and the assumption that "X" is an element of the grounding level of encodings, no further resources available for making "X" an encoding. Encodingism requires such a grounding level of encodings, so the encodingism assumption per se has yielded an incoherence: it assumes a grounding level of encoding elements, which cannot exist.

There are many other corollary arguments against encodingism (Bickhard 1993; Bickhard and Terveen 1995), but I will not pursue them here. The incoherence argument per se logically suffices to refute encodingism, but all I require for current purposes is a prima facie case that encodingism has serious problems — all I require is a motivation for considering an alternative model of representation.

1.1. *Representation as Isomorphism*

There are not only many corollary arguments against encodingism, which I will not pursue here, there are also many apparent rejoinders to the general critique of encodingism. For similar reasons of space, I cannot pursue most of those either (see Bickhard 1993; Bickhard and Terveen 1995), but there is one rejoinder that is inherent in one of the foundational frameworks for Artificial Intelligence — specifically, the Physical Symbol System Hypothesis — that I will address briefly.

Within the Physical Symbol System Hypothesis, symbols are defined in terms of pointers (*access*) to some other entity in the system (Newell 1980; Bickhard and Terveen 1995). This a strictly functional notion, strictly internal to a machine, and, as such, is unobjectionable. Problems emerge, however, when the attempt is made to extend the model to representation — in particular, to representation of external entities. The relationship between a symbol and what it is supposed to designate is fundamentally different in this external case — to construe that relationship in terms of pointer access is to presuppose what is to be explained. In particular, access is a primitive function built into a machine, but it is not primitive in the relationships between machine and the world. Representation *is*, in some sense, epistemic access, so access cannot be simply presupposed in a model of representation.

The alternative that is proposed is a relationship of the isomorphism of patterns: patterns in the world are what is designated, and they are designated by patterns in the system that are isomorphic to the designated external patterns (Newell 1980; Vera and Simon 1993). This is a richer concept than simple correspondence, but it, nevertheless, is still a version of encodingism, and is still logically unworkable.

Pattern isomorphy is still a version of correspondence — isomorphic, or structural, correspondence as well as single point to point correspondence. As such, it is subject to all of the problematics of encodingism:

- There is nothing about the internal pattern that carries knowledge of the fact of any such isomorphy, nor about what any such isomorphy might be with. There is no representational content specifying what the isomorphy is with — about the other end, the presumed represented end, of the isomorphic relationship.
- Isomorphy is multifarious: isomorphic correspondences can be defined ubiquitously, with almost anything. Which of these is the "representational" isomorphy?
- Isomorphy is transitively unbounded: any isomorphy with one pattern will also constitute an isomorphy with patterns causally prior to that one (as well as noncausal accidental isomorphisms all over the universe), and prior again, and so on. Which of *these* is the "representational" isomorphy?
- Isomorphy is transitive, but representation is *not* transitive: merely knowing the label of a map will not permit you to travel in the mapped territory — the label of the map does not represent the territory.
- Isomorphy is symmetric, but representation is not symmetric: the table does not represent your mental representation of the table.
- If representation is constituted as isomorphy, then, if the isomorphy exists, the representation exists, and it is correct. But if the isomorphy does not exist, then the representation does not exist, *and it cannot be incorrect*! Correspondence models, including isomorphy models, cannot account for representational error (Dretske 1988; Fodor 1987, 1990; Loewer and Rey 1991; Millikan 1984; B.C. Smith 1987).
- Still more deeply, correspondence models, including isomorphy models, cannot account for the possibility of representational error (error of such correspondence) that is *detectable by the system itself*. But, without system detectable representational error, representational learning, among other error guided processes, is not possible (Bickhard and Terveen 1995).

Isomorphy models of representation are versions of informational approaches to representation, of the assumption that the representational relationship is version of the informational relationship (Fodor 1987, 1990; Loewer and Rey 1991; Hanson 1990; B.C. Smith 1987). An informational relationship, in turn, is just one version of a correspondence relationship. Informational approaches to representation are certainly the dominant approaches today, but, if the arguments against encodingism are correct, they are ultimately unworkable. It is clear that no one knows today how to make such an approach work: "we haven't got a ghost of a Naturalistic theory about [encoding]." (Fodor 1987b: 81) "The project of constructing a representational theory of the mind is among the most interesting that empirical science has ever proposed. But I'm afraid we've gone about it all wrong." (Fodor 1994: 113). The Physical Symbol System Hypothesis, then, does not provide a solution to the problems of encodingism. More generally, isomorphic correspondences are no improvement over correspondences per se in attempting to model representation.[1]

2. Interactivism

For an agent interacting with its world, the ability to anticipate what interactions might be possible next would be a useful resource (Bickhard 1993). Anticipations would permit the agent to select among those possibilities those that are most suited to its current internal conditions, or to select those that are most to be avoided. Such possibilities for further interaction will vary as the situation of the agent varies, so some process for constructing and updating the anticipations would be required.

The critical property of such anticipations for my current purposes is that they might be wrong, and might be discoverable to be wrong by the system itself: if the system engages in an indicated possible interaction, and the interaction fails — fails to proceed as indicated — then the anticipation was in error. Anticipations — indications of possible interactions — can be false, and can be discovered to be false by the system. Anticipations have truth values, for the system. Possessing truth values is at least one fundamental property of representations.

I argue that these primitive truth values are in fact foundational to all representation. Indications of potential system interactions are the most primitive, the foundational, form of representation, out of which all other representation is constructed.

There are many aspects and promissory notes involved in this claim. I will address only two here:

1. Can this notion of anticipation be explicated in purely functional terms?
2. How could interactive anticipations account for such representations as those of objects?

Examples of other issues that I will *not* address here include: How could this notion of representation account for abstract representations, such as number? How could such a model of representation account for perception, for language? How could it be consistent with rational thought? And so on. These are all addressed elsewhere (Bickhard 1980, 1993; Bickhard and Richie 1983; Bickhard and Terveen 1995: Hooker 1995). For my current purposes, I need only a prima facie plausibility of interactive representation, not a demonstrated adequacy in all senses, because I am primarily aiming at the implications of such an interactive model of representation for issues of modularity. In particular, I will argue that, in the interactive model, issues of action and of motivation as action selection are most fundamentally intrinsic aspects of anticipatory interactive systems, not separate modules.

2.1. *Functional Anticipations*

Rendering the necessary notion of anticipation in functional terms is not difficult: pointers and subroutines suffice. In particular, a pointer to a subroutine can indicate the potentiality of the interactions that would be engaged in by that subroutine, while further pointers to internal outcomes should that subroutine be in fact executed constitute the anticipations that are detectable by the system. If the system does engage that subroutine and the internal outcome of the interaction is not one of those indicated, then the indications have been falsified — and falsified for the system itself. There are other architectural frameworks in which the requisite anticipations can be modeled (Bickhard and Terveen 1995), but demonstrating the adequacy of pointers and subroutines suffices to demonstrate that no non-functional notions are necessary.

Interactive anticipation yields the possibility of system detectable error. System detectable error, in turn, is necessary for error guided activities, such as goal directedness or learning. System detectable error, then, is a necessity for any but the most simple and primitive forms of life or artificial agents.

3. Representing Objects

Primitive interactive anticipations have truth values, and, thus, constitute primitive forms of representation. They implicitly predicate to the environment whatever interactive properties would support the indications of internal outcomes (Bickhard 1993). Such primitive representation, however, is appropriate primarily to simple organisms and simple artificial agents. More complex agents involve more complex representations, and that complexity must be accounted for.

There are two primary resources in the interactive model of representation that permit it to model complex representations, such as objects. These resources are conditional indications, and iterated indications.

All indications are conditional at least in the sense that they are evoked in certain internal system conditions, and not in others. That is, they are conditional on particular internal states of the system. In turn, interactions actually engaged in yield subsequent internal conditions as the internal outcomes of those interactions. This yields the possibility of iterated indications: interaction I_7 is possible given current system states and will yield outcomes O_1, O_2, or O_3, while, if O_1 is reached, then interactions I_{10}, I_{23}, and I_{34} will be possible, which would yield outcomes O_{88}, … and so on. Indications can branch and iterate, forming potentially complex webs of indicated interactive potentiality.

One possibility for such webs is that they might close into loops and other reciprocal indication relationships. A subweb might even exhibit the property of closure: all states in the web are reachable from all other states via some class of interactions that relate those states. With one additional property, I claim that we now have the necessary resources for modeling simple object representation. That additional property is invariance.

In particular, a typical object, say a child's toy block, offers many possible interactions — visual scans from various perspectives, multiple manipulations, dropping, chewing, and so on. Furthermore, every one of these possibilities indicates all the others, perhaps with necessary intervening interactions (for example, turning the object around to create the possibility of the visual scan from that angle). That is, the organization of the possibilities is closed. Still further, that overall pattern of possibilities, together with its closure, is invariant with respect to a large class of interactions. Clearly it is invariant with respect to each interaction in the web itself, but the interactive pattern of the block, for example, is also invariant under various throwings, locomotions of the infant, storing in the toy chest, and so on. It is *not* invariant, however, under burning, crushing, dissolving in acid, and so on.

Epistemically, objects just *are* closed invariant patterns of physical interaction. That's all they can be to infants and monkeys. Accounting for objects in terms of theories about objects, in terms of earth, air, fire, and water, for example, or in terms of atoms and molecules, is a much higher order accomplishment. Accounting for those higher order possibilities requires, among other things, addressing the representation of abstractions — something I will not undertake here.

The claim is that interactive representation is capable of accounting for representations of physical objects, in the generally Piagetian manner just outlined (Piaget 1954). The concluding claims of this section, then, are that interactive representation is renderable in strictly functional terms, and that it has at least a prima facie initial plausibility of serving as an adequate approach to all representation.

4. Pragmatics and Representation

Interactivism is a version of the general pragmatic approach to representation (Hookway 1985; Houser and Kloesel 1992; Rosenthal 1983, 1990; Thayer 1973). Such approaches share the assumption that representation is an emergent of action, and that classical correspondence, or "spectator," models of representation are inadequate (J.E. Smith 1987). Although pragmatism has to date had relatively little influence on studies of cognition and artificial intelligence, there are exceptions. Most of these derive their pragmatist influences from Jean Piaget (influenced by James Baldwin, who, in turn, was influenced by Peirce and James). Interactivism shares a general pragmatism with such approaches, and shares a more specific influence from Piaget. The differences, consequently, are more subtle than they are with encoding models — all genuine pragmatist approaches *share* a rejection of correspondence models of representation (or at least an attempted rejection. In a number of cases, I claim that the attempt to avoid encodingist presuppositions has in fact not been fully successful). Within Artificial Intelligence, perhaps the best known of pragmatist approaches is the work of Drescher (1986, 1991), so I turn now to a brief comparison between interactivism and Drescher's model.

4.1. *Drescher*

Drescher's model of representation is essentially that of Piaget. There are strong commonalities between the interactive model and Piaget's model (Bickhard and Campbell 1989), and, therefore, there are strong commonalities between the interactive model (Bickhard 1980, 1993; Bickhard and Richie 1983; Bickhard and Terveen 1995) and Drescher's (1986, 1991). Most of these commonalities follow from the basic framework of modeling representation as an emergent of action systems.

For example, if representation is construed as correspondence, there is a temptation to think that the world could impress itself into a passive but receptive mind, leaving behind representational correspondences. This is essentially Aristotle's model of perception, and is still with us in the technologically more sophisticated, but logically no better, notions of passive transduction and induction (Fodor and Pylyshyn 1981; Bickhard and Richie 1983). On the other hand, if representation is an emergent of action systems, there is no such temptation to think that action systems — interactively competent control structures — could possibly be impressed by the environment into a passive mind. If representation is emergent out of action, then perception and learning and development must all be active constructive processes (Bickhard 1992). Furthermore, since such knowledge constructions cannot be assured of being correct — if they could, then the knowledge would already be present — they must be subject to being tried out and eliminated if they fail. That is, an action framework for understanding representation forces a variation and selection constructivism, an evolutionary epistemology (D.T. Campbell 1974).

For another example, consider again just what it is that is most fundamentally being represented in an action based model. Representation is most fundamentally of future potentialities for further action and interaction. Pragmatic representation is future looking instead being backward looking down the stream of inputs coming into the organism — pragmatic models are not models of a spectator looking into the past of that input stream. In particular, pragmatic representation is intrinsically representation of *possibilities*. Pragmatic representation is intrinsically *modal*. This is one of many fundamental differences between pragmatic models and correspondence models (correspondence models, in fact, have in principle difficulties handling modal representation). It is worth noting that representation in children does not begin with strictly "actual" representation and then later add a layer of modality, as standard logic and encoding frameworks would suggest. Instead, children begin with representation that is intrinsically undifferentiated between various aspects of modality, and actuality, possibility, and necessity must be progressively differentiated in development over the course of some years (Bickhard 1988; Piaget 1986, 1987).

There are many more commonalities between interactivism and Piaget's genetic epistemology, and, therefore, with Drescher's model — striking commonalities, especially in the context of the contrary but dominant encoding orientations. There are also, however, important differences (Bickhard 1982, 1988b; Bickhard and Campbell 1989; Campbell and Bickhard 1986). Piaget's notion of representation is action based and modal, but it is still subtly a correspondence model. For Piaget, concepts are structures of potential actions that are isomorphic with structures of potentialities in the environment — a kind of modal isomorphism of patterns (Bickhard 1988b; Bickhard and Campbell 1989; Campbell and Bickhard 1986; Chapman 1988) — and Piaget's model of perception is straightforwardly an encoding model (Piaget 1969). This is action based, constructive, and modal — all different from standard approaches — but it is still a correspondence notion. There is still no way for those mental structures to pick out or to specify what they are supposed to represent — to provide or constitute knowledge that they are in isomorphism or what they are in isomorphism with. There is still no way for these mental structures to avoid the problems of encodingism.

These problematics carry over into Drescher's model. His is similarly action based, constructive, and modal. He recognizes the necessity of action feedback for the construction of representation, for learning — in particular, pragmatic error feedback of when an action does not work as anticipated. But he still construes representation itself — that which is learned — in terms of correspondences between "items" in the system and conditions in the world (cf. Dretske 1988). "Drescher has recognized the importance of pragmatic error for learning, but has not recognized the emergence of representational error, thus representational content, out of pragmatic error. In the interactive model, in contrast, representational error is *constituted* as a special kind of pragmatic error." (Bickhard and Terveen 1995: 281). Drescher's model is a momentous advance over standard approaches in the literature of artificial intelligence and cognitive science. I suggest, however, that there remain residual problems — problems that are avoided in the interactive model.

In presenting and discussing the interactive model, I have presented brief but focused contrastive discussions with the Physical Symbol System Hypothesis and with Drescher's model. Clearly there are innumerable additional models and approaches that could be examined, and deserve to be examined, such as Cyc, SOAR, PDP, machine learning, autonomous agents, dynamic systems approaches, and so on, as well as further cognitive issues that deserve attention, such perception, learning, language, rationality, instantiations in the central nervous system, and so on. I cannot address these here, but would suggest alternative sources to the interested reader (Bickhard 1991, 1995, In preparation; Bickhard and Terveen 1995; Hooker 1995). I will turn now from the interactive model of representation per se to one of its interesting consequences: an inherent integration of issues of representation, action, and motivation.

5. Representation, Action, and Motivation

Encoding representations represent in virtue of some correspondence between them and that which they represent. Typically, those correspondences are assumed to be constructed or invoked via some sort of processing of inputs (Fodor 1990; Newell 1980), but even that is not logically necessary: Representational correspondences are intrinsically atemporal. In particular, encodings do not require any agent in order to exist; they are not dependent on action — however much it may be that action is taken to be dependent on (interpreting) them.

Interactive representation, in contrast, is an emergent of certain forms of organization of an interactive agent. Interactive representation cannot exist in a passive system — a system with no outputs. Interactive representation is the anticipatory, the implicit predicational, aspect of interactive systems. Representation and interaction are differing functional aspects of one underlying system organization similarly to the sense in which a circle and a rectangle are differing visual aspects of one underlying cylinder. Action and representation are not, and cannot be, distinct modules.

A similar point holds for motivation too, but to see this requires a brief digression on motivation per se. Encoding representations are consistent with models of completely passive systems; correspondingly, the typical assumption is that the default condition of the system is inactivity. In such a view, the basic question of motivation is "What makes the system do something rather than nothing?" Motivation is a matter of pushing or pulling the system out of its default inactivity.

Living systems, however, are not passive. They are constantly in activity of some sort: to cease activity is to become dead. For living systems, then, the question of motivation is mis-stated: instead of "What makes the system do something rather than nothing?" the proper question of motivation is "What makes the system do this rather than that?" That is, the question of motivation is a question of action and interaction *selection*, not of action and interaction activation or stimulation. The system is always doing something, the question is what determines what it does (Bickhard and Terveen 1995).

In this form, however, motivation becomes a functional matter — the function of interaction selection. That function, in turn, is precisely what interaction indications are useful for. Indicated interactions with their indicated potential internal outcomes form a primary resource for the system to select what interactions to engage in next. That is, interactive indications and their associated internal outcomes not only implicitly predicate interactive properties of the environment, they also serve the motivational function of interaction selection. Motivation and representation are both aspects, along with interactive competency per se, of one underlying ontology of interactive system organization (Bickhard and Terveen 1995; Cherian and Troxell 1995a).

In complex organisms, and other complex systems, it is possible for relatively specialized and dedicated subsystems to develop that subserve complex functions of representation or of motivation. But, if the interactive model is correct, such specializations must arise out of, and on the foundation of, the basic interactive competence, interactive representation, and interactive motivational selection aspects of underlying interactive system organization. Certainly we do not find specialized such subsystems in simple organisms, only in complex organisms.

6. Conclusions

The three phenomena of action and interaction, representation, and motivation, then, do not form separate functional modules that can simply be pasted together in a more complicated system if a more complex design is desired, or if modeling those additional complexities in a natural agent is desired. To the contrary, neither interaction nor representation nor motivation can be correctly modeled without, at least implicitly, modeling all three.

Conversely, if such modularization *is* possible within some modeling approach, then that approach is almost certainly assuming or presupposing an encodingism toward representation. That is, if such modularization is possible in a modeling approach, then that approach is almost certainly founded on a logical incoherence.

Cognition, then, is *not* an autonomous subsystem — and any approach or programme that permits cognition to seem autonomous is foundationally flawed. Such a foundational incoherence, in turn, can have myriad and ramified pernicious consequences throughout the programme or programmes involved — and programmatic errors can be extremely difficult to diagnose and to avoid (Bickhard and Terveen 1995). Nevertheless, encodingism and its associated modularizations dominate contemporary artificial intelligence and cognitive science. It will not be possible to understand or to design beings with minds within such an approach — artificial intelligence and cognitive science are dominated by programmatic assumptions that make their own highest level programmatic aspirations impossible (Bickhard and Terveen 1995).

Notes

1. I should point out that, although there are many important and useful properties of connectionist nets, the trained correspondences that are supposed to constitute representations in standard connectionism are no improvement over the designed or isomorphic correspondences that are supposed to constitute representations in GOFAI (Bickhard and Terveen 1995).

References

Beer, R.D. 1990. *Intelligence as Adaptive Behavior*. New York: Academic.

Beer, R.D. 1995. Computational and dynamical languages for autonomous agents. In *Mind as Motion: Dynamics, Behavior, and Cognition,* R. Port and T.J. van Gelder (eds), Cambridge, MA: MIT Press.

Beer, R.D. In press. A dynamical systems perspective on agent-environment Interaction. *Artificial Intelligence*.

Beer, R.D., Chiel, H.J., and Sterling, L.S. 1990. A biological perspective on autonomous agent design. In *Designing Autonomous Agents,* P. Maes (ed), 169–186. Cambridge, MA: MIT Press.

Bickhard, M.H. 1980. *Cognition, Convention, and Communication*. New York: Praeger.

Bickhard, M.H. 1982. Automata theory, artificial intelligence, and genetic epistemology. *Revue Internationale de Philosophie* 36(142–143), 549–566.

Bickhard, M.H. 1988. The necessity of possibility and necessity: Review of piaget's possibility and necessity. *Harvard Educational Review* 58(4), 502–507.

Bickhard, M.H. 1988b. Piaget on variation and selection models: Structuralism, logical necessity, and interactivism. *Human Development* 31, 274–312. Also in *Jean Piaget: Critical Assessments,* L. Smith (ed), 83, 388–434, 1992. London: Routledge.

Bickhard, M.H. 1991. A pre-logical model of rationality. In *Epistemological Foundations of Mathematical Experience,* L. Steffe (ed), 68–77. New York: Springer-Verlag.

Bickhard, M.H. 1992. How does the environment affect the person? In *Children's Development within Social Context: Metatheory and Theory,* L.T. Winegar and J. Valsiner (eds), 63–92. Hillsdale, NJ: Erlbaum.

Bickhard, M.H. 1993. Representational content in humans and machines. *Journal of Experimental and Theoretical Artificial Intelligence* 5, 285–333.

Bickhard, M.H. 1995. Intrinsic constraints on language: Grammar and hermeneutics. *Journal of Pragmatics* 23, 541–554.

Bickhard, M.H. In preparation. Critical principles: On the negative side of rationality. In *Non-formal Approaches to Rationality*, C.A. Hooker and H.I. Brown (eds).

Bickhard, M.H. and Campbell, R.L. 1989. Interactivism and genetic epistemology. *Archives de Psychologie* 57(221), 99–121.

Bickhard, M.H. and Richie, D.M. 1983. *On the Nature of Representation: A Case Study of James J. Gibson's Theory of Perception.* New York: Praeger.

Bickhard, M.H. and Terveen, L. 1995. *Foundational Issues in Artificial Intelligence and Cognitive Science — Impasse and Solution.* Amsterdam: Elsevier Scientific.

Brooks, R.A. 1991a. New approaches to robotics. *Science* 253(5025), 1227–1232.

Brooks, R.A. 1991b. How to build complete creatures rather than isolated cognitive simulators. In *Architectures for Intelligence*, K. VanLehn (ed.), 225–239. Hillsdale, NJ: Erlbaum.

Brooks, R.A. 1991c. Challenges for complete creature architectures. In *From Animals to Animats*, J.-A. Meyer and S.W. Wilson (eds), 434–443. Cambridge, MA: MIT Press.

Campbell, D.T. 1974. Evolutionary epistemology. In *The Philosophy of Karl Popper*, P.A. Schilpp (ed), 413–463. LaSalle, IL: Open Court.

Campbell, R.L. and Bickhard, M.H. 1986. *Knowing Levels and Developmental Stages.* Basel: Karger.

Chapman, M. 1988. *Constructive Evolution: Origins and Development of Piaget's Thought.* Cambridge: Cambridge University Press.

Cherian, S. and Troxell, W.O. 1995a. Interactivism: A functional model of representation for behavior-based systems. In Morán, F., Moreno, A., Mereld, J.J., Chacón, P. *Advances in Artificial Life*, Proceedings of the Third European Conference on Artificial Life, Granada, Spain, 691–703. Berlin: Springer.

Cherian, S. and Troxell, W.O. 1995b. Intelligent Behavior in machines emerging from a collection of interactive control structures. *Computational Intelligence* 11, 565–592.

Churchland, P.M. 1989. *A Neurocomputational Perspective.* Cambridge, MA: MIT Press.

Clancey, W.J. 1991. The frame of reference problem in the design of intelligent machines. In *Architectures for Intelligence: The Twenty-Second Carnegie Symposium on Cognition*, K. VanLehn (ed), 357–423. Hillsdale, NJ: Lawrence Erlbaum.

Drescher, G.L. 1986. Genetic AI: Translating piaget into lisp. Cambridge, MA: MIT AI Memo No. 890.

Drescher, G.L. 1991. *Made-Up Minds*. Cambridge, MA: MIT Press.

Dretske, F.I. 1988. *Explaining Behavior*. Cambrdige, MA: MIT Press.

Fodor, J.A. 1981. The present status of the innateness controversy. In *RePresentations*, J. Fodor, 257–316. Cambridge, MA: MIT Press.

Fodor, J. A. 1983. *The Modularity of Mind: An Essay on Faculty Psychology*. Cambridge, MA: MIT Press.

Fodor, J.A. 1987. *Psychosemantics*. Cambridge, MA: MIT Press.

Fodor, J.A. 1987b. A Situated Grandmother? *Mind and Language* 2, 64–81.

Fodor, J.A. 1990. *A Theory of Content*. Cambridge, MA: MIT Press.

Fodor, J.A. 1994. Concepts: A potboiler. *Cognition* 50, 95–113.

Fodor, J.A. and Pylyshyn, Z. 1981. How direct is visual perception?: Some reflections on Gibson's ecological approach. *Cognition* 9, 139–196.

Hanson, P.P. 1990. *Information, Language, and Cognition*. Oxford: Oxford University Press.

Hooker, C.A. 1995. *Reason, Regulation, and Realism: Towards a Regulatory Systems Theory of Reason and Evolutionary Epistemology*. Albany: SUNY.

Hookway, C. 1985. *Peirce*. London: Routledge.

Houser, N. and Kloesel, C. 1992. *The Essential Peirce*. Vol. 1. Indianapolis: Indiana.

Loewer, B. and Rey, G. 1991. *Meaning in Mind: Fodor and his critics*. Oxford: Basil Blackwell.

Malcolm, C. A., Smithers, and T. and Hallam, J. 1989. An emerging paradigm in robot architecture. In *Proceedings of the Second Intelligent Autonomous Systems Conference*, T. Kanade, F.C.A. Groen, and L.O. Hertzberger (eds), 284–293. Amsterdam. 11–14 December 1989. Published by Stichting International Congress of Intelligent Autonomous Systems.

Millikan, R.G. 1984. *Language, Thought, and Other Biological Categories*. Cambridge, MA: MIT Press.

Newell, A. 1980. Physical symbol systems. *Cognitive Science* 4, 135–183.

Palmer, S.E. 1978. Fundamental aspects of cognitive representation. In *Cognition and categorization*, E. Rosch and B.B. Lloyd (eds), 259–303. Hillsdale, NJ: Erlbaum.

Piaget, J. 1954. *The Construction of Reality in the Child*. New York: Basic.

Piaget, J. 1969. *The Mechanisms of Perception*. New York: Basic.

Piaget, J. 1986. Essay on necessity. *Human Development* 29, 301–314.

Piaget, J. 1987. *Possibility and Necessity*. Vols. 1 and 2. Minneapolis: University of Minnesota Press.

Rosenthal, S.B. 1983. Meaning as habit: Some systematic implications of Peirce's pragmatism. In *The Relevance of Charles Peirce*, E. Freeman (ed.), 312–327. La Salle: Monist.

Rosenthal, S.B. 1990. *Speculative Pragmatism*. La Salle: Open Court.

Smith, B.C. 1987. *The Correspondence Continuum*. Stanford, CA: Center for the Study of Language and Information, CSLI-87–71.

Smith, J.E. 1987. The reconception of experience in Peirce, James, and Dewey. In *Pragmatism Considers Phenomenology*, R.S. Corrington, C. Hausman and T.M. Seebohm (eds), 73–91. Washington, DC: University Press.

Smithers, T. 1994. On behaviour as dissipative structures in agent-environment system interaction spaces. Presented at the meeting Prerational Intelligence: Phenomenology of Complexity in Systems of Simple Interacting Agents, November 22–26, 1993, part of the Research Group, Prerational Intelligence, Zentrum für Interdisziplinäre Forschung (ZiF), University of Bielefeld, Germany, 1993/94.

Thayer, H.S. 1973. *Meaning and Action*. Indianapolis: Bobbs-Merrill.

Vera, A.H. and Simon, H.A. 1993. Situated action: A symbolic interpretation. *Cognitive Science* 17(1), 7–48.

Part II: Epistemology and Methodology

Seán Ó Nualláin

The papers in this section focus on the following topics; symbol-grounding, the systematicity argument, functionalism and parallel distributed processing (PDP), the consequences for conventional CS of emotion's informational role (given that emotion of course involves embodiment) and finally, aspects of the dynamical systems approach.

MacDorman's paper is an appropriate successor to Bickhard's. It begins by supplying an excellent and conscientious account of the symbol-grounding problem. The possibility of grounding being implemented via a Fodorian language of thought is investigated before being rejected. The PDP approach is then outlined before the main theme is introduced; in order truly to ground cognitive categories in a biologically situated manner, they must be interrelated with sensorimotor expectations. Nonadaptive symbol-grounding mechanisms involve over-rigid categorization.

The way forward, then, must be through grounding cognitive categories in sensorimotor projections. An abstract model for this is given before the example of a fish in a simple aquatic environment is focussed on. MacDorman's careful final claim is remarkably Piagetian; abstract reasoning is guided by (initially) sensorimotor interaction between subject and environment. On the same grounding there, Mc Kevitt uses his "*droit d'éditeur*" to insert a previously published paper on a Chinese room extension. Were the participant's experience multi-modal, could we predicate understanding? As will become clear, much current computational work focusses on vision-language integration. At a guess, understanding requires also embodiment, particularly if the (currently favored) Merleau-Pontyan notion of meaning is correct.

In a book whose origin is a statement of scientific crisis, Aizawa's paper using as illustration the type for all future paradigm shifts (Ptolemaic to Copernican) is obviously of heightened interest. Ptolemaic astronomy could indeed take on board the movements of the Sun, Mercury and Venus; to put it in the correct terms, the phenomena could be "saved." However, the fact that

Mercury and Venus are never found in opposition to the Sun could be explained in any meaningful way only in the Copernican System. Aizawa uses this example in discussing whether PDP can explain systematicity. His PDP systems, characterized *qua* nodes and connections, can perhaps save any particular substitutability instance (construction of "X shoots Y" requiring that "Y shoots X" also be possible). However, it cannot explain it; nor, surprisingly, can the opposing classical viewpoint. It is concluded that, despite this failing, PDP may have an important role in psychological modelling and is of intrinsic mathematical and commercial interest. A lucid exposition of the role of tensors in Smolensky's Harmony Theory is given *en route*.

Pylyshyn's throughgoing attempt to found cognitive science had a functionalist ethos. Essentially, this hinges on a "multiple realizability" view of mental processes; their formal analysis, the analysis that mattered, was identical irrespective of whether they were implemented on a Mac computer or an array of beads held together by string. Berkeley addresses himself to the issue of whether PDP can be compatible with functionalism, which he characterizes as subtracting from each mental process everything except its causal role. The "and" gate is discussed at the implementational (hardware), symbolic and semantic levels (as Pylyshyn characterizes the latter). An equivalent connectionist network is then outlined. The conclusion from these last two papers that the Fodor/Pylyshyn critiques of PDP are lacking in some respect is inescapable.

De Lancey's paper stresses, as we have seen, that emotion has enormous negative consequences for classical cognitivism. The notion of emotion as being in some sense rational has a rich code of philosophical acceptance from Hume to our contemporary de Sousa. Details of relevant psychological experimentation are included.

Finally, many researchers like Thelen and Smith propose abandoning all classical CS discourse in favor of the dynamical systems vocabulary of strange attractors, trajectories in state-space (more familiar to us), etc. Dautenhahn and Christaller first outline the methodology used for autonomous robots and then propose how remembering (*qua* Bartlett), rehearsal and empathy might be captured. Functionality is assumed to emerge from the engagement of primitive symbols with the world; dynamical symbols can arise from the type of trajectories just noted. The work is interesting, and may be of great consequence when combined with the type of analysis Hoffman puts forward in the last paper in this book.

How to Ground Symbols Adaptively

K.F. MacDorman
Computer Laboratory
Cambridge University

1. Introduction

In the aftermath of the cognitive revolution, psychologists and AI researchers have generally modeled intelligent behavior in terms of internal symbols and rules for their combination. This has had its advantages. People can use rules and symbols to construct formal systems. Not only are these systems amenable to programming and, hence, computer simulation, but people can also readily interpret and explain what they do. However, human behavioral evidence can only support *external* use of symbols and rules — for example, in languages — their internal use being largely inferred from this;[1] phenomenological evidence suggests that we can at least think about objects wordlessly and in a manner more suggestive of sensorimotor rehearsal than the recombination of symbols (Rosenfield 1992; Kosslyn 1994); neurophysiological evidence is insufficient to conjecture — even functionally — the systematic deployment of symbols in the brain (Feldman and Ballard 1982; Rumelhart and McClelland 1986; Smolensky 1988).

From an evolutionary standpoint, symbol use has appeared very recently indeed, and only in the last few thousand years has it been formalized in lexicons, grammars, and axiomatic systems. It is only because of technological developments in the last few decades that human beings have been able to develop artifacts able to manipulate symbols efficiently. In sum, evolution and social and technological development have comparatively recently facilitated the systematic use of symbols and rules. Undoubtedly they are *end-products* of human cognition. Nevertheless, instead of viewing them *as* end-products — as aspects of the phenomena to be explained — cognitive theorists often instead view them as the primary units in terms of which their theories are constructed.

They have sometimes justified their theories' symbol-based ontology by arguing that this is what is appropriate for the simulation of higher-level (usually conscious) cognition and behavior. It is undeniably true, however, that all human behavior requires and manifests nonsymbolic processing. Purely symbolic human behavior cannot exist — even typing symbols on a keyboard involves sensorimotor coordination. Behavior's symbolic aspects are often so tightly bound with the nonsymbolic that we cannot make sense of the symbolic without taking the nonsymbolic into account. Even when considering ordinary verbal behavior — the epitome of symbol use — a transcript of a conversation can lead one to miss the conversation's point entirely (see Cowley and MacDorman 1995).

Although, of necessary, any outward expression has an analogue form, on the basis of human rationality and language, Fodor (1975), Chomsky (1965, 1986), and others have posited that, internally at some deep level, human beings systematically use rules. Yet one need only observe people interacting to see inconsistencies in both grammar and argumentation. Far from being inborn, the ability to think logically is largely culturally mediated and taught. Formal schooling promotes the ability to reason from the logical, content-independent form of a statement instead of from practical experience concerning its subject matter (Wertsch 1985, 1991; Scribner and Cole 1981: 126–128; Luria 1976). Thus, it may be argued that the origins of rationality are more sociocultural than biological.[2] Many other signs converge on the conclusion that, contra Newell (1980), minds are not symbol systems.[3] Whereas reifying a concrete problem makes it easier for a symbol system to solve, redescribing an abstract problem in a concrete setting can make the problem easier for people to solve (Wason 1981). Instead of getting in the way, the trappings of real life contribute to finding a solution. Could the reason for this be because representations are not grounded in abstract systems but sensorimotor experience?

Taking the symbol system as a starting point leads to various sticky dilemmas. Chief among them is the following: If people are symbol systems, then how could their internal symbols ever be causally connected with external objects? Trivially equating symbols with mental concepts brings in still more unresolved problems, such as the mind-body problem, further muddying the waters with its long prescientific history of debate. Researchers can avoid these dilemmas by focusing on questions that are biologically fundamental: How can organisms reliably act to exploit patterns in their sensory projections? How can they learn to recognize and handle things never before experienced by their genetic predecessors? It is worthwhile to have at least a tentative explanation of these abilities before inquiring into how human beings are able to solve logic puzzles or to communicate symbolically. This is because our higher cognitive abilities presuppose and utilize more basic abilities — in particular the ability to become sensitized to salient sensorimotor patterning.

Intelligent creatures appear to model the world — not by means of categorical rules — but rather through sensorimotor expectations — both conscious and nonconscious. Although these expectations may sometimes appear rule-like, unlike rules they develop inductively from sensorimotor projections.[4] This process may occur because of the direct influence of experience, the indirect influence of natural selection, or their combined influence whereby experience enables evolved predispositions (see Hinde 1987, e.g. on fear of snakes). As stated, from an evolutionary standpoint, symbols and rules appeared very late. The only place we are certain to find their systematic use is inside a digital computer or similar contrivance. It is no surprise then that it was inspiration drawn from the power of computers that lead Newell to equate the mind with a symbol system. However, even in computers, if we examine our programs carefully, seldom are we to find the orderly semantics of a symbol system (Smith, forthcoming). Nevertheless, computers have at least shown us that symbol systems can be implemented and that their symbols can be causally connected to the external world — either through an outside interpreter or through the sensing equipment of purpose-built systems. This paper is intended to explore how this grounding can be developed adaptively, that is, how a robot can learn to interact directly with its environment without an intervening user and without the need for its programmer to anticipate the nature of its relationship to that environment. In this sense its approach is ecological (Gibson 1979), sensorimotor, and adaptive.

What is missing in the symbol system account is an explanation of how symbol systems could learn to detect new objects that cannot be decomposed into already recognizable feature categories. Presumably new categories must be developed by natural selection or inductively through learning. However, one of the criteria of a symbol system is that it be semantically interpretable, which is to say that it be possible to assign its symbols a meaning. This precludes its formal rules from being applied to categories before they are fully fledged in the sense of corresponding to 'natural kinds' of things. What room is there then for modifying categories? Semantic interpretability has sometimes been taken to imply logical consistency because self-contradictory statements like *The king is and is not bald* are considered meaningless and, therefore, uninterpretable. However, it appears impossible for a symbol system to be logically consistent when the very symbol categories on which it is based face, as the result of selective pressures and new experience, continual selective and inductive modification.

1.1. *Outline*

Section two provides some philosophical background to the symbol grounding problem. The lack of adaptability of symbol systems reveals itself in the limitations of robotic implementations based on feature detectors. Attempts at making symbol systems more adaptable fail because (1) they make naive assumptions about the structure of reality and the ease with which it may be perceived, (2) they rely on the introduction of a homunculus — a teacher or innate knowledge — that exceeds the capabilities of the system that are to be explained, or (3) their modularity prevents behavior-based skills from playing any role in planning.

Section three places the problem of grounding cognitive categories in a behavioral and biological context. To be biologically plausible and behaviorally efficacious, the symbol categories of a symbol system would need at least to be based on and preceded by developed sensorimotor expectations. Memory-based models can be used to learn sensorimotor relationships and have proved to be more flexible, robust, and biologically plausible than conventional models. It is shown how one particular model, which has been applied to learning hand-eye relationships, can be used to develop sensorimotor expectations.

Section four is intended to offer an empirical, bottom-up approach to grounding cognitive categories in sensorimotor projections. A preliminary simulation suggests that an expectational model can directly map tactile, chemoreceptive, and motor information and their expected physiological consequences. Not only can an agent use the model proposed to discover salient aspects of its environment (aspects we often identify with objects and features of objects) but also to direct both reactive behavior and longer-term planning.

It is hoped that by abandoning the demand that a cognitive theory be, *at its foundations*, formally rule-governed, it may be possible to chart a path from sensorimotor coordination to the use of symbols in abstract reasoning and interindividual communication.

2. Philosophical Background: The Debate over Perceptual Analysis

Working in such disparate fields as psychology, neuroscience, philosophy of mind, linguistics, and artificial intelligence, proponents of symbolic representation have set themselves a highly ambitious task: the explanation or simulation of thought and its physical manifestations. To this end many of them have exploited a powerful tool and metaphor. It has even been called the only game in town. It emerged in part from synergy between propositional logic and the technological marvel of our day, the digital computer. I am referring here to the symbol system. Most attempts at operationalizing thinking share much in common with it. Its symbolic representations are often said to reflect a mental realm of concepts and relations which serve to represent actual objects and events. In stronger versions these representations are expressed formally in what Fodor (1975) calls a *language of thought* (LOT). Harnad (1990) reconstructs from Fodor (1975), Newell (1980), Pylyshyn (1980) and others the following definition of a symbol system:

> A symbol system is (1) a set of arbitrary *physical tokens* (scratches on paper, holes on a tape, events in a digital computer, etc.) that are (2) manipulated on the basis of *explicit rules* that are (3) likewise physical tokens and *strings* of tokens. The rule-governed symbol-token manipulation is based (4) purely on the *shape* of the symbol tokens (not their "meaning"), i.e. it is purely *syntactic*, and consists of (5) *rulefully combining* and recombining symbol tokens. There are (6) primitive *atomic* symbol tokens and (7) *composite* symbol-token strings. The entire system and all its parts — the atomic tokens, the composite tokens, the syntactic manipulations (both actual and possible) and the rules — are all (8) *semantically interpretable*: The syntax can be *systematically* assigned a meaning (e.g. as standing for objects, as describing states of affairs). (p. 336)

Like Descartes' pineal gland, an intended purpose of a symbol system is to link the mental and the material (Stoutland 1988). It is meant to offer an intermediate and functional level of explanation between physics and the beliefs, goals, and desires of folk psychology (Russell's *propositional attitudes*). Analogies have been drawn with computer software because a symbol system's operation is conceived of as being independent of any particular physical realization, be it a digital computer or a brain. Because under this theory mental states can be implemented in limitless ways, Fodor and Pylyshyn do not generally suppose that mental categories can be reduced to brain categories.

A symbol system is reliant on a separate perceptual system maintaining consistency between its active symbolic representations and the states of affairs they are supposed to represent in the physical world.[5] This is what Harnad (1990) calls the *symbol grounding problem*. Harnad likens the activity of a symbol system to someone who does not know Chinese (or any other language) trying to learn Chinese from a Chinese/Chinese dictionary. You want to know what a symbol refers to, so you look it up, only to find more meaningless symbols. You move from symbol to symbol without ever arriving at the actual thing that the symbol represents. Thus, to escape infinite regress, it is argued that the meaning of some symbols must not be derived from other symbols.

Fries drew similar conclusions concerning scientific statements and what he called the *predilection for proofs*. He argued that the call for all statements to be logically justified by other statements leads to infinite regress. He concluded that the *mediate* knowledge represented in the symbols of a language must be justified by the *immediate* knowledge of perceptual experience. Haugeland (1985) discusses infinite regress in the derivation of symbol meaning in relation to AI programs whose only connection to the world is through textual interactions with their users. Only in the mind of their human user are these program's symbols connected to the things they represent.

The idea of a language of thought may be traced back to antiquity as most certainly may its critiques. It appears in the Enlightenment in the work of Leibniz (1950). Inspired by Hobbes' view of all reasoning being mere calculation, in *De Arte Combinitoria* Leibniz sought to develop a logical calculus on an idea from his schooldays that all complex concepts were but combinations of a few fundamental concepts. By contrast, in Plato's Cratylus Socrates argued that knowledge of things cannot be derived from symbols but must be derived through an investigation of interrelations between things.[6] Kant in his *Critique of Pure Reason* pointed out that a formal definition of a dog by itself is not sufficient to recognize a particular dog.

Interestingly enough, the language of thought hypothesis is in line with the empiricist tradition of Hume insofar as representations corresponding to *complex ideas* are composed of those corresponding to *simple ideas* as the result of experience. As Hume first put forth in his *Treatise of Human Nature* (1739), "The idea of a substance is nothing but a collection of simple ideas, that are united by the imagination, and have a particular name assigned to them, by which we are able to recall, either to ourselves or others, that collection" (1978: 16). In his *Enquiries* Hume asserts that we can conceive of a virtuous horse because we can conceive of virtue and unite that with the familiar figure of a horse (1975: 19). This is not unlike Fodor's suggestion that the concept *airplane* may be composed of the concepts *flying* and *machine* on the basis of experience (p. 96) or Harnad's suggestion that *zebra* may be composed of *horse* and *stripes*.

But Fodor goes further than Hume. He argues that, to account for the potentially infinite number of belief states we can entertain and our systematic capacity for understanding sentences, it would appear that concepts must be expressed in an internal language that is both *productive* and *systematic* (1981: 147–149).[7] Elementary representations are combined to form complex representations according to formal syntactic rules. Harnad likewise highlights the utility of symbol systems being systematic: in order that propositions of grounded symbols may be assigned a semantic interpretation (p. 343). What distinguishes Fodor's claim about airplanes from Hume's about virtuous horses is that Fodor argues that we use trivially simple definitions based on a doctrine of 'natural kinds' of things.

2.1. *The Language of Thought Nativist Position*

Where Harnad and Fodor differ is on the question of how symbols are to be grounded. Whereas Harnad favors LOT empiricism whereby objects are rendered identifiable by associating their invariant features, Fodor favors LOT nativism whereby they trigger concepts that are already latent in the mind. This controversy, long predating LOT, has raged in philosophy since at least the time of Locke and was later to infect psychology.

Both Fodor and Maze are critical of LOT empiricism. They consider it to beg the question: if it is possible for perception to organize symbols into hypotheses like *the robin is on the lawn* by association, "where do 'robin' and 'lawn' and, for that matter, 'being on,' come from? It must be obvious that the acquisition of the background knowledge presents just the same difficulties as the interpretation of current sensory information." (Maze 1991: 173) As Fodor (1975) points out, "If, in short, there are elementary concepts in terms of which all the others can be specified, then only the former needs to be assumed to be unlearned" (p. 96).

Fodor proposes a nativist solution to the problem of grounding the concepts expressed in a language of thought. He posits a passive mechanism of perceptual analysis that derives symbolic representations of real objects from sensory data (the raw unstructured readings of physical parameters). *Demons* each sensitive to a single physical property shriek *yes* or *no* depending on whether a hypothesis is present or absent in the environment. These demons activate innate elementary concepts which, once properly combined, are used to reason formally about the world.

Part of the controversy surrounding perceptual analysis has resulted from its apparent support for the existence of a 'grandmother cell' (a neuron which fires whenever you recognize your grandmother). However, similar architectures have won supporters in the neurosciences. Marr (1982) proposed a theory of stereoscopic vision based on *feature detectors* analogous to Fodor's shrieking demons.[8] The results of experiments on feature recognition in monkeys (e.g. Fujita *et al.* 1992) have actually suggested to some that visual memories may be written in an alphabet of iconic figures (Stryker 1992) not unlike Leibniz's "alphabet of human thoughts" (1951: 20). This interpretation contrasts with the view that recognition and recall are evoked by highly distributed brain activity (Rumelhart and Norman 1981; Rumelhart and McClelland 1986; Smolensky 1988).[9] This distributed account has lead others to conclude that it may be impossible to translate between thoughts and brain activity, for example, by means of an intervening cognitive level such as a language of thought (Stern 1991). It is unlikely that Fodor and Pylyshyn ever saw this as the purpose of LOT, as they argued for the autonomy of psychology and the irreducibility of intentional states.

By proposing an innate method of transducing elementary concepts from sensory projections, Fodor's nativism, exemplified by Marr, does not constitute a radical departure from the empiricist tradition. Although Hume (1975), its quintessential figure, believed that our ideas were not innate, he claimed that our impressions were (p. 22).[10] And as our simple ideas, according to Hume, are just copied from and caused by these impressions, it takes only a small step to conclude, as Fodor does, that simple ideas are already latent in the mind.

While Maze and Fodor agree that symbols cannot ultimately be grounded empirically, Maze also attacks Fodor's nativist view:

> By elementary concepts he seems to mean natural kinds, which, he points out, cannot be subjected to 'definitial elimination' without loss of meaning. Thus, the innate language must be provided with terms for every natural kind which we can potentially identify, which would include, just for a start, every one of the millions of species of animals, birds, fish, insects, vegetable life and so on with which the world is stocked. (1991: 173)

Interpreted in this way, Fodor's nativism implies that we are hard-coded to recognize specific things, like Antarctic penguins and DNA molecules, that our ancestors have never before seen. This would require a miraculous feat of evolution.[11] Maze concludes that the LOT theory of the mind must be false because (1) its symbols would be solipsist if ungrounded and (2) its symbols cannot be grounded, neither empirically nor innately. His first point may be arguably true, although Fodor (1980) appears to maintain that cognition is solipsist and that this is unfortunate but must be accepted. Maze's second point remains an open question — one that can be answered empirically. A preliminary chemoreceptive simulation in section three suggests that representation *can* be grounded on the basis of empirical and evolved adaptation.

Traditional academic robotics is fully compatible with LOT nativism, although few in the field would exhort this philosophical stance. The robot

Shakey designed at SRI is a good (though rather hackneyed) example of this design methodology (Nilsson 1984).[12] Its programmers furnished it with a stock of symbols and operator rules. The robot used the symbols to compose propositions. These propositions represented states of affairs such as relations between nearby objects. The robot scanned its environment in a perceptual stage to determine which propositions to include in its internal description of its surroundings. The presence of a box at a doorway, for example, might be represented as *at* (*box*, *doorway*). Solving a problem involved finding a chain of operator rules whose application would transform the propositions in the robot's current state to those in its goal state. Each rule (e.g., for moving a box) had certain preconditions (like the robot being at the box) and resulted in certain additions and deletions to the robot's world description (the location of both the robot and the box having changed). The chain of operator rules served as an action plan for the robot to carry out. This sensorimotor process involved matching internal symbols against external objects so that the objects could be located and moved.

The trouble with this approach is that it demanded that Shakey be stocked with symbols for every object and relation it could possibly be required to handle. Otherwise, whenever it needed a new symbol to represent a new elementary concept or relation, its programmer would have to add it. So long as Shakey remained within the confines of the simple environment of boxes and platforms expressly set up for it, the robot's symbols appeared to be grounded. However, it is important to note that Shakey had to rely on its programmer to set up a causal relation between its internal symbols and the objects it detected. The adaptability of robots possessing this kind of architecture could not begin to approach that of most vertebrates, let alone human beings, *unless* we include their programmers in our overall conception of what constitutes the robot.

However, it would be unacceptable to include the programmer as part of any robotic system that is intended to explain how the mind works. The reason this is a problem is because it introduces a homunculus, the programmer, who entirely duplicates the talents of the mind the robot's workings were meant to help explain (Dennett 1979: 123). A different sort of explanation of how objects originate is needed (for discussion, see Smith 1995).

It may seem unfair to criticize Shakey, a robot developed almost thirty years ago. However, very up-to-date robotics systems like SOMASS, designed by researchers who write about symbol grounding, not only exhibit Shakey's vices but try to make a virtue of them (Malcolm 1995). In SOMASS the intention was to make the clearest possible separation between the classical symbolic planning module and the behavior-based plan execution module so that the purity of symbolic planning would not be tarnished by the implementation-dependent details of sensorimotor coordination. In the planning module, a particular instance of an object is represented solely by types and combinations of types (instantiated by feature detectorsJ. Since types by their very nature abstract away instance-dependent detail, information crucial to sensorimotor coordination — such as information about an object's contours — is lost. This information is important for manipulating objects of varying shapes and sizes. In SOMASS, its handling is kept down in the plan execution module.

There are at least three reasons why *not* to place an artificial rift between planning and sensorimotor coordination, especially if the architecture is to serve as a cognitive model for symbol grounding. (1J The robot is dependent on *a priori* symbols and cannot learn new elementary symbols inductively. (2J Making plans about the movement of concrete objects requires a consideration of their analogue features. For example, it is necessary to take care in angling an object while moving it through a narrow passageway. (3J Analogue features of particular past episodes appear to influence thinking without having been typecast in advance. Higher-level thought may not so much depend on the ability to draw on ever more abstract categories as the ability to draw on particular instances that appear superficially remote but bear an abstract relation to the matter at hand.

2.2. *Can Neural Nets Redeem Perceptual Analysis?*

The connectionism versus symbolic representation debate has seen a history of rival claims made about the adequacy of each methodology in modeling mind, brain, and behavior.[13] Harnad (1990J suggests, however, that a successful theory may be required to capitalize on the advantages of both. He proposes that neural networks or other statistical learning mechanisms might be able to form basic categorical representations from invariant features in the environment. These representations could serve as grounded symbols, and a symbol system could be developed from them in a bottom up fashion.

Specifically, neural networks would create iconic and categorical representations. *Iconic representations* are analogue copies of "the proximal sensory projections of distal objects and events [preserving their] 'shape' faithfully" (p. 335J. "In the case of horses (and visionJ, they would be analogs of the many shapes that horses cast on our retinas." They allow us to discriminate between horses by "superimposing icons and registering their degree of disparity" (p. 342J. *Categorical representations* are "learned and innate feature detectors that pick out the invariant features of objects and event categories from their sensory projections" (p. 335). "They are icons that have been selectively filtered to preserve only some of the features of the shape of the sensory projection: those that reliably distinguish members from nonmembers of a category." They allow us, for example, to identify a horse as a horse and not "a mule or a donkey (or a giraffe, or a stone)" (p. 342). They serve as the grounded *elementary symbols* out of which, Harnad contends, a symbol system ought to be constructed.

Harnad advances the hypothesis that internal symbols could develop causal links with sensory projections through "the acquired internal changes that result from a history of behavioral interactions" with the distal objects they represent (p. 343). Nevertheless, despite his acknowledgment of the importance of a sensorimotor grounding, he appears to accept that a purely sensory grounding could develop independently of it (p. 341). Is this possible?

Perceptual analysis is usually thought of as setting up a type-token relationship so that an instance of a particular type activates its corresponding symbol. In this way the sensory projections of a robin might instantiate the *robin* symbol. Methods of learning this kind of input-output mapping divide according to those that are *supervised* and, hence, require at least some information about what their correct output should be and those that are *unsupervised* and, hence, require no additional information.[14] Unsupervised learning methods develop categories on the basis of similarities in their input. Without augmentation, however, these similarities are based solely on correlations in sensory projections. There is no reason to believe that these correlations would coincide with what we would perceive as being objects or with the behavioral interactions the environment could potentially afford the individual. In order to do that, the categories would need to take into account the influence of motor signals on affective and physiological variables. Even if unsupervised learning happened to causally connect a robin symbol with the sensory projections of a particular robin in a particular placement and posture as viewed from a particular perspective, since the link would depend solely on the current physical structure of sensory projections, there is no reason to suppose that it would remain in place across changes in placement, posture, perspective, etc. In a note concerning object constancy in an earlier paper (1987: 561), Harnad acknowledges this weakness of unsupervised learning, and this may be one reason why he does not propose its use for symbol grounding. A further danger with unsupervised methods is that, by accentuating the gross appearance of the sensory projections, potentially important detail is filtered out. This is precisely the opposite of what is needed.

Supervised learning methods require either reinforcements or a so-called teacher that gives the right answer. Harnad apparently had a teacher in mind when he proposed that "icons, paired with feedback indicating their names, could be processed by a connectionist network that learns to identify" them correctly (p. 344). However, to give correct instruction a teacher must have access to the same categories that are to be learned. If the teacher or reinforcer were *innate*, there would be no fundamental sense in which the categories could be developed *empirically*. But only in humans, where there *are* teachers, could any explicit instruction be taken from an *external* source. Moreover, as Socrates pointed out in the Cratylus, there is an aetiological problem with relying on a teacher to name objects: "If things are only to be known through names, how can we suppose that the givers of names had knowledge ... before there were names at all, and therefore before they could have known them?" (Plato 1953: 104).

Harnad's approach has several drawbacks. It would appear that, against its own claim, categories are not developed from the bottom up but learned from a teacher. This assumes *a priori* symbol categories, not categories that are developed from experience. The introduction has already argued that a symbol system could not develop elementary categories empirically and retain systematicity and semantic interpretability (Harnad's eighth requirement). Furthermore, a system that learns in this manner introduces a teacher-homunculus that exceeds the cognitive capacities of the system it is meant to help explain.

From reinforcements it is possible to learn without explicit instruction. Unfortunately, reinforcement learning only addresses how a decision function to select the best action in a given state may be learned by trial and error. Its convergence rests on the Markov property that the identification of a state is sufficient to determine what action to take. This means that each state must be recognizable by those sensory features that are relevant to the prediction of state-transitions and rewards (see Watkins 1989; Watkins and Dayan 1992). In practice, the burden of ascertaining what features should determine the state is placed squarely on the designer who hand-codes an *a priori* state detection mechanism.

Reinforcements are, of course, available both to humans and to creatures in the wild but (in lieu of an innate categorizer) they must be appraised in terms of the consequences of action. If categorical representations of states are to be learned from reinforcements (as they potentially could be), we must grant that a sensorimotor grounding is a prerequisite for a sensory grounding. In this case, perceptual analysis could not be considered a wholly passive process. Any passively learned categories would require scaffolding from either innate or developed sensorimotor categories. An account of how this could feasibly be accomplished is outlined in the next two sections.

3. Developing Sensorimotor Expectations: A Foundation for Perceptual Categories

This section considers the shortcomings of nonadaptive approaches to grounding perceptual categories. By developing a fuller, more biological notion of grounding, it attempts to go beyond the minimal requirement that they be causally linked with actual objects, events, and relations. At their foundations, representations are seen as part of an organism's sensorimotor loop with its environment and, through an empirical process of adaptation, they mediate its activity. In developing this idea, the section advances a theory of sensorimotor expectations and explores computational techniques for implementing it.

It would be difficult for proponents of symbolic representation to explain its evolutionary origin without granting that it, at some point, must have conferred on its bearer selective advantage.[15] Symbolic representation can offer no benefit, however, unless symbolic categories capture features of the surrounding world that are relevant to an organism behaving so as to enhance its reproductive success. Therefore, to achieve minimal biological grounding, an internal category must stand in some relation to sensorimotor coordination.

Genes are passed on when an organism is sufficiently able to discriminate sensorimotor patterning relevant to its reproductive success. The essential kind of invariance in this patterning, the kind that motivates an organism at least to behave as though it categorizes different instances, with their disparate sensory projections, as similar, is that those different instances can be dealt with in similar ways to produce similar physiological effects. This is not to deny that why we learn what we learn is intimately tied up with affect and is only ultimately determined by natural selection. However, we may for the purposes of this section safely put aside the complex affective and social components of behavior, for in the following discussion it is only necessary to consider the simplest forms of adaptation to understand the drawbacks of perceptual analysis (perception based on innate feature detectors).

3.1. *Evidence from Vision for Sensorimotor Integration*

The brain's processing of visual input is highly complex, and much occurs even before this information is integrated with sensorimotor information from other modalities. Hence, in humans and other complex organisms, visual information cannot be mapped directly onto sensorimotor categories. Nevertheless, sensorimotor integration must come at some point in order to ground perceptual categories in sensorimotor activity, and I would argue that this grounding comes prior to and serves as the foundation for more abstract, systematic modes of behavior.

Vision's dependence on coordination is well acknowledged. The manner and degree of this dependence has been contested by neuroscientists for more than 40 years — ever since it became clear that whether the brain attributes a movement on the retinal image to the viewer or to an object depends on whether the motion was self-induced (see Gyr *et al.* 1979). It is only by taking self-induced movements into account, for example, that the brain can maintain the stability of the visual scene. Kohler (1964) showed that such movements were necessary for subjects to adapt to seeing the world through reversing prisms. The importance of movement is illustrated by the fact that if the eye is held focused at a fixed point in a static scene, once neurons have habituated the viewer even fails to see. Distance perception may depend more heavily on movement parallax than binocular disparity, and evidence suggests that infants become sensitive to it first (Slater 1989). However, Marr's stereoscopic theory of vision echoes perceptual analysis in its neglect of sensorimotor coordination. However, there has been a growing realization in the 1990s that computational models need to be extended to account for the role of action in vision (Ballard 1991; Blake and Yuille 1992).

A consideration of the shortcomings in Marr's attempt at perceptual analysis will illustrate the importance of relating internal categories to sensorimotor coordination. Marr proposed that different neurons in a network could be attuned to varying degrees of binocular disparity in the retinal image. In this manner a network could extract a viewer's distance from a perceived object. However, to be grounded, this distance measure would need to be related to the viewer's behavior, such as the movements necessary to grasp the object. Otherwise, the measure could not serve any behavioral ends nor could a selective process be involved directly in its evolution.

This observation is acutely evident with body variables. For example, although there is a correspondence between the topography of the retina and arrays of cells in various cortical regions, the output from these cells does not capture any spatial relationship.[16] To accomplish this, further representations would be required to relate the retinal image to the individual's motor activities such as the eye movements that would be required to foveate a peripheral object. To be effectual and, therefore, evolutionarily explicable, symbols must be grounded in sensorimotor categories that are relevant to coordination. Otherwise, they are as useless as a cockpit furnished with unlabeled and undimensionalized instruments in the hands of a pilot lacking instructions as to their bearing on the flight of the aircraft.

3.2. *A Theory of Developed Sensorimotor Expectations*

The aim of this paper is to propose a possible method of modeling the development of behaviorally-relevant categories. The concept of *sensorimotor representation* is substituted for *symbolic representation*. It denotes adaptations that take place within an individual on the basis of past sensorimotor projections, and it depends on, among other things, the individual's particular body, affective systems, and life history. When an agent is using an internal representation, what makes that representation a representation is not the fact that it looks like one to us (that we can interpret it) but that it can function as one for the agent. What this means for our purposes is that the organism's sensorimotor processes that need to be appropriately related to objects, events, or relations can be appropriately related to their internal representations.

There is likely to be a close relationship between how representations function and how they develop and change. With this in mind, Sommerhoff and MacDorman (1994, §5) proposed an approach to modeling sensorimotor relationships based on neurophysiological expectancies. They also discussed evidence for these expectancies from brain research and suggested one way they might be represented neurally. This expectational approach shall be further developed here from the standpoint of behavioral simulation.

An individual develops expectations on the basis of spatial, temporal, and affective correlations in sensorimotor projections. These expectations need not be conscious. They may concern, for example, the consequent effect of self-induced movement on sensory projections from internal and external sense organs. An individual may also develop them passively. They may, therefore, concern likely trends in physiological, affective, and external sensory projections that can occur either without motor involvement or independently of it. An individual's anticipation of consequences is always contingent on that individual acting in particular environmental circumstances. Therefore, sensorimotor projections elicit, revise, and sustain an individual's developed expectations, and only a subset of them will be active at any given moment. They are in this sense conditional expectations.

Once activated, expectations prepare the individual for the kinds of sensorimotor projections that typically ensue. They do this by expediting *anticipatory responses*. Unexpected sensorimotor projections initiate *orienting responses*. They stimulate the revision of old expectations and the development of new ones in order to account for the unexpected projections. Sensorimotor expectations are able to model the relationship between individual and environment because they are developed by means of sensorimotor projections from actual bodily interactions. Other individuals are an important part of that environment and certainly the most complex (see Cowley and MacDorman, 1995).

We shall next consider how this theory of expectations can be used to adaptively ground an *a priori* symbol system. Memory-based techniques provide one possible route to its implementation.

3.3. *Grounding an A Priori Symbol System*

The propositions of a symbol system resemble the disembodied knowledge found in books. This contrasts with an individual's knowledge which is full of affective variables and relates to body and personal history. As with the words in a book, the symbols of a symbol system are ungrounded insofar as they have been *abstracted* from their relation to a particular person. Moreover, the symbols manipulated by most AI programs are not part of an autonomous system (e.g. a robot). Hence, they are not required to mediate between the program's own sensory projections and motor signals. This is because the program has neither. The symbols the program generates are only intended for human consumption, and their meaning is determined by a user's interpretation. Although these programs cannot interact with the world through bodies under their direct control, their output influences their user's behavior; and people have successfully applied them to a variety of endeavors from mineral exploration to medical diagnosis.

A robot, by contrast, must rely on its own sensors and actuators. If Fodor claims that internal symbols can be functionally analogous to intentional states, in order to relate these symbols to the outside world, they ought to be abstracted from and grounded in the robot's own sensorimotor coordination. If we consider the difference between an agent acting on the presence of objects, events, and relations by recognizing them for itself or by being given a description of them (perhaps by its own perceptual system), intuitively, the difference may seem trivial. This is because we can simply perceive a world populated by objects, and much of the brain activity involved in recognizing and responding to them is introspectively opaque. Being subconscious it can only be inferred.[17] We further take for granted that our similar bodies, sense organs, affective systems, language, culture, etc., makes possible intersubjective agreement on what constitutes an object. But far from being trivial, grounding symbols in behavior has proved unexpectedly difficult. For example, it appears to be much harder to automate the task of moving chess pieces about a board than that of deciding what moves to make.

Very few programs including programs that play chess could be said to have the clear semantics of a symbol system. Internal symbols often do not stand for external states of affairs, and indeed their semantics can be rather convoluted (Smith, forthcoming). Nevertheless, by Harnad's eight criteria (1990: 336), we may give a symbol-system interpretation to significant portions of a chess-playing program. The game has explicit rules for the abstract movement of pieces (as opposed to their physical movement on a board). The manipulation of internal representations is purely syntactic, and a significant subset of these representations may be interpreted semantically — for example, as standing for chess pieces, board positions, and chess moves.

To illustrate how learning can provide a well-adapted sensorimotor grounding to the symbols shared by a chess program and a piece locating system, let us consider in more detail an aspect of the hand–eye coordination task of moving chess pieces. A robot must move its six-jointed arm so that its hand is positioned to grasp a piece. For eyes the robot has two cameras, both trained on the chess board. If either its hand or a chess piece becomes visible, feature detectors activate a proposition that contains the corresponding symbol and its location in camera-based coordinates. For example, the proposition *at (rook, cam$_1$loc (497, 325), cam$_2$loc (890, 194))* may signify that there is a castle centered at the position $x = 497$, $y = 325$ in the first camera's image plane and $x = 890$, $y = 194$ in the second camera's image plane. To give the *rook* symbol a sensorimotor grounding, the robot must be able to pick up the castle and move it across the board. The robot could move its hand into position by constantly monitoring the visual consequences of small changes in its joint angles, and using this negative feedback to correct for errors in the manner of a servo-mechanism. This method would be comparatively slow, however. If only the robot were able to know which joint angles corresponded to the object's visual location, it could make a beeline through the coordinate space of joint angles, thus positioning its hand in one swift and graceful movement.

Technically, the robot needs to be able to make a *phase space transformation* from points in the 4–D visual coordinate space of the camera's image planes to points in the 6–D proprioceptive coordinate space of the robot arm's joint angles. In humans some neuroscientists have assigned this task to the cerebellum.[18] People who lack a working cerebellum cannot make rapid well-coordinated movements and must apparently rely on constant conscious monitoring of small motions. In robotics the standard approach has been to compute the transformation according to a mathematical function that has been analytically derived from a set of equations describing the geometry of the robot's cameras and the kinematics of its arm (Paul 1981). This kind of *a priori* mapping leads to robots whose behavior lacks resilience to mishaps, such as a joint sticking or a camera being knocked askew. Omohundro (1990) contrasts this rigidity with the adaptability found in nature:

> Biology must deal with the control of limbs which change in size and shape during an organism's lifetime. An important component of the biological solution is to build up the mapping between sensory domains by learning. A baby flails its arms about and sees the visual consequences of its motor actions. Such an approach can be much more robust than the analytical one. (p. 310)

Considering the speed at which neural signals travel, servo-mechanical control is not fast enough to account for the fluidity of human movement. Taking a swing at a golf ball and many other kinds of movement are simply too rapid to rely on it. This is especially true of ballistic movements like throwing a ball (see Carpenter 1990: ch. 9). Were these movements to exploit any kind of feedback, it could *only* come from an internal mapping. This is because external feedback concerning goal attainment (where the ball has landed) arrives after control of the object has ceased (the hand has already released the ball).

Furthermore, innate motor patterns are not flexible enough to provide a sensorimotor mapping for bodies that grow and change in sometimes unpredictable ways. Following any significant alteration in kinematics, perseverative changes in a creature's nervous system are needed to effect adaptations in sensorimotor mappings. This is why, *contra* Harnad (1990: 341), sensory feedback and innate motor patterns probably play a relatively minor role in grounding perceptual categories as compared to patterns of behavior developed in light of sensorimotor experience. We should not be surprised if all perceptual information is, to some extent, filtered through developed sensorimotor mappings in which case sensorimotor learning must be a prerequisite for any kind of passive learning.

3.4. *Memory-based Implementations for Learning Sensorimotor Mappings*

There is a broad class of algorithms for learning nonlinear mappings like those needed for sensorimotor coordination, including, for example, feed-forward neural networks that learn connection weights by back-propagation (Rumelhart and McClelland 1986). Indeed, connectionism has already been applied to learning in sensorimotor domains (e.g. Mell 1988). However, Omohundro (1987, 1990) has instead advocated the application of closest point learning algorithms to geometric problems. They are easier to analyze and, for this kind of application, learn far more quickly and accurately (at least when implemented on ordinary computers). Methods abound for finding closest points, and they shall not be discussed here (see Friedman *et al.* 1977; Tamminen 1982; Ramasubramanian and Paliwal 1992; Micó, Oncina and Vidal 1994). There are even efficient methods of finding closest points in dissimilarity spaces where, as is the case for phoneme recognition, the dimensionality of the instance and the categorical representation are likely to differ and the computational cost of comparing them is high. For simplicity's sake, however, discussion in this paper will be limited to Euclidean distance measures in ordinary vector spaces.

Clocksin and Moore (1989) successfully applied a closest point algorithm to learning stereoscopic hand-eye coordination like that required for moving chess pieces. Clutching a pen light (to simplify the recognition of its hand), their robot moved its arm about. It recorded in a composite phase space the locations it visited in its visual and proprioceptive subspaces. If nearby points in its visual subspace had been reached by more than one set of joint angles, it would only remember the set that resulted in the least contortion to its arm. During a 'dreaming' period, the robot was disconnected from its arm and recorded further points by making linear interpolations from neighboring points. To approximate the joint angles corresponding to a new visual location, the robot consulted the joint angles for the closest point recorded in the visual part of its phase space. Thus, previous sensory projections acted as categorical representations to divide up the phase space into Voronoi hyperpolyhedra (a special class of multidimensional convex hulls where the hyperplane boundaries that quantize the phase space are equidistant from the closest two points). If an object were recognized, the image plane coordinates of its location would fall into one of these hulls, and the corresponding representation would serve as the robot's expectation about the movements required to approach it. In a comparatively short period of time the robot learned to move its hand to within a centimeter of a given goal point without the use of an algorithm for either stereoscopic vision (like that proposed by Marr) or arm kinematics.

Even more significantly, phase space learning methods (of which the closest point method is only one example) are highly general. There are only two assumptions implicit in the above model, and both of them generalize to other geometric problems. The first is *continuity* which indicates that nearby points in the domain of a mapping are also nearby in its range. The second is *smoothness* which indicates that the mapping can be approximated by local interpolation. The shortcoming with this example is that a chess board shares little in common with the complexity of the real world. However, the learning techniques used here generalize to more intricate environments. As Clocksin and Moore note, the same class of techniques could have been used if the robot were controlling wings or fins or viewing the world through reversing prisms.

3.5. *Improving the Implementation in Light of the Expectational Theory*

Sommerhoff and MacDorman's high-level description of how an individual could develop expectations (beginning with representations about sensorimotor relationships) does not immediately suggest a particular implementation. Nevertheless, we may appraise the suitability of an implementation in terms of a description such as theirs. We shall presently consider Clocksin and Moore's hand-eye coordinating robot. The suitability of their implementation will, of course, depend on many things that must be left out of the high-level description, for example, what computing elements are available and the intended sensorimotor domain. Unsurprisingly, implementations that model qualitatively different kinds of patterning will require some degree of differentiation and modularity. Nevertheless, given some of the broad adaptive features of biological systems considered in this section, we may use Sommerhoff and MacDorman's description to appraise Clocksin and Moore's implementation and to suggest how it may be improved.

We can do this by assigning the functional particulars of the implementation an *intertheoretic interpretation* in terms of the description. In this way, we can evaluate the plausibility of the implementation *vis-à-vis* the description. Indeed, in addition to implementations this process can be applied to representational frameworks such as Fodor's language of thought. The more detailed the description, the more deficiencies it can potentially expose in a theory or implementation, so ideally it should be very detailed (containing as much low-level information as possible), and cognitive scientists should all be able to agree on it. This is, unfortunately, not realistic given how controversial representational theories in this field are. In what follows we shall analyze, in the context of our chess game, Clocksin and Moore's implementation of hand-eye coordination in terms of Sommerhoff and MacDorman's high-level account of how an individual represents sensorimotor relationships (§3 and 5, 1994).

The proposition about the rook's location on the robot's image planes qualifies as a sensorimotor representation because the sensorimotor processes that need to be appropriately related to the manipulation of the rook are appropriately related to the proposition. Specifically, the processes that determine where to move the robot hand rely on the proposition's image plane location values for the rook. Also, the robot derives lasting benefit from sensing the visual and proprioceptive consequences of its arm movements insofar as they leave their mark in the form of corresponding sensorimotor expectations about the movements required to reach visual locations. Indeed, the core of the robot's behaviorally grounded representations consists of these developed expectations. Hence, Clocksin and Moore's implementation certainly reflects the spirit of Sommerhoff and MacDorman's high-level description and, up to a point, conforms to it. The robot, however, does not exhibit *orienting responses* to unexpected sensorimotor projections. They are generally accompanied by a shift in attention from ongoing activity to the unanticipated perception and, according to the present approach, lead to the development of new expectations or the modification of insufficiently accurate expectations.

To account for orienting responses, the robot must be able to shift emphasis between developing, correcting, and using its phase space model as circumstances dictate. If the robot's expectations are sufficiently accurate for it to position its hand so as to move a piece, the robot need not augment or modify them. If not, the robot can, in the final stages of positioning its hand, resort to positioning it servo-mechanically by exploiting negative feedback from vision. (This appears to be what a person does in extending an arm to place a finger on a spot. The arm movement slows right before the spot is reached.) Once suitably positioned, the robot can then store in its phase space the new visual and proprioceptive position of its hand so as to fill in the gap in its sensorimotor model. It is also important that the robot be able to distinguish between a lack of expectations and expectations that are no longer sufficiently accurate because of bodily or environmental change. A lack of expectations can be corrected simply by further exploration so that expectations may be developed from previously-unencountered sensorimotor projections. However, insufficiently accurate expectations must be modified, amended, or forgotten. If the robot's arm or cameras are thrown into misalignment, this may necessitate replacing all the expectations that are related to hand-eye coordination, and it may be more efficient simply to forget them all and to start afresh by relearning the mapping.

Once amended Clocksin and Moore's implementation conforms to Sommerhoff and MacDorman's high-level description at least insofar as it concerns the robot's sensorimotor coordination in mapping from visual locations to corresponding arm movements. (Piece recognition is another matter.) All the points in the robot's phase space constitute developed conditional expectations. Active expectations — those that have been elicited, revised, or sustained by sensorimotor projections — prepare the robot with anticipatory responses for sensorimotor projections that are likely to follow. In the present example, the recognition of a particular chess piece triggers the anticipatory response of the robot's arm movement; in turn, the arm movement triggers the anticipatory response of grasping the rook.[19] Unexpected projections initiate orienting responses in order to develop new expectations or modify insufficiently accurate expectations. If the arm is not positioned close enough to the rook, the robot must resort to positioning it servo-mechanically and then augment or modify its expectations.

3.6. *Implications and Extensions*

One important feature of Clocksin and Moore's memory-based approach to hand-eye coordination is that no intermediary form of representation was necessary for the robot to map from vision to proprioception. The robot was able to integrate information from these two modalities without, for example, computing a depth map from binocular disparity. A depth map by itself is only grounded insofar as it serves (or could potentially serve) to coordinate the robot's activity. It would have required considerably more computation than their memory-based approach, and it is unclear what it could have contributed because values for depths would still need to be mapped onto arm movements. The mapping is more straightforwardly performed directly from the image planes of the two cameras. If the depth values were to be represented in some objective unit of measure, the robot would need to have a means of calibrating itself accordingly. A depth map would also probably make it harder for the robot to adapt to unanticipated change. Even were it to choose its units subjectively for its own convenience, if its cameras were knocked out of alignment, it would be difficult for it to compensate for the change in its camera geometry. Incorrect values would require it to relearn the sensorimotor mapping from scratch, thus offering no advantage over Clocksin and Moore's approach.

This is not to deny that intermediary forms of representation do at some point become useful or even necessary. However, there is little point in developing representational models without first considering how they are to mediate between sensation and action. Otherwise, we might develop models based on our own introspection that, in practical robotics problems, would prove unworkable or superfluous.

In the hand-eye coordination example, sensorimotor projections from vision elicit expectations about proprioception. This kind of mapping is also possible within the same modality. By the phase space methods described, for example, a robot can learn to map between movement parallax and binocular disparity, or binocular disparity and vergence, or vergence and changes in apparent size, etc., or any combination of these. Information from these sources can also be used to set up expectations about the movements required to reach an object. Bodily movements can likewise be used to set up expectations related to vision. As mentioned at the beginning of the section, this has been observed and studied extensively in humans. A person walking without vision from one location to another automatically updates the relative direction of surrounding objects (see Rieser, Guth and Hill 1986). From past experience an individual develops a feeling about which way to turn.

This too can be modeled expectationally. For example, a robot could learn to update, among other things, expectations about the direction it would have to turn to reach objects solely on the basis of proprioceptive information. In this way, a robot can use one source of sensory information to elicit expectations that are more directly related to a second source and, thereby, to compensate for a lack of information from that second source. In cases like this, the expectations would probably be less accurate; however, phase space models at least offer graceful degradation in predictive accuracy when fewer dimensions are available. (In the simplest closest point model, this means fewer terms in the Euclidean distance measure.)

Expectations about how bodily movements transform a viewpoint-dependent representation of an object (or configuration of objects) can in fact serve as a viewpoint-independent representation of that object. *By themselves*, the sensory projections of an object as viewed from a particular perspective can say little about either the projections of that object as viewed from another perspective or how motor actions could be related to that object. However, a viewpoint-dependent representation *plus* expectations about how motor actions transform its projections make it possible to predict the appearance of the object from other angles and distances. Thus, they together provide a viewpoint-independent representation.[20] From this we can see that sensorimotor expectations concerning how motor actions transform an object's sensory projections indeed facilitate the development of an allocentric (object-centered) frame of reference from an egocentric frame of reference (see Feldman 1985, for background). This is because expectations concerning transformations caused by self-induced movements can be used to compensate for those movements, for example, in visualizing activity from an allocentric perspective.

As mappings develop between and within different sensory modalities, it is possible to see how expectations can be elicited by ever more indirect means of contact. For example, a robot might initially be obliged to follow the contours of objects with its hands and eyes to discern their shape and how they behave. Gradually, sensorimotor expectations could develop from it having touched and foveated points along the contours of objects. These expectations map the topography of the image plane to eye saccades so that it is no longer necessary to follow the contours of an object with hand or eye to elicit expectations about its shape. Instead its shape is recognizable from the still image. Eventually, sensorimotor expectations develop not only through self-induced movement but also by passively watching things happen.

So far we have only increased the adaptability of the robot's hand-eye coordination. Yet although the robot can now adaptively adjust to changes in its visual geometry and arm kinematics, its feature detecting mechanisms do not improve with practice. Moreover, it can only classify objects as belonging to a finite number of specified categories. This is no disadvantage in playing chess because there are only six kinds of pieces. However, the real world is more complex. To adapt to environmental change it may be necessary to become sensitized to new kinds of sensorimotor patterning or face extinction. The next section is intended to illustrate how this could be accomplished.

As we can see, there are a number of ways in which playing chess is not like getting about in a natural environment. On the one hand, chess players can agree on the fundamental objects of chess (its ontology). These objects are the game's (abstract) pieces, board positions, legal moves, etc. Chess players may, of course, disagree about other matters such as whether the pieces are situated in a particular strategic arrangement, the significance of that arrangement, and what term should best describe it. But, on the other hand, it would be dubious to suppose that what exists for each individual in the surrounding natural (or social) environment is just the same as what exists for other individuals (or other species) and can be readily mapped onto internal representations. What that individual notices or responds to (or can notice or can respond to) depends in part on such things as that individual's body, interests, and life history.

There are other limitations about the domain of chess that, unlike physical environments, make the game amenable to symbol systems. Unlike real environments, chess only requires the recognition of a few objects which are known in advance: six kinds of chess pieces of two different colors, and their position on the chess board. Likewise, abstract piece movement is constrained by the rules of play which do not change. Furthermore, a chess program can get away with being solipsistic. A human can directly interpret the meaning of its output so the program does not need to connect its symbols to actual pieces and board positions. However, even in a highly constrained domain like chess, if a chess program had to recognize and move actual chess pieces, an *a priori* grounding of the kind provided by feature detectors may not be flexible enough to recognize all the different kinds of chess sets it might potentially confront.

4. Adaptively Grounding a Symbol System from the Bottom Up

We began, in the above discussion, with a pre-existing chess playing symbol system and a pre-existing piece locating perceptual module. The point was to consider a method of grounding their shared symbols adaptively in sensorimotor coordination. So far we have taken for granted the task of recognizing objects. Although innate detectors may recognize low-level features of objects and even whole objects where they are particularly common and important (like faces and hands), there are unlikely to be innate detectors for even a fraction of the objects that people recognize in their daily lives. So, left unanswered are questions of how behaviorally relevant categories could be discovered, how they could serve as the elementary symbols of a symbol system, and how a symbol system might develop from the bottom up. I should like to address them shortly through the simulation of a fish in a simple aquatic environment, but first shall give a more abstract introduction of the model I will be using.

4.1. *A Phase Space Model for Integrating Multimodal Sensorimotor Information*

Of the five senses, perceptual research has tended to concentrate on the last to evolve: vision. This sense appears to rely on highly evolved special-purpose processes for recognition. Unfortunately, this fact has tended to distract attention from the importance of learning to integrate multimodal sensorimotor information. The purpose of the following expectational model is to demonstrate how it is possible to directly map sensorimotor information from more basic senses like smell and touch to cognitive categories that can be used not only for reactive behavior but also higher-level planning. It may be possible to adapt similar principles to vision after a certain amount of bootstrapping from innate feature detectors.

We may represent the sensory stimulation that an agent receives from *external* sources as a point **e** $(e_1, e_2, \ldots e_n)$ in a sensory subspace E. The stimulation may arrive either directly from its sensors or pass through an intervening preprocessing stage. (Preprocessing in Clocksin and Moore's penlight-holding robot, for example, reduced two large matrices of light intensity values from the two camera's image planes to a point in a visual phase space of just four dimensions.) Likewise, we may represent the sensory stimulation that an agent receives from *internal* sources as a point **i** in a sensory subspace I. For a very simple agent the point may be determined by a single scalar value such as a reinforcement from a reward function. In a more complex agent, it may reflect numerous physiological, affective, or higher-level cognitive variables. Finally, we may also represent the motor signals the agent sends to its actuators as a point **m** in a motor subspace M.

The agent can represent the consequence of a self-induced action by storing points in a *sensorimotor phase space* composed of its motor subspace and its sensory subspaces before and after carrying out the action. We will assume for the moment discreet time steps where the sensory effects of an action are available by the next time step. Thus, at time t, $(\mathbf{m}_t, \mathbf{e}_t, \mathbf{e}_{t+1}, \mathbf{i}_t, \mathbf{i}_{t+1})$ determine the location of the point $\mathbf{p}_t$ in a composite phase space. The internal changes $\Delta\mathbf{i}_t$ correlated with $(\mathbf{m}_t, \mathbf{i}_t, \mathbf{e}_t)$ are simply $\mathbf{i}_{t+1} - \mathbf{i}_t$, and the external changes $\Delta\mathbf{e}_t$ are $\mathbf{e}_{t+1} - \mathbf{e}_t$.

As long as there is some degree of consistency across time and space in what the agent senses, it can exploit the points in its sensorimotor phase space to predict the consequences of its actions based on what has happened to it in the past. Once again, this may be achieve by the closest point method. Thus, the agent may approximate the likely affect of a motor signal **m** on its sensory stimulation $(\mathbf{i}_t, \mathbf{e}_t)$ by finding the point $\mathbf{p}_x$ $(\boldsymbol{m}_x, \mathbf{i}_x, \mathbf{e}_x, \mathbf{i}_{x+1}, \mathbf{e}_{x+1})$ such that $(\mathbf{m}_x, \mathbf{i}_x, \mathbf{e}_x)$ is the closest point to $(\mathbf{m}, \mathbf{i}_t, \mathbf{e}_t)$ in that subspace. The predicted values for sensory stimulation in the next time step are $(\mathbf{i}_{x+1}, \mathbf{e}_{x+1})$. It follows that the predicted change in the agent's internal sensory stimulation is $\mathbf{i}_{x+1} - \mathbf{i}_t$, thus reflecting expected change in its physiological state as sensed and its affective state; the predicted change in the agent's external sensory stimulation is $\mathbf{e}_{x+1} - \mathbf{e}_t$. Thus, points contained within the sensorimotor phase space may serve as *expectations* about the physiological, affective, and ecological consequences of possible actions. The agent may select the action it deems best on the basis of these expectations or, in order to fill in gaps in the model, an action with relatively unknown consequences.

This method upholds the property of graceful degradation found in artificial neural networks without sacrificing semantic interpretability. If we consider the case where the closest point is selected by means of an Euclidean distance measure, estimates will be less accurate if the squared difference between the current external sensory stimulation and all the previously recorded external sensory points ($\mathbf{e}_1$ to $\mathbf{e}_t$) cannot be calculated for every relevant dimension. However, the agent's ability to find a closest point would be impaired only if the lack of contribution from those dimensions would cause a different point to be selected. That point may still be acceptable. Just as someone who is blinded may still be able to perform a task, so may the agent despite input from one or more sensory modalities or sources of sensory information being incomplete or missing.

The closest point method may be put to use calculating many other kinds of approximate mappings than those discussed here. For example, it is possible to fill in lost data from a sensor with estimates by finding the closest point in the external sensory subspace based on the remaining data.

Importantly, to an outside observer, the semantics of the system should be relatively clear. By peering into the phase space we can always explain why a certain prediction was made by tracing it to the particular sensorimotor projections on which it is grounded. Of course, this does not mean that an observer would be able to see what the agent 'sees.' The agent's ontology — what there is to it — will depend on what it senses and, therefore, its particular sensors and how it appraises their output in terms of its developed expectations.

It would be useful for the agent to be able to develop the most accurate sensorimotor model possible in the least amount of time. Ideally, the closest point to the agent's current sensory stimulation in phase space will indeed provide an expectation that offers the best prediction. But just what should the agent be trying to predict? That depends on what we want the agent to *do*. In nature selective pressures would favor models that offer a clear choice between those actions that are likely to enhance reproductive success and those that are not. As an animal cannot relive its life to see what works best, it can only estimate indirectly what action is most likely to enhance its reproductive success. The expected future proximity of physiological and affective variables to their optimum provides a measure of its expected future well being (and, as we shall discuss, can be estimated using Q-learning or other methods of learning from delayed rewards). Exactly how physiological and affective variables should best be weighted can be left to evolution. Hence, it is indirectly, through these variables, that evolution influences perception and behavior. (Of course, instinct, which is not discussed in this paper, also plays a role in behavior.)

There are numerous ways an agent can improve its model once it has indicators in terms of which it can evaluate its model's predictive power.

Statistical methods can be used to estimate the linear or nonlinear correlation between the indicators and the other sensorimotor dimensions so that dimensions can be weighted according to their likely impact on the indicators. (This method can be used to eliminate irrelevant sensorimotor dimensions.) However, often their relative influence will vary from one region of the phase space to another. For ease of illustration, we may consider a two-segmented arm being monitored by a camera that is fixed perpendicular to the plane in which the arm has mobility. When the arm is extended, a change in the angle between the base and the first segment of the arm will have a much greater influence over the location of the arm's end in camera-based coordinates than when the arm is contracted. (You may easily demonstrate this by moving your own arm.)

In the arm example, a better method would be local linear interpolation. We are trying to approximate a mapping from the visual to the proprioceptive subspace, so that we can approximate the proprioceptive coordinates of some new point **p** given only its visual coordinates. If the proprioceptive subspace has k dimensions, we need $k + 1$ nearby points in the visual subspace in order to interpolate (or extrapolate) the proprioceptive coordinates of **p**. An added restriction is that the points not fall on the same hyperplane of dimensionality $k - 1$. A weighted sum of vectors emanating from one of the $k + 1$ points to the k remaining points describes the location of **p** in the visual subspace, which, of course, is known from the camera's image plane. This provides a system of k linear equations with k unknowns. A straight-forward algorithmic solution provides k weights. The $k + 1$ points in visual subspace correspond to $k + 1$ points in the proprioceptive subspace. Hence the k vectors in terms of which the new point was originally described in the visual subspace have k corresponding vectors in the proprioceptive subspace. The weights may be applied to these vectors, providing an estimate of the new point in the proprioceptive subspace.

4.2. *Forming Behaviorally Relevant Categories by Adaptive Quantization*

Staying in constant motion a fish swims about a shallow pond. As the fish sways back and forth, its receptors sample the concentrations of different chemicals in the water.[21] It controls its behavior through two decision-making systems, one which controls its oral cavity and the other which controls its navigation. The former is only activated when the fish bumps into something, perhaps another fish or a piece of debris. When this happens, the fish selects one of four motor responses. It (1) ignores it, (2) ingests it, (3) takes it up in its mouth, or (4) empty the contents of its mouth. Its choice of action at time t is represented by a point $\mathbf{m}_t$ in the motor subspace of its *sensorimotor phase space* ($\mathbf{m}_t$, $\mathbf{e}_t$, $\mathbf{e}_{t+d}$, $\mathbf{i}_t$, $\mathbf{i}_{t+e}$, $\Delta\mathbf{i}_t$). The levels of stimulation to its different kinds of chemoreceptors from waterborne substances determine its external sensory projection. Before performing an action they are represented by the point $\mathbf{e}_t$ and the point $\mathbf{e}_{t+d}$ afterwards. Before performing an action, its essential physiological variables are represented by the point $\mathbf{i}_t$. After the results of the action have had time to take effect, it is represented by the point $\mathbf{i}_{t+e}$. We shall assume that the dimension of $\mathbf{i}_t$ and $\mathbf{i}_{t+e}$ have already been weighted so that the fish's health is roughly proportional to the proximity of $\mathbf{i}_t$ to a known optimum. The change in the fish's health correlated with the action taken and the current external sensory projections ($\mathbf{m}_t$, $\mathbf{e}_t$) are represented by the point $\Delta\mathbf{i}_t = \mathbf{i}_{t+e} - \mathbf{i}_t$.

The fish develops expectations about the effects of its motor actions on its physiological variables by recording unexpected sensorimotor projections in its phase space. In a new situation the fish approximately predicts the effect of alternative actions on its physiological state by finding the closest point in sensorimotor phase space given its current levels of chemoreceptive stimulation. (We shall delay for a moment discussion of what we mean by *closest*.) Therefore, in this example, points corresponding to unexpected sensorimotor projections quantize the phase space into convex hulls to set up expectations about the consequences of future actions.

How do we quantify what makes a sensorimotor projection unexpected in determining when it is appropriate for the fish to form a corresponding expectation? If we are to be very strict then any occurrence that deviates at all from the fish's current expectations is unexpected. In this case, except in the unlikely event that its expectation were spot on, the fish should simply commit all its experiences to memory. Economy speaks against this, because the fish would be learning detail that is of marginal utility. We note that the fish's receptors are of limited accuracy and also that some of the stimulation they receive would not be coming from whatever the fish were currently trying to distinguish. It would be a waste of the fish's precious resources to learn what is for the most part noise in the system. There is another reason that speaks against remembering every occurrence. The better able the fish is to treat behaviorally equivalent things as the same, the better able it is to survive and reproduce. This implies that the fish should not only ignore extraneous detail but also take determined measures to filter out any form of misleading information.

In a complex and rapidly changing world it is important to grasp that there are certain things worth discriminating rather finely. If two different species of fish, for example, produce nearly identical sensory projections ($\mathbf{e}_x \approx \mathbf{e}_y$) but radically different physiological changes when ingested ($\Delta\mathbf{i}_x \neq \Delta\mathbf{i}_y$) — if, for example, one is poisonous — then priority must be given to detecting any telltale differences in their sensory projections. However, if two different species of fish produce somewhat different sensory projections ($\mathbf{e}_x \neq \mathbf{e}_z$) but nearly identical physiological changes ($\Delta\mathbf{i}_x \approx \Delta\mathbf{i}_y$), then they can be lumped together in the same behaviorally equivalent category. We may capture this intuition by first weighting the dimensions of the phase space so that each variable receives proper emphasis. If, for example, slight changes in a physiological variable spell life or death for the fish but the level of stimulation to a particular kind of chemoreceptor is relatively unindicative of what is in its environment, then the weighting should reflect this. The weighting may be performed by standard linear last-squares. Here we are trying to predict the change in the fish's overall health $|\Delta\mathbf{i}_t|$ in terms of its sensorimotor projection $\mathbf{p} = (\mathbf{m}_t, \mathbf{e}_t, \mathbf{e}_{t+d}, \mathbf{i}_t, \mathbf{i}_{t+e}, \Delta\mathbf{i}_t)$. If we have a set of expectations $\mathbf{p}_1, \mathbf{p}_2, \ldots \mathbf{p}_r$, each k dimensional, then we are looking for weight vector $\mathbf{w} = w_1, w_2, \ldots w_k$ that minimizes the linear squared error of the following equations:

$$
\begin{array}{llll}
w_1\, p_{1,1} & + w_2\, p_{1,2} + \ldots + w_k\, p_{1,k} & & = |\Delta\mathbf{i}_1| \\
w_1\, p_{2,1} & + w_2\, p_{2,2} + \ldots + w_k\, p_{2,k} & & = |\Delta\mathbf{i}_2| \\
 & \vdots & & \\
w_1\, \mathrm{p}_{r,1} & + w_2\, p_{r,2} + \ldots + w_k\, p_{r,k} & & = |\Delta\mathbf{i}_r|
\end{array}
$$

Once properly weighted, the phase space may be quantized.

The granularity of the quantization will vary according to the distance threshold used. The threshold determines when an experience is sufficiently different from nearby categorical representations to warrant the formation of a new category. The categories thus formed will reflect to varying degrees behavioral and structural similarities depending on trade-offs made in the weighting. The most appropriate granularity at which this quantization should be performed may be discovered empirically. The granularity is roughly analogous to vigilance in adaptive resonance theory (Grossberg 1988) or the number of units whose weights are to be adjusted in competitive learning (Rumelhart and Zipser 1985) and, indeed, we see that the quantization itself need not have been performed by the closest point approach but could have been performed by these or other methods of unsupervised learning.

As mentioned earlier, ideally the phase space should be quantized so as to minimize the misleading affect of chemicals from more distant sources in discriminating what is at hand. This can be approached in the following manner. Note that each behaviorally relevant category is determined by the placement of its categorical representation and those of surrounding categories. Therefore, the collection of all these representations quantizes the phase space. At first, each sufficiently distinct sensorimotor projection acts as a representation. However, when a second sensorimotor projection falls in the same category, the representation is refined by intersecting it with the new sensorimotor projection and resetting it to the result (see Figure 1). In our specific example, for each kind of chemoreceptor (that is, each dimension), the future value of the representation will be assigned the minimum of its present value and the value of the new sensorimotor projection. To use Harnad's terms, the purpose of a categorical representation is to "reliably distinguish members from nonmembers of a category" (p. 342). It should not, therefore, be a prototypical expectation but rather the invariance that is shared by the members of a category but not shared by nonmembers.

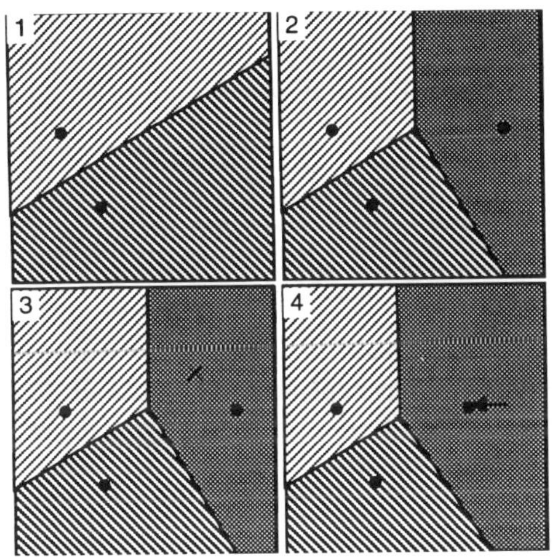

Figure 1. (1) Two categorical representations quantize two dimensions of the sensorimotor phase space. (2) A sufficiently dissimilar sensorimotor projection forms a new representation. (3) A sensorimotor projection (×) is categorized by a representation. (4) It is intersected with the representation thus refining it.

Apart from the decision-making system for controlling its oral cavity, the fish has a second decision-making system for navigation. The two systems are autonomous except for the fact the second system exploits the categories formed by the first. Based on the categorical representations that the oral cavity decision making system has so far formed, the navigational system ascertains which corresponding sources of chemicals are present in its vicinity. It does this by trying to account for the level of stimulation to the fish's different kinds of chemoreceptors in terms of its categorical representations by means of a modified linear least-squares.

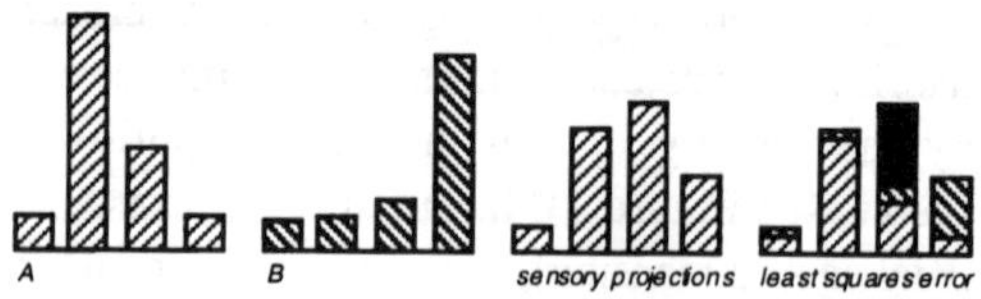

Figure 2. This simple illustration shows how least-squares can be used to account for a sensory projection in terms of the known categorical representations A and B. Both A and to a lesser degree B appear to be present. However, there appears also to be an unexplained source of the third chemical transmitter, shown here in black. Potentially the fish could actively seek to uncover unknown chemical sources.

Notice that here the weight coefficients $a_1, a_2, \ldots a_r$ determine the likely contribution of the external sensory subspace of each categorical representation $\mathbf{e}_1, \mathbf{e}_2, \ldots \mathbf{e}_r$ to the current external sensory projection $\mathbf{e}_t$, and do not vary from one dimension to another of the same projection:

$$\begin{array}{llllll} a_1 e_{1,1} & + a_2 e_{2,1} & + \ldots & + a_r e_{r,1} & = e_{t,1} \\ a_1 e_{1,2} & + a_2 e_{2,2} & + \ldots & + a_r e_{r,2} & = e_{t,2} \\ & \vdots & & & \\ a_1 \mathrm{e}_{1,l} & + a_2 \mathrm{e}_{2,l} & + \ldots & + a_r \mathrm{e}_{r,l} & = \mathrm{e}_{t,l} \end{array}$$

(The computations for this modified least-squares were made by means of singular value decomposition as explained in Press *et al.* 1992.) There are two additional restrictions. None of the weight coefficients are permitted to be negative. This is, of course, because a substance cannot have a negative presence in the water. Also, the weighted sum of the representations cannot exceed the sensory projection for any one chemical transmitter. This is because such an overshoot could never be accounted for by an additional but unknown representation.

Of the categorical representations whose corresponding sources are apparently present (above a certain threshold), the fish selects the one which could potentially, given the most appropriate action, bring its current physiological variables closest to their optimum. To climb the gradient of chemicals associated with that categorical representation, it uses klinotaxis (the successive comparison of concentrations; see Jones 1992) as follows. We assume that the concentrations of chemicals will in general exponentially increase as the fish swims directly towards their source. As the fish moves ahead, it may either veer to the left or right. It will continue veering in the same direction as long as the concentrations accounted for by the selected categorical representation are accelerating. Otherwise, it will try veering in the other direction.

So what have we achieved so far? We have shown that perceptual information can be expressed *directly* in terms of sensorimotor categories. However, by *sensory* we must now include *internal* feedback from physiological or affective variables. In our example we have developed a simulated fish that learns to interact with its environment so as to optimize its physiological variables. It learns for example, what to eat and what to avoid eating. From the raw sensory projections of its different kinds of chemoreceptors it learns behaviorally relevant categories that progressively better approximate behaviorally similar aspects of its environment. Based on the expectations it has developed about the consequences of its actions from its past experiences, it follows a course of action that keeps its essential variables within their normal ranges for survival.

In extending this model, there are many avenues that may be pursued. Animals become social early on in evolution and must compete for mates and distinguish friend from foe. The navigational decision making system could be enlarged so that the fish not only homes in on fish that it can profitably interact with but also avoids predators. Also further variables analogous to hormone levels could be added to its sensorimotor phase space. To keep them in their normal ranges, actions governing courtship and mating behavior would need be taken.

4.3. *Developing Categories into a Symbol System*

At this point the fish has learned behaviorally relevant categories. Conceivably we could simply connect these categories to the symbols of an already existing symbol system. To obtain a description of the current state of the world, a symbol can be instantiated if its corresponding categorical representation is estimated to be present in its surroundings above a certain threshold. This provides a listing of all the chemical producers within a certain region near the fish, and may be calculated in the usual manner by accounting for the current sensory projections in terms of the fish's discovered categorical representations by least-squares minimization. The symbol system part would then need to be programmed with goals and operator rules that could be applied to achieve those goals. For example, if the fish had the goal of taking care of its young, then it might have an operator rule telling it that, among other things, bringing certain kinds of food to its offspring would help it do this. There would be other rules telling it how to get the food: for example, by searching for it, taking it into its mouth, searching for its young, and then spewing out the food. The symbol system could be implemented using a symbolic planner like STRIPS, SMLP, or

SOAR. What would be unsatisfying about this is that the system would still require domain-specific programming.

However, because the fish has already developed behaviorally relevant categories that it is able to exploit, it is in a good position to assemble simple interactions into more complex patterns of behavior. A method akin to learning from delayed rewards would provide an adequate means of doing this, for example, the temporal differencing of Sutton's (1988) adaptive critics or Watkin's (1989) Q-learning. The fish's state would be determined according to the categorization of its sensory projections when it bumps into something. If in these states the fish happened to perform actions in an order that produced some kind of internal reward (which could be learned by a genetic algorithm), then this reward would be credited across the last few states (or state-action pairs). Eventually it would learn to produce rewardable sequences of actions in the correct circumstances. So, for example, it would learn to take the correct food to its offspring. Within the constraints of how we define state and after a sufficiently long period of trial and error, the fish would be behaving as if it had an *a priori* symbol system.

It is possible to use this model to approximate symbol system behavior more closely than this. With slight modification, the sensorimotor phase space can be made to serve as a rudimentary model of the world allowing the fish to plan sequences of actions it has never attempted before. At present the fish can use the phase space to map from external sensory projections and desirable internal physiological changes to appropriate motor actions. However, the phase space also tells the fish what effect its actions are likely to have on succeeding external sensory projections. Therefore, if it has a certain projection as its goal, it can backwards chain to the current projection in the same manner that a symbolic planner backwards chains on operator rules. Then the fish can attempt to perform the chain in its correct order. Each motor action in the chain has the precondition of the fish being exposed to the external sensory projection matching the planned action. Of course, in a nondeterministic environment, provisions must be made for the fish to retry actions or take corrective actions when the action first attempted fails to produce the desired change in the external sensory projection (i.e. closed loop control).

4.4. *Vanishing Intersections*

Although it was possible to refine categorical representations in the fish example to capture invariant aspects of chemoreception, this approach may not generalize, for example, to vision. As Harnad (1990) has pointed out, it has many critics:

> It has been claimed that one cannot find invariant features in the sensory projections because they simply do not exist: The intersection of all the projections of the members of a category such as "horse is empty. The British empiricists have been criticized for thinking otherwise... The problem of vanishing intersections (together with Chomsky's "poverty of the stimulus argument") has even been cited by thinkers such as Fodor as a justification for extreme nativism." (p. 344)

If one assumes that the intersections are taken solely at the level of raw sensory projections, then for vision intersections may well vanish. The fact that they do not vanish for chemoreception would be just a peculiarity of that domain. However, extreme nativism does not solve the problem of vanishing intersections; innate feature detectors would still need to exploit the invariance associated with different types to categorize particular instances according to their sensory projections. The only difference is that there would be no need to *learn* how to do this. (Presumably the features detectors would already have evolved specifically for it.)

Arguably intersections do occur but at higher levels of abstraction. Horses do not exist as recognizable entities at the lowest levels of visual processing. At this level invariance facilitates the detection of simpler and more universally applicable categories, and here it may be appropriate to speak in terms of Marr's innate detectors of edge segments, blobs, boundaries, and orientations. Higher levels are sensitive to horse-indicative invariance in lower levels, and no doubt this invariance is integrated from multiple sensorimotor modalities. Yet here it may be more fruitful to think in terms of *feature selection* than feature detection. The global 'interpretation' of a figure may place constraints on lower-level processing thus permitting a categorization to have multiple equilibria as the Necker cube, Rubin's (1915) vase-face, and Jastrow's (1900) duck-rabbit figures exemplify (see Wittgenstein 1958, further examples in Gregory 1970, and the Gestalt literature).

To exploit physical invariance in sensorimotor projections, the *same* invariant features need not be present for every instance of a category. Any one of a disjunctive set of invariant features, sampled from any number of sensory modalities, can serve to indicate sensorimotor invariance at a more abstract level. And it is at more abstract levels that invariance is most significant. Either consciously or nonconsciously, an individual needs to be able to recognize *abstract* invariance — what something is, the ways a particular individual can interact with it, and the likely outcome of those interactions. At more concrete levels *variance* can be crucial — not the abstract properties an instances shares with other members of a class but its particular shape, etc.

At this point a word of caution is in order. The variant-invariant dichotomy presented here is very much one of our own making, as is the metaphor about levels of abstraction. For our purposes the term *abstract* could be especially misleading. It is not intended to refer to a cognitive category that has been abstracted from any particular individual, for such could not exist. This is because the perception of an object's behavioral possibilities depends on who is interacting with it — that individual's particular body, life history, etc. In the present discussion one category is referred to as being more abstract than another only because it is further removed from the physical structure of the sensory projection. This usually means that later brain processing is involved and its recognition is more likely to depend on developed expectations than hardwired detectors. This is also why we should doubt whether the iconic and categorical representations that Harnad refers to are separate and distinct types of representation. In the current approach, a representation may serve a role that is more iconic or more categorical, but it is still the same point in phase space. Planning how to move an object through a narrow passageway is more iconic. Referring to an object by name is more categorical. The latter may seem archtypically categorical; however, a different name may be used to refer to an object depending on the particular surfaces of that object that you happen to be trying to manipulate.

In the fish example the sensorimotor phase space is limited to only one level, and our quantization of it is based on the structure of the phase space, or specifically, on the distribution of the categorical representations within it. Nevertheless, once sensory projections have been scaled with respect to alternative actions and their likely physiological consequences, the structure thus uncovered can no longer be equated merely with patterns in external sensory projections. This is because, in certain cases, the same category will be activated by structurally dissimilar sensory projections and, in other cases, different categories will be activated by structurally similar sensory projections. The former occurs when the same action produces a similar physiological effect; the latter when structurally minor details can be exploited to discriminate things that appear similar but produce widely different physiological effects.

5. Conclusion

In the simulated fish example, we saw how an adaptive model could ground internal categories in sensorimotor invariance. These categories served as expectations concerning the consequences of motor actions. The fish exploited its sensorimotor categories to maintain its physiological variables near their optimum values. The effectiveness with which this is accomplished may be increased by learning from delayed feedback so as to better estimate the future effect of motor actions on those variables. By these methods the agent brings its past — both successes and failures — to bear on its present. The use of these simple sensorimotor categories may be extended to activities at which symbol systems were once considered uniquely proficient. For example, an agent can use them to plan — regardless of whether its goals are expressed in terms of internal variables or external observations.

However, the adaptive model outlined in this paper is arguably not a symbol system, although it can to some extent mimic the behavior of one. This is because its sensorimotor categories are not imported from a human-imposed ontology and, hence, will not necessarily be logically consistent or semantically interpretable. Instead, they have been developed through the interaction between the agent's own motivational system, sensors, and actuators. It is necessary to give up the requirement of semantic interpretability because a traditional symbol system cannot be grounded in the sense of it being able to learn new symbols from scratch. This is because in these systems the connection between symbol and referent must be either inborn or learned under the supervision of a teacher. Unfortunately, neither possibility can serve as an adequate explanation of human cognitive abilities. They both indefinitely postpone the question of how the connection between symbol and referent could have originated. Supervised learning additionally fails to explain how those lacking language can also perceive a world populated by objects.

Furthermore, an exclusive reliance on innate categories is evolutionarily implausible; it fails to explain how animals are able to perceive sensorimotor patterning that cannot be decomposed into categories inherited from their ancestors. This is not to deny that evolution has endowed creatures with feature detectors. Evidence suggests that they contribute to low-level sensory processing and even the recognition of faces and hands. It would appear that they are ideally placed to help bootstrap empirically developed sensorimotor categories. However, they cannot replace them. Taken by themselves feature detectors are insufficient to ground the vast numbers of symbols required to represent all the different kinds of potentially recognizable things.

In robotics the common procedure of designing a symbol system and then trying to develop a perceptual module to link its symbols to the objects they represent is likewise unsatisfying because it places the same constraints on the system's adaptability. Since a robot developed according to this procedure can only detect the features that it was programmed to detect, it will never be able to cope with a situation if its proper handling depends on the discovery of new features (and not merely new combinations of existing features). This is not only a problem for symbol systems, but also for methods of reinforcement learning that are based on an *a priori* mapping from sensory projections to perceived states.

While no-one disputes that sensorimotor dynamics are clearly implementation dependent, some have falsely presumed that planning can or should operate on abstract categories that are wholly independent of body and experience and devoid of analogue accompaniments. Attempts to isolate planning from sensorimotor control may make the robot programmer's job easier, but it will not make the robot itself better able to adapt to environmental change. A potentially more flexible approach is to plan in terms of developed sensorimotor categories at whatever level of abstraction.

Cognition serves affect (Piaget and Inhelder 1969; Zajonc 1980). According to the model proposed here, developed sensorimotor expectation also serve affect while they serve as a basis for cognition. This is because cognition is applied to a world model — a phase space of sensorimotor expectations — that has been developed empirically. Nevertheless, model learning, reinforcement learning, and the simple kind of problem solving discussed here cannot explain how children learn languages as quickly and as systematically as they do nor how people can successfully make plans in novel domains without prior learning. No doubt, an explanation of the latter will require models that come to grips with abstract analogical aspects of discernment. But that does not invalidate the central point here: that abstract reasoning is grounded in adaptive ecological activity and not an *a priori* symbol system.

Acknowledgments

I would like to thank William Clocksin, Stephen Cowley, Charles Elkan, and Bruce Mangan for their helpful comments on earlier drafts of this paper.

Notes

1. Clearly, there are rule-like regularities in how we string words together; they are symbols insofar as they stand for things without resembling them. In this sense the relationship between word and thing represented is arbitrary with perhaps the rare exception of onomatopoeia.
2. Although our educational system may try to enforce logical argumentation from first principles, even academics do not manage to put aside *ad hominem* arguments when discussing the work of their colleagues.
3. For a clarification of what a symbol system is, see Harnad's definition in the first section.
4. Of course, this is not to deny that creatures are under selective pressure to maintain a certain degree of consistency in their internal models.
5. Point (*8*) merely asserts that the symbols are grounded. It is unclear whether this means the symbols are required to be interpretable to the agent of which they are a part or merely to an outside observer (see Dennett 1987).
6. "How real existence is to be studied or discovered is, I suspect, beyond you and me. We must rest content with the admission that knowledge of things is not to be derived from names. No; they must be studied and investigated in their connection with one another." (Plato 1953: 104–105)
7. This point was first made by Chomsky (1965, 1986). However, Chomsky did not argue that people's concepts could be functionally identical. Rather he argued that language should be modeled according to universal principles, such as the rules that the ideal speaker might have for combining symbols.
8. Starting from the bottom up, he first proposed algorithms for extracting edge segments, blobs, boundaries, and orientations from a static scene and then for discriminating groups of these primitives according to their size, orientation, and spatial arrangement. These computations yielded three levels of representation: the primal, 2 $^1/_2$–D, and 3–D sketch.
9. The difference between the notion of a grandmother cell and distributed connectionist representation may only be a matter of degree. It is quantified in Valiant's (1994) mathematical theory of representation and retrieval in the brain.
10. Hume argued that ideas were not innate because the blind cannot form any idea of colors nor the deaf of sounds as long as neither had experienced the corresponding impressions (1975: 20). Hume's term *impression* includes affect.
11. The anthropic principle has indeed been invoked to justify it. The argument, imported from physics, is that if the universe were not the way it is, we would not be here to witness it. It is not a denial of evolution, but it does permit highly improbable evolutionary occurrences. In physics it is used only a last resort.
12. Shakey's planning system is based on STRIPS (see Fikes and Nilsson 1971). Brooks (1991a, b) has been one of the strongest critics of the application of symbol systems to robotics.

13. On the connectionist side see Smolensky (1988) and Dreyfus and Dreyfus (1988). On the symbolist side see Pinker and Prince (1988) and Minsky and Papert (1988).
14. Examples of unsupervised learning include statistical cluster analysis, vector quantization (Swaszek 1985), Kohonen nets (Kohonen 1984), competitive learning (Rumelhart and Zipser 1985), and adaptive resonance theory (Grossberg 1988). Examples of learning from a teacher include the backpropagation of errors (Rumelhart, Hinton and Williams 1986) and the generalized delta rule, and from reinforcements include temporal differences (Sutton 1988) and Q-learning (Watkins and Dayan 1992).
15. The argument has been put forward that a language system developed like a spandrel in a cathedral, that is, because of the demands of surrounding architecture and not because of anything it contributed in its own right. This argument could also be extended to symbolic representation. Yet considering the complexity of both language systems and symbol systems and the lack of empirical evidence that surrounding architecture would have created a demand for them, it seem unlikely that either could be spandrels.
16. It is easy to look on the nerve cells' pattern of activation as a representation of an outside reality. There are two reasons why this cannot be so. (1) The form of the pattern is ecologically mediated: it depends on such things as the organism's particular sense organs. Even in perceiving the same object from the same angle, different internal patterns will result depending on the physiology of the organism. (2) What is more germane here, the pattern is only meaningful because we are able to interpret it. In fact, we can imagine the nerve cells displaced so that their pattern of activation appears completely random but produces the same effect, just as we can imagine the knots of a net tangled up without their interconnections having been interfered with.
17. It has been noted that we cannot explain how we recognized someone's face with the same certainty that we can explain, for example, how we solved a mathematics problem.
18. Pellionisz and Llinás (1979) conjectured that Purkinje cells in the cerebellum were responsible for making coordinate space transformations by acting as a metric tensor.
19. These responses are rather simple and uninteresting because they do not involve interaction. Nevertheless, expectations may elicit anticipatory responses when two or more agents must coordinate their activity — for example, if one must catch a ball thrown by another during a game.
20. This representation need not rely on any intermediary form of representation, such as a representation of the object's shape in allocentric (object-centered) coordinates.
21. Many vertebrates exploit olfaction for feeding, navigation, reproduction, and the avoidance of predators (Stoddart 1980). Evidently even birds like the petrel will assess the direction of an odor source by swaying their heads from side to side (p. 192). Fish will swim up the gradient of chemicals emitted by food to its source. They make use of klinotaxis, successive comparisons of concentrations, and tropotaxis, simultaneous comparisons among distinct chemosensory organs (Jones 1992: 292). Although fish have innate preferences for certain odors and tastes, experience can change their response to chemical cues (Jones 1992: 300).

References

Ashby, W.R. 1952. *Design for a Brain*. London: Chapman and Hall.

Ballard, D.H. 1991. Animate vision. *Artificial Intelligence* 48(1), 57–86.

Blake, A. and Yuille, A. 1992. *Active Vision*. Cambridge, MA: MIT Press.

Brooks, R.A. 1991a. Intelligence without reason. *IJCAI-91: Proceedings of the 12th International Joint Conference on Artificial Intelligence*, 569–595. Sidney, Australia (Vol. 1). San Mateo, CA: Morgan Kaufmann.

Brooks, R.A. 1991b. Intelligence without representation. *Artificial Intelligence* 47, 139–159.

Carpenter, R.H.S. 1990. *Neurophysiology* (2nd ed.). London: Edward Arnold.

Chomsky, N. 1965. *Aspects of the Theory of Syntax*. Cambridge, MA: MIT Press.

Chomsky, N. 1986. *Knowledge and Language: Its Nature, Origin, and Use*. New York: Praeger.

Clocksin, W.F. and Moore, A.W. 1989. Experiments in adaptive state-space robotics. *Proceedings of the 7th Conference of the Society for Artificial Intelligence and Simulation of Behavior*, 115–125.

Cowley, S. J. and MacDorman, K. F. (1995). Simulating conversations: The communion game. *AI and Society* 9(3).

Dennett, D.C. 1979. *Brainstorms*. Hassocks: Harvester Press.

Dennett, D.C. 1987. *The Intentional Stance*. Cambridge, MA: MIT Press.

Dreyfus, H.L. and Dreyfus, S.E. 1988. Making a mind versus modeling the brain: Artificial intelligence back at a branchpoint. *Daedalus* 117(1), 15–43.

Feldman, J.A. 1985. Four frames suffice: A provisional model of vision and space. *Behavioral and Brain Sciences* 8(2), 265–288.

Feldman, J.A. and Ballard, D.H. 1982. Connectionist models and their properties. *Cognitive Science* 6, 205–254.

Fikes, R. and Nilsson, N. 1971. STRIPS: A new approach to the application of theorem proving to problem solving. *Artificial Intelligence* 2(3–4), 189–208.

Fodor, J.A. 1975. *The Language of Thought*. Cambridge, MA: Harvard University Press.

Fodor, J.A. 1980. Methodological solipsism considered as a research strategy in cognitive psychology. *Behavioral and Brain Sciences* 3(63), 63–110.

Fodor, J.A. 1981. *Representations: Philosophical Essays on the Foundations of Cognitive Science*. Brighton, UK: Harvester.

Friedman, J.H., Bentley, J.L. and Finkel, R.A. 1977. An algorithm for finding best matches in logarithmic expected time. *ACM Transactions in Mathematical Software* 3(3), 209–226.

Fujita, I., Tanaka, K., Ito, M. and Cheng, K. 1992. Columns of visual features of objects in monkey inferotempral cortex. *Nature* 360(6402), 343–346.

Gibson, J.J. 1979. *The Ecological Approach to Visual Perception*. Boston, MA: Houghton Mifflin.

Grossberg, S. 1988. *Neural Networks and Natural Intelligence*. Cambridge, MA: MIT Press.

Gyr, J., Wiley, R. and Henry, A. 1979. Motor sensory feedback and geometry of visual space: A replication. *Behavioral and Brain Sciences* 2, 59–64.

Harnad, S. 1987. Category induction and representation. In *Categorical Perception: The Groundwork of Cognition*, S. Harnad (ed). Cambridge, UK: Cambridge University Press.

Harnad, S. 1990. The symbol grounding problem. *Physica D* 42(1–3), 335–346.

Haugeland, J. 1985. *Artificial Intelligence: The Very Idea*. Cambridge, MA: MIT Press.

Hinde, R.A. 1987. *Individuals, Relationships and Culture: Links between Ethology and the Social Sciences*. Cambridge: Cambridge University Press.

Hume, D. 1975. *Enquiries Concerning Human Understanding and Concerning the Principles of Morals*. Oxford: Oxford University Press.

Hume, D. 1978. *A Treatise of Human Nature*. Oxford: Oxford University Press.

Jastrow, J. 1900. *Fact and Fable in Psychology*. Boston: Houghton, Mifflin and Co.

Jones, K.A. 1992. Food search behavior in fish and the use of chemical lures in commercial and sports fishing. In *Fish Chemoreception*, T.J. Hara (ed). London: Chapman and Hall.

Kohler, I. 1964. The formation and transformation of the perceptual world (Fiss, trans.). *Psychological Issues* 3, 1–173.

Kohonen, T. 1984. *Self-Organization and Associative Memory*. Berlin: Springer-Verlag.

Kosslyn, S.M. 1994. *The Resolution of the Imagery Debate*. Cambridge, MA: MIT Press.

Leibniz, G. 1951. De arte combinatoria. In *Selections*. New York: Schribner.

Luria, A.R. 1976. *Cognitive Development: Its Cultural and Social Foundatons*. Cambridge, MA: Harvard University Press.

Marr, D. 1982. *Vision*. New York: Freeman.

Malcolm, C.M. 1995. The SOMASS system: A hybrid symbolic and behavior-based system to plan and execute assemblies by robot. In *Hybrid Problems, Hybrid Solutions*, J. Hallam, *et al.* (eds), 157–168. Oxford: ISO Press.

Maze, J.R. 1991. Representationalism, realism and the redundancy of 'mentalese'. *Theory and Psychology* 1(2), 163–185.

Mell, B.W. 1988. Building and using mental models in a sensory-motor domain: A connectionist approach. *Proceedings of the Fifth International Conference on Machine Learning*, 207–213.

Micó, M.L., Oncina, J. and Vidal, E. 1994. A new version of the Nearest-Neighbour Approximating and Eliminating Search Algorithm (AESA) with linear preprocessing time and memory requirements. *Pattern Recognition Letters* 15, 9–17.

Minsky, M.L. and Papert, S.A. 1988. *Perceptrons* (expanded ed.). Cambridge, MA: MIT Press.

Newell, A. 1980. Physical symbol systems. *Cognitive Science* 4, 135–183.

Nilsson, N.J. 1984. Shakey the robot. Technical Report No. 323, SRI AI Center, Menlo Park, CA.

Omohundro, S.M. 1990. Geometric learning algorithms. *Physica D*, 42(1–3), 307–321.

Paul, R.P. 1981. *Robot Manipulators, Mathematics, Programming and Control.* Cambridge, MA: MIT Press.

Pellionisz, A. and Llinás, R. 1979. Brain modeling by tensor network theory and computer simulation. The cerebellum: Distributed processor for predictive coordination. *Neuroscience* 4, 323–348.

Piaget, J. and Inhelder, B. 1969. *The Psychology of the Child.* New York: Basic Books.

Plato. 1953. *The Dialogues of Plato: Cratylus*, (Vol. 3), B. Jowett (Trans). Oxford: Oxford University Press.

Press, W.H., Teukolsky, S.A., Vetterling, W.T. and Flannery, B.P. 1992. *Numerical Recipes in C: The Art of Scientific Programming* (2nd ed). Cambridge, UK: Cambridge University Press.

Prince, A. and Pinker, S. 1988. On language and connectionism: Analysis of a parallel distributed-processing model of language-acquisition. *Cognition* 28(1–2), 73–193.

Pylyshyn, Z. 1980. Computation and cognition: Issues in the foundations of cognitive science. *Behavioral and Brain Sciences* 3, 111–169.

Ramasubramanian, V. and Paliwal, K. K. 1992. An efficient approximation-elimination algorithm for fast nearest-neighbour search based on a spherical distance coordinate formulation. *Pattern Recognition Letters* 13, 471–480.

Rieser, J.J., Guth, D.A. and Hill, E.W. 1986. Sensitivity to perceptive structure while walking without vision. *Perception* 15, 173–188.

Rosenfield, I. 1992. *The Strange, Familiar, and Forgotten: An Anatomy of Consciousness.* New York: Knopf.

Rubin, E. 1915/1958. Figure and Ground (trans.). In *Readings in perception*, D.C. Beardslee and M. Wertheimer (eds). Princeton: Van Nostrand.

Rumelhart, D.E., Hinton, G.E. and Williams, R.J. 1986. Learning internal representations by error propagation. In *Parallel Distributed Processing: Exlorations in the Microstructure of Cognintion, Volume 1: Foundations,* Rumelhart and McClelland. Cambridge, MA: MIT Press.

Rumelhart, D.E. and McClelland. 1986. *Parallel Distributed Processing: Explorations in the Microstructure of Cognition, Volume 1: Foundations.* Cambridge, MA: MIT Press.

Rumelhart, D.E. and Norman, D.A. 1981. A comparison of models. In *Parallel Models of Associative Memory*, G.E. Hinton and J.A. Anderson (eds). Hillsdale, NJ: Lawrence Erlbaum Associates.

Rumelhart, D.E. and Zipser, D. 1985. Feature discovery by competitive learning. *Cognitive Science* 9(1), 75–112.

Scribner, S. and Cole, M. 1981. *The Psychology of Literacy.* Cambridge, MA: Harvard University Press.

Selfridge, O.G. 1959. Pandemonium: A paradigm for learning. *National Physical Laboratory, Symposium No 10: Mechanisation of Thought Processes*, Her Majesty's Stationery Office, London, Vol. 1. 513–531.

Slater, A. 1989. Visual memory and perception in early infancy. In *Infant Development*, A. Slater and G. Bremner (eds). London: Lawrence Erlbaum.

Smith, B.C. 1995. *On the Origin of Objects.* Cambridge, MA: MIT Press.

Smith, B.C. *The Middle Distance* (Vols. 1–5). (Forthcoming.)

Smolensky, P. 1988. On the proper treatment of connectionism. *Behavioral and Brain Sciences*, 11, 1–74.

Sommerhoff, G. and MacDorman, K.F. 1994. An account of consciousness in physical and functional terms: A target for research in the neurosciences. *Integrative Physiological and Behavioral Science* 29(2), 151–181.

Stern, D.G. 1991. Models of memory: Wittgenstein and cognitive science. *Philosophical Psychology* 4(2), 203–218.

Stoddart, M.D. 1980. *The Ecology of Vertibrate Olfaction.* London: Chapman and Hall.

Stoutland, F. 1988. On not being a behaviourist. In *Perspectives on Human Conduct*, L. Hertzberg and J. Pietarinen (eds). Leiden: E.J. Brill.

Stryker, M.P. 1992. Elements of visual perception. *Nature* 360(6402), 301.

Sutton, R.S. 1988. Learning to predict by the method of temporal differences. *Machine Learning*, 3(1), 9–44.

Swaszek, P.F. 1985. *Quantization.* New York: van Nostrand Reinhold.

Tamminen, M. 1982. The extendible cell method for closest point problems. *BIT* 22, 27–41.

Valiant, L.G. 1994. *Circuits of the Mind.* Oxford: Oxford University Press.

Wason, P.C. 1981. Understanding the limits of formal thinking. In *Meaning and Understanding*, H. Parret and J. Bouveresse (eds). Berlin: Walther de Gruyter.

Watkins, C.J.C.H. 1989. *Learning from Delayed Rewards.* Unpublished doctoral paper. King's College, Cambridge.

Watkins, C.J.C.H., and Dayan, P. 1992. Q-learning. *Machine Learning* 8(3–4), 279–292.

Wertsch, J.V. 1985. *Vygotsky and the Social Formulation of Mind.* Cambridge, MA: Harvard University Press.

Wertsch, J.V. 1991. *Voices of the Mind.* London: Harvester.

Wittgenstein, L. 1958. *Philosophical Investigations* (2nd ed.). Oxford: Blackwell.

Zajonc, R.B. 1980. Feeling and thinking: Preferences need no inferences. *American Psychologist* 35, 151–175.

From Chinese Rooms to Irish Rooms

New Words on Visions for Language[1,2]

Paul Mc Kevitt & *Chengming Guo*
Institute of Electronic Systems — *Computer Science Department*
Aalborg University — *Tsinghua University*

1. Introduction

Today's dictionaries are sorely lacking in information that people have had in their heads for years. If one thinks of *a dog, a cat, parties, love, hate, sex, loud, bang, greasy, furry, running, jumping, swooning, ice cream*, etc. then one has a picture in one's head of these objects, emotions, sounds, feelings, and actions or some situation where they occurred in past personal history or in a film. Such pictures and sounds, and their manifestation in the symbols of written language itself were a major part of the emphasis of the writings of Joyce (1922, 1939) and others. For example, Joyce (1922) uses letters in English to produce the sounds of the waves as they come rushing towards the seashore on Dollymount Strand.

Today's dictionaries such as Longman's Dictionary of Contemporary English (LDOCE) (see Procter 1978), Collins' COBUILD (see Sinclair 1987) and Webster's, whether in text form or in electronic form do not contain much pictorial information; they typically encode words in symbolic natural language form with symbolic natural language descriptions. Encyclopedias do contain pictures but they do not contain definitions of words, rather knowledge about words and specifically objects in the world. It is not clear to us why dictionaries have had this bias towards symbolic natural language but it certainly seems very strange behavior.

Although there has been much progress in developing theories, models and systems in the areas of Natural Language Processing (NLP) and Vision Processing (VP) there has been little progress on integrating these two subareas of Artificial Intelligence (AI). Although in the beginning the general aim of the field was to build integrated language and vision systems, few were done, and there quickly became two subfields. It is not clear why there has not already been much activity in integrating NLP and VP. Is it because of the long-time reductionist trend in science up until the recent emphasis on chaos theory, non-linear systems, and emergent behavior? Or, is it because the people who have tended to work on NLP tend to be in other Departments, or of a different ilk, to those who have worked on VP? There has been a recent trend towards the integration of NLP and VP (see Denis and Carfantan 1993; Dennett 1991; Mc Kevitt 1992, 1994a; Pentland 1993; and Wilks and Okada in press).

Dennett (1991: 57–58) says:

> Surely a major source of the widespread skepticism about 'machine understanding' of natural language is that such systems almost never avail themselves of anything like a visual workspace in which to parse or analyze the input. If they did, the sense that they were actually understanding what they processed would be greatly heightened (whether or not it would still be, as some insist, an illusion). As it is, if a computer says, "I see what you mean" in response to input, there is a strong temptation to dismiss the assertion as an obvious fraud.

There has been fallout from the field of AI having become split into subfields. Many questions of ambiguity of words, sentences and paragraphs in NLP go away if one has some other perceptual source such as sounds or visual input feeding in to solve the ambiguity problem. People in NLP talked of solving ambiguities solely in terms of symbolic natural language frameworks and these debates sometimes wound round in circles as other perceptual inputs were forgotten. The dictionaries and knowledge bases for NLP suffered from two limitations: (1) the split from vision processing, and (2) our history of having symbolic natural language dictionaries. In vision processing many systems were built which attempted to uncover objects solely on the basis of visual recognition whereas natural language and other perceptual inputs could have helped to solve many of these problems.

There have been two problems that have worried us in the field of NLP for years: (1) Where are symbolic semantic primitive meanings in NLP programs grounded? and (2) How come some words in dictionaries have circular definitions so that words end up defining each other? We believe that these two problems were caused in part by the fact that people were thinking of NLP without taking other perceptual sources into account. So, it would seem that we have caused problems for ourselves by our reductionist tendencies. We show here how the problems can be solved in part by resorting to integrated representations for at least language and vision.

2. Integrated Lexicons

There has been much work in AI and in NLP on defining dictionaries for use in large intelligent systems. Examples are the Machine Tractable Dictionaries (MTDs) produced from Machine Readable Dictionaries (MRDs) such as LDOCE (see Guo 1995; Guthrie *et al.* 1991) and COBUILD (see Sinclair 1987). In fact, the idea of processing MRDs to obtain lexicons for NLP systems has become one of the largest research areas in NLP. There has also been work on encoding large encyclopedias for AI systems (see Lenat and Guha 1989). However, we argue that such dictionaries and encyclopedias are limited as they have carried over bad habits from existing MRDs in that they only contain symbolic natural language definitions for words and do not contain the necessary knowledge for solving intelligent problems. We call here for spatial and visual representations to be added to the lexicons and knowledge bases for AI systems.

Partridge (1995) points out that language and vision have a single perceptual mechanism where he shows that independent work on a cognitive model of visual perception and of perception of lexical items reveals a common framework underlying the two sets of cognitive mechanisms. Marconi (1996) says that NLP systems have not been able to *understand* because they have no vision component in which symbols representing natural language concepts are grounded. Wilks (1996) discusses the relationship between language, vision and metaphor and argues that visual processing can embody structural ambiguity but not anything analogous to metaphor. He says that metaphor is connected with the extension of sense and only symbols can have senses.

In the recent moves towards constructing integrated AI systems integrated lexicons have been developed. Many of these use spatial representations for words. An interesting venture has been that of developing animations for describing primitive concepts in natural language. Narayanan *et al.* (1994) discuss the possibility of developing dynamic visual primitives for language primitives. Language primitives which themselves represent dynamic processes such as entering a building can be dynamically visualized. They map the Conceptual Dependency primitives of Schank (1972) such as PTRANS (which is the physical act of transferring an object from one location to another) into pictures and show how simple stories can be mapped into picture sequences. Beardon (1995) has developed a visual language for iconic communication which he argues will be easier to use than language in a computer environment. He has developed a prototype system called CD-Icon which is also based on Schank's Conceptual Dependency representation.

Another area which has seen the rise of integrated lexicons is that of document processing. Rajagopalan (1994) defines a picture semantics for specifying the connection between patterns, colors, and brush types used in diagrams and a domain such as urban scenes. Srihari (1994) and Srihari and Burchans (1994) define a lexicon with spatial semantics for the problem of document understanding where textual captions are used as collateral information in the interpretation of corresponding photographs. Their system called PICTION uses a syntactic/semantic lexical database which is constructed from LDOCE, OALD (Oxford Advanced Learner's Dictionary), WordNet (see Beckwith *et al.* 1991) and manual augmentation with ontological information and visual semantics. The entire lexicon is represented as a LOOM knowledge base.[3] The lexicon captures fixed, visual properties of objects, such as size and color as well as procedural information such as the scalar and shape properties of the object *hat*, and location *above head*, for sentences like "the person wearing the hat." Nakatani and Itoh (1994) have developed semantic representations with primitives for colors and words that qualify colors and use this for an image retrieval system that accepts natural language.

Many of the integrated representations for language and vision are spatial in nature. Olivier and Tsujii (1994) describe a quantitative object model and qualitative spatial and perceptual model for prepositions such as *in front of*. The meanings of prepositions in his system called WIP are automatically tuned in accordance with reference to objects. Gapp and Maaß (1994) describe spatial models for route descriptions given through verbal utterances. Reyero-Sans and Tsujii (1994) describe a visual interlingua of spatial descriptions which is used to represent linguistic meaning and is motivated by the problem of machine translation for English and Spanish. Chakravarthy (1994) describes a perceptually-based semantics of action verbs such as *hit* and *walk* where lexical entries for verbs can be built up from perceptual predicates by inheritance from existing more primitive lexical entries. The perceptual semantics contains state descriptions for verbs and would give, for example, for *cut*, the object, instrument, trajectory and duration of cutting. Dolan (1994) describes a framework for vision processing which makes use of a large lexical database which has been automatically derived from MRDs. He suggests that MRDs contain much of the information needed about the physical and common-sense properties of objects in vision processing. Meini and Paternoster (1996) discuss a lexical structure where each lexical entry has two pointers, one to a knowledge base and the other to a catalogue of shapes containing structural descriptions of objects denoted by the lexical items.

There are also full blown multi-modal lexicons under development. Young (1983) and Wright and Young (1990) describe a knowledge representation called Cognitive Modalities (CM) for neural networks which is a cognitive, non-verbal representation of information. The CM system uses a cognitive lexicon of some 8,000 modality specific elements which are grouped according to the sub-modal categories of perception and non-sensorimotor events. It is hoped the representation will be used for machine translation and information retrieval. Modalities such as touch and time are encoded in CM. Phonological information is currently being encoded into CM and of course CM includes the traditional pragmatic, semantic and syntactic information encoded in traditional lexicons.

Hence, we can see that there are now moves afoot to developing integrated lexicons and a quick review of them shows that central themes are (1) animation and iconization of primitives, (2) spatial relations and (3) multi-modal forms.

3. Word Problems

One of the suggested solutions to problems of NLP over the years has been to reduce word and sentence representations to primitives (see Wilks 1977, 1978). Schank defined 14 of such primitives for Conceptual Dependency (see Schank 1972, 1973, 1975) and Wilks had some 80 in his Preference Semantics system (see Wilks 1973, 1975a, 1975b, 1975c, 1977). However, all this reductionist work had difficulties because of at least three reasons: (1) general inference rules applied to primitives don't follow through when applied to specific situations or contexts, (2) *grounding*: how are the primitives grounded in the world? i.e. what gives them their meaning? (3) *circularity*: some words are defined in terms of primitives but those primitives are defined in terms of the original words. We shall focus on the second and third problems here. First, the grounding problem.

3.1. *Primitive Grounding*

The problem with dictionaries such as LDOCE and MTDs for years has been that words are defined in terms of a defining vocabulary of 2000 words but that there is no definition of the defining vocabulary itself. Also, in systems like Schanks' and Wilks' where are the primitives grounded? Harnad (1990, 1993) has brought the grounding problem further and said that symbolic processing systems have, in general a problem with grounding their symbols and that this problem can only be freed up by using other perceptual sources such as visual input. This is his answer to Searle's Chinese Room problem where Searle argues that a machine cannot "understand" the symbols it represents but just passes them around with no feeling for their meaning (see Searle 1980, 1984, 1990).[4]

Figure 1. The Chinese Room

The Chinese Room is shown in Figure 1 where the sentence "What are the directions of lexical research?" in Chinese is being passed to John Searle in the Chinese Room (note that John doesn't get the English translation which we have placed there for the reader). In some sense what Searle is arguing is that the computer behaves like a hypertext system does, encoding text and being able to manipulate and move it around but having no sense of its meaning. Jackson and Sharkey (1996) argue that connectionist architectures are necessary for grounding perceptual systems or at least that such grounding will be easier with such architectures. Wilks (1995) points out that primitives in natural language do not have any obvious visual analogues and that no definition of primitives is necessary because they are explained by the procedural role they play in language as a whole. Katz (1972) claims that linguistic primitives play the role that neutrinos play in science.

We argue here that defining vocabularies and primitives can be defined in terms of spatial or pictorial representations to obtain meanings. So, for example, taking a primitive for the concept of abstract transfer (called ATRANS by Schank) we can have a picture showing two agents with an object being transferred between the two. This picture could be animated as demonstrated by Beardon (1995) and Narayanan *et al.* (1994) and could be shown to a user on demand. Furthermore, a definition of the changes in spatial relations with respect to ATRANS could be represented. For example, this would detail the object, instrument, trajectory and duration of transfer as defined in the perceptual semantics of Chakravarthy (1994). Hence, now we have an Irish Room, going even further than Joyce where he tried to bring perception into written symbols.

Figure 2. The Irish Room

The Irish Room is shown in Figure 2 where the sentence "What are the directions of lexical research?" in English is being passed to Seán the Leprechaun in the Irish Room (note that this time the input is annotated with icons). The Irish

Room is one where a Leprechaun who cannot understand English is locked in a room and has the task of using a Gaelic rule book for manipulating English words. Each English word has an icon or picture sequence attached to it. Then, to an outside observer, Seán the Leprechaun appears to be able to understand English just as a computer program which manipulates symbols could appear to do so. However, this time Seán begins to understand the words because he/she has reference to their meaning. Sounds, smells and even touch can be added in later! This solution to the Chinese Room problem has also been suggested by Harnad (1990, 1993), and discussed at length by Marconi (1996) and Meini and Paternoster (1996).

3.2. *The Irish Room in a Chinese Room*

In order to test whether the Irish Room idea works one of us (Paul Mc Kevitt) tried an experiment at Tsinghua University in Beijing, China on August 20th, 1994 at 9.00–10.00 PM. I asked Jun-Ping Gong, Jin Zhang and Yan-Qing Yu to write a few sentences in Chinese and to annotate them with icons to see if I could guess what the sentences were. The examples are shown in Figure 3 (note that I didn't get the English translation which we have placed there for the reader). First I was given the sentence "You are very strong" in Chinese annotated with icons. I guessed the sentence meant "Some person SOMETHING big" which isn't too bad! Then, it became interesting because I was given another sentence: "A cat is strong" in Chinese with one more icon (a cat) and recognized the ending characters were the same as those used at the end of the previous example. I guessed it right first time as the icon for cat was obvious! We only worked here with simple icons and the system was working quite well. We predict that with more fine tuned icons, videos, sounds and other sensory information the results would be much better. Next, the circularity problem.

Figure 3. The Irish Room in China

3.3. *Breaking the Circle*

How come in some dictionaries you look up a word like ‘gorse’ and find that the definition of that word involves ‘furze’ and when you look up ‘furze’ its definition uses ‘gorse’? In LDOCE, the primitive for ‘disease’ is defined to be ‘disorder’ or ‘illness’ and these in turn are defined as ‘disease.’ This has been a problem for dictionaries for years.

Again we propose a solution here where one defines a word by using a definition that uses other words but also spatial and visual structures. These structures would give partial definitions of words so that there would only be at most partial circularity in definitions. The result is partial circularity or no circularity at all.

To sum up we have argued that spatial and pictorial or animated picture sequences can be used to ground symbolic natural language specifications of the meanings of words and reduce circularity in their definitions. Our solution has not happened by accident. It has been argued for years (see Barnden 1990;

Lakoff 1986) that much of language use, such as metaphor, is based on spatial relationships and mental analogical mappings.

4. Learning Words

Another problem that dictionaries have had for years has been how to acquire the knowledge in them. Techniques have been developed for automatically gleaning information from MRDs for NLP (see Guthrie *et al.* 1991) and also this has been done by hand (see Guo 1995). A discussion of how new words can arise from an existing dictionary pool is given in Rowe and Mc Kevitt (1991). We believe that similar techniques will be needed for gleaning spatial and visual information automatically for integrated lexicons which encode both symbolic and spatial information. Just like people learn the word for a dog by looking at, hearing and even touching lots of prototype dogs we believe that eventually computers will need to be able to do this if they are to be considered intelligent. At least, computers should have visual as well as symbolic representations for dogs but should also be able to learn what zebras are by putting pictures of horses together with stripes to get them. Such pictorial information is missing from today's dictionaries. The ability to develop and learn new words such as metaphors is to a large extent based on spatial and pictorial mappings. Our systems of the future will need to be able to apply algorithms for such mappings to existing dictionaries to derive new ones. And, of course, Wittgenstein (1963: 42) pointed out already that "It is only in the normal cases that the use of a word is clearly prescribed."

5. Intentions

A theory of intention analysis (see Mc Kevitt 1991b) has been proposed as a model, in part, of the coherence of natural-language dialogue. A central principle of the theory is that coherence of natural-language dialogue can be modeled by analyzing sequences of intention. The theory has been incorporated within a computational model in the form of a computer program called the Operating System CONsultant (OSCON) (see Guthrie *et al.* 1989; Mc Kevitt 1986, 1991a, 1991b; Mc Kevitt and Wilks 1987; and Mc Kevitt *et al.* 1992a, 1992b, 1992c, 1992d). OSCON, which is written in Quintus Prolog, understands, and answers in English, English queries about computer operating systems.

The computational model has the ability to analyse sequences of intention. The *analysis of intention* has at least two properties: (1) that it is possible to recognize intention, and (2) that it is possible to represent intention. The syntax, semantics and pragmatics of natural-language utterances can be used for intention recognition. Intention sequences in natural-language dialogue can be represented by what we call *intention graphs*. Intention graphs represent frequencies of occurrence of intention pairs in a given natural-language dialogue. An ordering of intentions based on *satisfaction* exists, and when used in conjunction with intention sequences, indicates the local and *global* degree of expertise of a speaker in a dialogue.[5]

The architecture of the OSCON system consists of six basic modules and two extension modules. There are at least two arguments for modularizing any system: (1) it is much easier to update the system at any point, and (2) it is easier to map the system over to another domain. The six basic modules in OSCON are as follows: (1) ParseCon: natural-language syntactic grammar parser which detects query-type, (2) MeanCon: a natural-language semantic grammar (see Brown *et al.* 1975; and Burton 1976) which determines query meaning, (3) KnowCon: a knowledge representation, containing information on natural-language verbs, for understanding, (4) DataCon: a knowledge representation for containing information about operating system commands, (5) SolveCon: a solver for resolving query representations against knowledge base representations, and (6) GenCon: a natural-language generator for generating answers in English. These six modules are satisfactory if user queries are treated independently, or in a context-free manner. However, the following two extension modules are necessary for dialogue-modeling and user-modeling: (1) DialCon: a dialogue modeling component which uses an intention matrix to track intention sequences in a dialogue, and (2) UCon: a user-modeler which computes levels of user-satisfaction from the intention matrix and provides information for both context-sensitive and user-sensitive natural-language generation.

It has been pointed out recently by Schank and Fano (1995) that in order to perform tasks in the world *understanding* is a question of relating visual and linguistic input to the intentions (goals, plans and beliefs) derived from the task. They point out that expectations are a large part of understanding and say "We need to be able to reason about the things we can sense and the situations in which we will be able to detect them. Reminders to ourselves such as strings around fingers, notes on doors, alarm clocks, etc. all betray an underlying model of what we will notice (e.g. strings, sounds, notes) as well as the situations in which we will notice them (e.g. we are all likely to see our finger, pass through a door before leaving, and hear an alarm clock next to our beds."

We agree with Schank and Fano and believe that our own work in intention modeling can only be fulfilled by incorporating the analysis of visual scenes as well as symbolic natural language. In particular, our beliefs about people before they say anything at all are based on body language, clothes, looks, makeup, style and so on and work on modeling beliefs in language (see Ballim and Wilks 1990, 1991; and Wilks and Ballim 1987) will need to be augmented and integrated with work on determining beliefs from visual input. Indeed, work has already begun on determining intentions from vision and language (see Gapp and Maaß 1994; Herzog and Wazinski 1994; and Maaß 1994).

6. Conclusion

We conclude here by pointing out that integrated lexicons are needed for language and vision processing where such lexicons provide extra structures which describe objects and actions rather than just having the flat symbolic representation which we have been used to.

We believe that these extra structures will involve spatial and pictorial animated representations of objects and actions and that these representations will enable systems to conduct better processes such as analogical reasoning over lexicons. Integrated lexicons will cause many of the problems of symbol grounding and semantic primitives to disappear and the lexicons will become grounded in many forms of perceptual input. Searle's Chinese Room Problem will go away as machines will have more of a feel for the meanings of the words they know in the form of an Irish Room.

The analysis of intentions is not only important for interpreting the actions of agents in visual environments but also for determining what agents mean when they use words. That is, words have meanings which people intend them to have. Cognitive Science (CS) and Computer Science (CS) are converging on Information, Intentions and Integration and we propose the following formula for future development:

$$\mathrm{CS} = \mathrm{I} \times \mathrm{I} \times \mathrm{I} = I^3$$

Lexicons of the future will have in conjunction with flat semantic representations for word senses, spatial representations, pictures and sounds and these will all be used in computer systems for AI and for multimedia interfaces. Metaphors and new uses will easily be derived from analogical mappings between spatial relations and pictures. We believe the computer will change the whole meaning of lexicon and lexicons of the future will not be constrained by the symbolic language descriptions of the past. Such are our words on visions for lexicons.

Notes

1. Paul Mc Kevitt is currently funded for five years on a British Engineering and Physical Sciences Research Council (EPSRC) Advanced Fellowship under grant B/94/AF/1833 for the Integration of Natural Language, Speech and Vision Processing.
2. This paper is a reprint of that already published in *Artificial Intelligence Review*, Vol. 10(1–2), 1996 and *Integration of Natural Language and Vision Processing (Vol. III): Theory and Grounding Representations*, Mc Kevitt, Paul (ed), 1996. Dordrecht, The Netherlands: Kluwer-Academic Publishers. © 1996 Kluwer Academic Publishers.
3. LOOM (see ISX 1991) is a high level object-oriented programming language and environment for constructing knowledge based systems.
4. Searle asked us to imagine a Chinese Room where a person who cannot understand Chinese is locked in the room and has the task of using an English rule book for manipulating Chinese symbols. Then, to an outside observer, the person appears to be able to understand Chinese just as a computer program which manipulates symbols could appear to do so (see Searle 1984: 32–33).
5. By local expertise we wish to stress the fact that sometimes experts can act as novices on areas of a domain which they do not know well.

References

Ballim, Afzal, and Wilks, Yorick. 1990. Stereotypical belief and dynamic agent modeling. *Proceedings of the Second International Conference on User Modeling*, University of Hawaii at Manoa, Honolulu, HA.

Ballim, Afzal and Wilks, Yorick. 1991. *Artificial Believers*. Hillsdale, NJ: Lawrence Erlbaum Associates.

Barnden, John A. 1990. Naive metaphysics: A metaphor-based approach to propositional attitude representation (unabridged version). Memorandum in Computer and Cognitive Science, MCCS-90–174, Computing Research Laboratory, Dept. 3CRL, New Mexico State University, Las Cruces, NM.

Beardon, Colin. 1995. Discourse structures in iconic communication. In *Artificial Intelligence Review, Special Volume on Integration of Natural Language and Vision Processing*, Paul Mc Kevitt (ed), Vol. 9, Nos. 2–3, 189–203.

Beckwith, R., Fellbaum, C., Gross, D. and Miller, G.A. 1991. WordNet: A lexical database organized on psycholinguistic primitives. *Lexicons: Using Online Resources to Build a Lexicon*. Hillsdale, NJ: Erlbaum.

Brown, John Seely, Burton, Richard R. and Bell, Alan G. 1975. SOPHIE: A step towards creating a reactive learning environment. *International Journal of Man-Machine Studies* 7, 675–696.

Burton, R. 1976. Semantic grammar: An engineering technique for constructing natural-language understanding systems. BBN Report No. 3453, Bolt, Beranek and Newman, Cambridge, MA.

Chakravarthy, Anil. 1994. Towards a perceptually-based semantics of action verbs. In *Proceedings of the Workshop on Integration of Natural Language and Vision Processing*, Twelfth American National Conference on Artificial Intelligence (AAAI-94), Paul Mc Kevitt (ed), 161–164. Seattle, WA.

Denis, M. and Carfantan, M. (eds). 1993. Images et Langages: Multimodalité et Modelisation Cognitive. Actes du Colloque Interdisciplinaire du Comité National de la Recherche Scientifique, Salle des Conférences, Siège du CNRS, Paris, April.

Dennett, Daniel. 1991. *Consciousness Explained.* Harmondsworth: Penguin.

Dolan, William B. 1994. Exploiting lexical information for visual processing. In *Proceedings of the Workshop on Integration of Natural Language and Vision Processing*, Twelfth American National Conference on Artificial Intelligence (AAAI-94), Paul Mc Kevitt (ed), 185–188. Seattle, WA.

Gapp, Klaus-Peter and Maaß, Wolfgang. 1994. Spatial layout identification and incremental descriptions. In *Proceedings of the Workshop on Integration of Natural Language and Vision Processing*, Twelfth American National Conference on Artificial Intelligence (AAAI-94), Paul Mc Kevitt (ed), 145–152. Seattle, WA.

Guo, Chengming. 1995. *Machine Tractable Dictionaries*. Norwood, NJ: Ablex.

Guthrie, Louise, Mc Kevitt, Paul and Wilks, Yorick 1989. OSCON: An operating system consultant. *Proceedings of the Fourth Annual Rocky Mountain Conference on Artificial Intelligence (RMCAI-89), Subtitled, 'Augmenting Human Intellect By Computer,'* 103–113. Registry Hotel, Denver, CO.

Guthrie, Joe, Guthrie, Louise, Wilks, Yorick and Aidinejad, H. 1991. Subject-dependent co-occurrence and word sense disambiguation. *Proceedings of the 29th Annual Meeting of the Association for Computational Linguistics*, Berkeley, CA. Also Memoranda in Computer and Cognitive Science, MCCS-91–206, 146–152. CRL, NMSU.

Harnad, S. 1990. The symbol grounding problem. *Physica D*, 335–346.

Harnad, S. 1993. Grounding symbols in the analog world with neural nets: A hybrid model. *Think* 2, 12–20.

Herzog, Gerd and Wazinski, Peter. 1994. VIsual TRAnslator: Linking perceptions and natural language descriptions. In *Artificial Intelligence Review, Special Volume on Integration of Natural Language and Vision Processing*. Vol. 8, Nos. 2–3, Paul Mc Kevitt (ed), 175–187.

ISX (1991). ISX Corporation LOOM Users Guide, version 4.1 edition.

Jackson, Stuart and Sharkey, Noel. 1996. Grounding computational engines. In *Artificial Intelligence Review, Special Volume on Integration of Natural Language and Vision Processing.* Vol. 10, Nos. 1–2, Paul Mc Kevitt (ed), 65–82.

Joyce, James. 1922. *Ulysses.* London: Faber and Faber.

Joyce, James. 1939. *Finnegans Wake.* London: Faber and Faber.

Katz, J. 1972. *Semantic Theory.* New York: Harper Row.

Lakoff, G. 1986. *Women, Fire and Dangerous Things.* Chicago, IL: The University of Chicago Press.

Lenat, Doug and Guha, R.V. 1989. *Building Large Knowledge-Based Systems: Representation and Inference in the Cyc Project.* Reading, MA: Addison-Wesley.

Marconi, Diego. 1996. On the referential competence of some machines. In *Artificial Intelligence Review, Special Volume on Integration of Natural Language and Vision Processing.* Vol. 10, Nos. 1–2, Paul Mc Kevitt (ed), 21–35.

Maaß, Wolfgang. 1994. From vision to multimodal communication: Incremental route descriptions. In *Artificial Intelligence Review, Special Volume on Integration of Natural Language and Vision Processing.* Vol. 8, Nos. 2–3, Paul Mc Kevitt (ed), 159–174.

Mc Kevitt, Paul. 1986. Building embedded representations of queries about UNIX Memorandum in Computer and Cognitive Science, MCCS-86–72, Computing Research Laboratory, Dept. 3CRL, New Mexico State University, Las Cruces, NM.

Mc Kevitt, Paul. 1991a. Principles and practice in an operating system consultant. In *Artificial Intelligence and Software Engineering, Vol. 1 Chapter on 'AI Mechanisms and techniques in practical software,'* Derek Partridge (ed). New York: Ablex Publishing Corporation.

Mc Kevitt, Paul. 1991b. Analyzing coherence of intention in natural-language dialogue. Ph.D. Thesis, Department of Computer Science, University of Exeter, Exeter, UK.

Mc Kevitt, P. (ed). 1992. Natural language processing. *Artificial Intelligence Review* 6(4), 327–332.

Mc Kevitt, P. (ed). 1994a. *Proceedings of the Workshop on Integration of Natural Language and Vision processing Twelfth American National Conference on Artificial Intelligence* (AAAI-94). Seattle, WA.

Mc Kevitt, P. (ed). 1994b. *Proceedings of the Workshop on Integration of Natural Language and Speech Processing Twelfth American National Conference on Artificial Intelligence* (AAAI-94). Seattle, WA.

Mc Kevitt, Paul and Wilks, Yorick. 1987. Transfer semantics in an operating system consultant: The formalization of actions involving object transfer. In *Proceedings of the Tenth International Joint Conference on Artificial Intelligence (IJCAI-87)* Vol. 1, 569–575. Milan, Italy.

Mc Kevitt, Paul, Partridge. Derek, and Wilks, Yorick. 1992a. Approaches to natural language discourse processing. *Artificial Intelligence Review* 6(4), 333–364.

Mc Kevitt, Paul, Partridge, Derek, and Wilks, Yorick. 1992b. Analyzing coherence of intention in natural language dialogue. Technical Report 227, Department of Computer Science, University of Exeter, Exeter.

Mc Kevitt, Paul, Partridge, Derek, and Wilks, Yorick. 1992c. Why machines should analyse intention in natural language dialogue. Technical Report 233, Department of Computer Science, University of Exeter, Exeter.

Mc Kevitt, Paul, Partridge, Derek, and Wilks, Yorick. 1992d. Experimenting with intention in natural language dialogue. Technical Report 234, Department of Computer Science, University of Exeter, Exeter.

Meini, Cristina and Paternoster, Alfredo. 1996. Understanding language through vision. In *Artificial Intelligence Review, Special Volume on Integration of Natural Language and Vision Processing*. Vol. 10, Nos. 1–2, Paul Mc Kevitt (ed), 37–48.

Nakatani, Hiromasa and Itoh, Yukihiro. 1994. An image retrieval system that accepts natural language. In *Proceedings of Workshop on Integration of Natural Language and Vision Processing*, Twelfth American National Conference on Artificial Intelligence (AAAI-94), Paul Mc Kevitt (ed), 7–13. Seattle, WA: AAAI Press.

Narayanan, A., Ford, L., Manuel, D., Tallis, D. and Yazdani, M. 1994. Language animation. In *Proceedings of Workshop on Integration of Natural Language and Vision Processing*, Twelfth American National Conference on Artificial Intelligence (AAAI-94), Paul Mc Kevitt (ed), 58–65. Seattle, WA: AAAI Press.

Olivier, Patrick, and Tsujii, Jun-ichi. 1994. Prepositional semantics in the WIP system. In *Proceedings of the Workshop on Integration of Natural Language and Vision Processing*, Twelfth American National Conference on Artificial Intelligence (AAAI-94), Paul Mc Kevitt (ed), 139–144. Seattle, WA: AAAI Press.

Partridge, Derek. 1995. Language and vision: A single perceptual mechanism? In *Artificial Intelligence Review, Special Volume on Integration of Natural Language and Vision Processing* Vol. 9, Nos. 4–5, Paul Mc Kevitt (ed), 291–303.

Pentland, Alex (ed). 1993. Looking at people: Recognition and interpretation of human action IJCAI-93 Workshop (W28) at The 13th International Conference on Artificial Intelligence (IJCAI-93), Chambéry, France.

Procter, P. 1978. *Longman Dictionary of Contemporary English*. London: Longman.

Rajagopalan, Raman. 1994. Integrating text and graphical input to a knowledge base. In *Proceedings of the Workshop on Integration of Natural Language and Vision Processing*, Twelfth American National Conference on Artificial Intelligence (AAAI-94), Paul Mc Kevitt (ed), 14–21. Seattle, WA: AAAI Press.

Reyero-Sans, Irina and Tsujii, Jun-ichi. 1994. A cognitive approach to interlingual representation of spatial descriptions. In *Proceedings of Workshop on Integration of Natural Language and Vision Processing*, Twelfth American National Conference on Artificial Intelligence (AAAI-94), Paul Mc Kevitt (ed), 122–130. Seattle, WA: AAAI Press.

Rowe, Jon and Mc Kevitt, Paul. 1991. An emergent computation approach to natural language processing. *Proceedings of the Fourth Irish Conference on Artificial Intelligence and Cognitive Science,* University College Cork, Ireland.

Schank, Roger C. 1972. Conceptual dependency: A theory of natural language understanding. *Cognitive Psychology* 3(4), 552–631.

Schank, Roger C. 1973. Identification and conceptualizations underlying natural language. In *Computer Models of Thought and Language*, R. Schank and K. Kolby (eds). San Francisco, CA: W. Freeman and Co.

Schank, Roger C. 1975. Conceptual information processing. Fundamental studies in computer science, 3. Amsterdam: North-Holland.

Schank, Roger and Fano, Andrew. 1995. Memory and expectations in learning, language and visual understanding. In *Artificial Intelligence Review, Special Volume on Integration of Natural Language and Vision Processing.* Vol. 9, Nos. 4–5, Paul Mc Kevitt (ed), 261–271.

Searle, J.R. 1980. Minds, brains and programs. *Behavior and Brain Sciences* 3, 417–424.

Searle, J.R. 1984. *Minds, Brains and Science.* London: Penguin Books.

Searle, J.R. 1990. Is the brain's mind a computer program? *Scientific American* 262, 26–31.

Sinclair, John (ed). 1987. *Looking Up: An Account of the COBUILD Project in Lexical Computing.* London: Collins.

Srihari, Rohini. 1994. Photo understanding using visual constraints generated from accompanying text. In *Proceedings of the Workshop on Integration of Natural Language and Vision Processing*, Twelfth American National Conference on Artificial Intelligence (AAAI-94), Paul Mc Kevitt (ed), 22–29. Seattle, WA: AAAI Press.

Srihari, Rohini and Burchans, Debra. 1994. Visual semantics: Extracting visual information from text accompanying pictures. *Proceedings of the Twelfth American National Conference on Artificial Intelligence (AAAI-94)*, 793–798. Seattle, WA: AAAI Press.

Wilks, Yorick. 1973. An artificial intelligence approach to machine translation. In *Computer Models of Thought and Language,* R. Schank and K. Kolby (eds). San Francisco, CA: W. Freeman and Co.

Wilks, Yorick. 1975a. Preference semantics. In *Formal Semantics of Natural Language,* Edward Keenan (ed). Cambridge, UK: Cambridge University Press. Also as Memo AIM-206, July 1993. Artificial Intelligence Laboratory, Stanford University, Stanford, CA.

Wilks, Yorick. 1975b. An intelligent analyzer and understander of English. *Communications of the ACM* 18(5), 264–274. Also in Barbara Grosz, Karen Sparck Jones and Bonnie Webber (eds). 1986. *Readings in Natural Language Processing*, 193–204. Los Altos, CA: Morgan Kaufmann.

Wilks, Yorick. 1975c. A preferential, pattern-seeking semantics for natural language inference. *Artificial Intelligence* 6, 53–74.

Wilks, Yorick. 1977. Good and bad arguments about semantic primitives. *Communication and Cognition* 10(3/4), 181–221.

Wilks, Yorick. 1978. Semantic primitives in language and vision. *Proceedings of the Second Conference on Theoretical Issues in Natural Language Processing*. Champaign-Urbana, IL.

Wilks, Yorick. 1995. Language, vision and metaphor. In *Artificial Intelligence Review, Special Volume on Integration of Natural Language and Vision Processing*. Vol. 9, Nos. 4–5, Paul Mc Kevitt (ed).

Wilks, Yorick and Ballim, Afzal. 1987. Multiple agents and the heuristic ascription of belief. *Proceedings of the 10th International Joint Conference on Artificial Intelligence (IJCAI-87)* 119–124. Milan, Italy.

Wilks, Yorick, and Okada, N. (eds). In press. *Computer Language & Vision Across the Pacific*. Norwood, NJ: Ablex.

Wittgenstein, Ludwig. 1963. *Philosophical Investigations* (translated by G.E. Anscombe). Oxford: Blackwell.

Wright, Ron, E. and Young, D.A. 1990. The cognitive modalities (CM) system of knowledge representation and its potential application in C31 systems. *Proceedings of The 1990 Symposium on Command and Control Research SAIC Report SAIC-90/1508,* 373–381.

Young, D.A. 1983. Interactive modal meaning as the mental basis of syntax. *Medical Hypotheses* 10, 5.

The Role of the Systematicity Argument in Classicism and Connectionism

Kenneth Aizawa
Department of Philosophy
Centenary College

Despite the prominence of the systematicity argument in the debate between Classicists and Connectionists, there is extremely widespread misunderstanding of the nature of the argument. For example, Matthews (1994), has argued that the systematicity argument is a kind of trick, where Niklasson and van Gelder (1994), have claimed that it is obscure. More surprisingly, once one examines the argument carefully, one finds that Fodor, Pylyshyn, and McLaughlin, themselves have not fully understood it.[1] In part as a result of this, many Connectionists who have tried to meet the challenge of explaining the systematicity of thought have been misled about what this challenge involves (e.g. Pollack 1990; Smolensky 1990; Niklasson and van Gelder 1994).

I have five principal objectives in this paper. First, I want to respond to those who believe that the systematicity argument is mere obscurity by providing a clear presentation of it. Second, and at the same time, I want to respond to those who believe the argument is mere trickery by showing it to be an instance of a rather familiar form of legitimate scientific reasoning. Third, having seen what sort of reasoning is involved in the argument, I wish to indicate how certain attempts to meet the challenge of systematicity are inadequate. Fourth, having seen the explanatory standard set by the systematicity argument, I want to indicate how even Fodor, Pylyshyn, and McLaughlin, fail to see how Classicism fails to meet the standard. Fifth, and finally, I wish to indicate what consequences the foregoing considerations have for the prospects of Classicism and Connectionism.

1. The Structure of the Systematicity Argument

The systematicity argument is an instance of a very common pattern of scientific reasoning. The best illustration of it I can find is in the positional astronomy of Ptolemy and Copernicus.[2] The astronomers of Ptolemy's day had observed that if one charts the motions of Mercury and Venus through the heavens over the course of several years, one finds that these planets are never found in opposition to the Sun. That is, when one looks into the sky, one does not find the Sun in one direction and Mercury and Venus 180° away. By contrast, the other known planets, Mars, Jupiter, and Saturn, *can* be found in opposition to the Sun. The fact to be explained, therefore, is why Mercury and Venus are never found in opposition to the Sun.

Both the Ptolemaic and Copernican theories could provide accounts of the phenomena involving Mercury and Venus. Copernicus supposed that all the planets orbit the Sun on deferents and epicycles in the order, Mercury, Venus, Earth, Mars, Jupiter, and Saturn. (See Figure 1, in which the Copernican

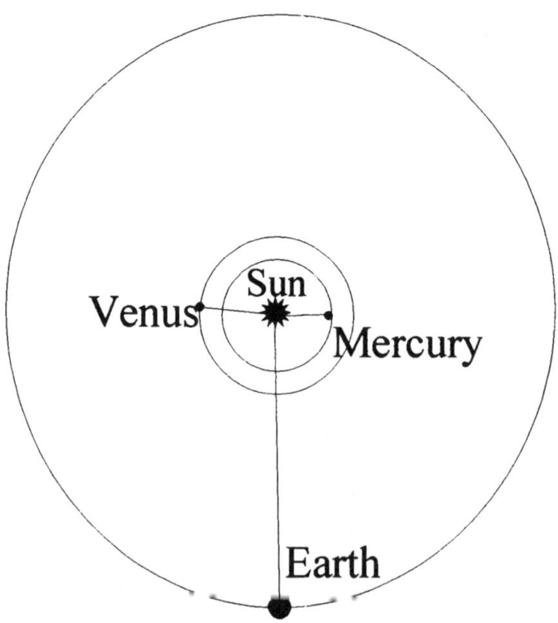

Figure 1. Copernican System

epicycles are omitted for the sake of clarity.) Given this hypothesis, it follows that Mercury and Venus can never appear in opposition to the Sun. In Ptolemaic astronomy, on the other hand, all the "planets" orbit the Earth on deferents and epicycles in the order Mercury, Venus, the Sun, Mars, Jupiter, and Saturn. This, however, is not all there is to the Ptolemaic account of the phenomena. In addition, one must suppose that the deferents of Mercury, Venus, and the Sun are locked together, that they are collinear, while the epicycles of Mercury and Venus carry them in circles back and forth before (or behind) the Sun. (See Figure 2.) According to Ptolemaic astronomy, this is why Venus and Mercury are never found in opposition to the Sun.

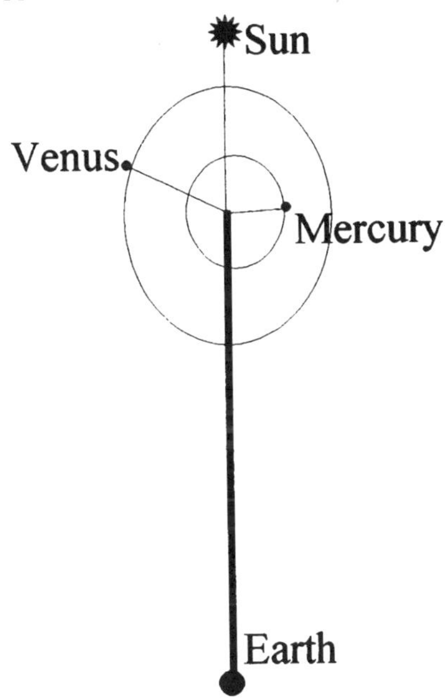

Figure 2. Ptolemaic system

Notice that both Copernicus and Ptolemy have an account of the phenomena — some story to tell about what is going on. They can both save the phenomena, if you will. But, Copernicus has a better explanation of the phenomena than does Ptolemy.[3] The reason does not have anything to do with the Ptolemaic theory using epicycles, where Copernican theory does not. Both theories postulated epicycles. Rather, the difference in explanatory power has to do with the way in which they save the phenomena. The Ptolemaic account must add the arbitrary hypothesis concerning the collinearity of the deferents of Mercury, Venus, and the Sun. Given the basic Ptolemaic framework of geocentrism, the Ptolemaic theory allows the deferents of Mercury, Venus, and the Sun to be collinear, but it also allows them to be non-collinear. From the perspective of the Ptolemaic theory, the collinearity hypothesis could go either way. By contrast, the Copernican account does not admit of this sort of arbitrariness. Were Jerry Fodor commenting on this case, he might offer the following diagnosis:

> The problem that the motions of Mercury and Venus pose for Ptolemaic astronomy is not to show that these motions are *possible* given the basic assumptions of Ptolemaic astronomy, but to explain how they could be *necessary* given those assumptions. No doubt it is possible for a Ptolemaic astronomer to rig the deferents of Mercury, Venus, and the Sun so that they make the motions of the planets come out right. The trouble is that, although the basic Ptolemaic assumptions permit this, they equally permit the Ptolemaic astronomer to rig the deferents of Mercury and Venus so that they wander arbitrarily far from the Sun or that they never wander far from Jupiter for that matter. The basic Ptolemaic architecture would appear to be absolutely indifferent as among these options. (cf. Fodor and McLaughlin 1990: 202)

The upshot of this is that we have some reason to prefer the Copernican theory over the Ptolemaic, namely, that it provides an explanation of the fact that Mercury and Venus are never in opposition to the Sun, where the Ptolemaic theory does not. This reason does not guarantee that the Copernican theory is true or that the Ptolemaic theory is false, but in various cases throughout the history of science, the failure to save the phenomena in the right way has been an indication that a theory, such as the Ptolemaic theory, is fundamentally misguided.

It seems quite clear that the systematicity argument is meant to have essentially the same form as the preceding argument for the Copernican theory over the Ptolemaic theory. The phenomenon to be explained in the systematicity argument is the systematicity of thought. The two competing theories are Classicism and Connectionism. Classicism is supposed to provide a better explanation of systematicity than does Connectionism for essentially the same reason that the Copernican theory provides a better explanation of the motions of Mercury and Venus than does the Ptolemaic theory. Ptolemaic astronomy and Connectionism are both supposed to rely on an arbitrary hypothesis in their accounts of the phenomena to be explained, hence offer inferior accounts of the phenomena. Hence, we have some defeasible reason to prefer Classicism to Connectionism. Notice that, if this analogy holds up, then we have some reason to think that, contrary to what Matthews (1994) claims, that the systematicity argument is not a trick. Moreover, we have gone some way to meeting Niklasson and van Gelder's complaints that the systematicity argument is obscure. So, let us consider this analogy in more detail.

Consider the phenomenon: thought is systematic. The thoughts that can occur to a normal cognitive agent are contentfully related so that, for example, if a cognitive agent can think that John loves Mary, then that cognitive agent can think that Mary loves John. Thus, a set of all and only the thoughts that might occur to a normal cognitive agent might be the following,

John loves John	Mary hates John
John loves Mary	Mary hates Mary
John loves Jane	Mary hates Jane
Mary loves John	Mary hates John
Mary loves Mary	Mary hates Mary
Mary loves Jane	Mary hates Jane
Jane loves John	Mary hates John
Jane loves Mary	Mary hates Mary
Jane loves Jane	Mary hates Jane

One does not, however, find normal cognitive agents who can think all and only the thoughts

John loves Mary	Bears hibernate
Einstein studied physics	Cats chase mice and other rodents.

Sets of possible thoughts such as this last one are evidently what Fodor and Pylyshyn had in mind when they wrote:

> Just as you don't find linguistic capacities that consist of the ability to understand sixty-seven unrelated sentences, so too you don't find cognitive capacities that consist of the ability to think seventy-four unrelated thoughts. (Fodor and Pylyshyn 1988: 40)

It is perhaps worth emphasizing here that this description of the systematicity of thought makes no presuppositions concerning the existence of mental representations. Nor does it even presuppose much about the nature of thought. For example, the systematicity of thought is consistent with the view that there are no mental representations and that an agent's having the thought that John loves Mary consists of nothing more than someone's taking a particular stance toward the agent. Recall that the systematicity of thought merely amounts to a normal cognitive agent having the ability to think sets of thoughts of a particular kind, whatever the ability to think a set of thoughts comes to. The point here is that the description of the systematicity of thought begs no questions in favor of mental representations or Classicism. So, again, Classicism has no tricks up its sleeve here.

As Fodor and Pylyshyn set up the argument, both Classicism and Connectionism assume that there are mental representations.[4] Where Classicism and Connectionism are supposed to differ is in their theories of the structure of mental representations.[5] Classicism maintains that (a) an agent's representation of, for example, John's loving Mary consists of three parts, "John," "loves," and "Mary," (b) that these parts mean John, loves, and Mary, respectively, and (c) and that the whole "John loves Mary" means what it does in virtue of the meanings of its parts and the way in which those parts are put together. Classicism postulates syntactic and semantic combinatorial structure. Connectionism, on the other hand, is supposed to maintain that the agent's representation of John's loving Mary lacks internal structure; it consists of a single atomic symbol "John-loves-Mary" that means John loves Mary. Classicism, then, attempts to explain the systematicity of thought as follows,

> [Classicism] says that having a thought is being related to a structured array of representations; and, presumably, to have the thought that John loves Mary is ipso facto to have access to the same representations, and the same representational structures, that you need to have the thought that Mary loves John. So *of course* anybody who is in a position to have one of these thoughts is ipso facto in a position to have the other. [Classicism] explains the systematicity of thought. (cf. Fodor 1987: 151)

Should Connectionists, for their part, produce a network in which the set of possible representations is systematic, Fodor and McLaughlin (1990: 202), have this to say:

> No doubt it is possible for [a connectionist] to wire a network so that it supports a vector that represents aRb if and only if it support a vector that represents bRa. … The trouble is that, although the architecture permits this, it equally permits [a connectionist] to wire a network so that it supports a vector that represents aRb if and only if it supports a vector that represents zSq; or, for that matter, if and only if it supports a vector that represents The Last of the Mohicans. The architecture would appear to be absolutely indifferent as among these options.

To make the point more concrete, consider a simple two-layer net with, say, five output nodes. One might have the output vectors represent this way:

(0.24 0.43 0.33 0.81 0.98)	represents	John is tall
(0.11 0.31 0.45 0.97 0.34)	represents	John is smart
(0.42 0.42 0.83 0.33 0.22)	represents	John is hairy
(0.93 0.44 0.53 0.21 0.03)	represents	Mary is tall
(0.45 0.12 0.33 0.02 0.91)	represents	Mary is smart
(0.72 0.12 0.99 0.97 0.82)	represents	Mary is hairy
(0.55 0.21 0.43 0.76 0.71)	represents	Jane is tall
(0.21 0.17 0.13 0.74 0.87)	represents	Jane is smart
(0.89 0.55 0.67 0.21 0.99)	represents	Jane is hairy,

with other output vectors not representing at all. Here we have a systematic set of representations; the meanings of all the representations are related. Alternatively, one can have output vectors represent this way:

(0.33 0.41 0.56 0.45 0.72)	represents	John loves Mary
(0.99 0.23 0.08 0.78 0.34)	represents	Einstein studied physics
(0.33 0.45 0.76 0.55 0.55)	represents	Bears hibernate
(0.23 0.37 0.88 0.93 0.74)	represents	Cats chase mice and other rodents,

where the other output vectors represent nothing. This second set of representations, however, is non-systematic; the meanings of the representations are unrelated.[6] To reiterate Fodor and Pylyshyn's point, then, Connectionist models fail to meet the challenge of explaining systematicity in just the way that Ptolemaic astronomy failed to meet the challenge of explaining the motions of Mercury and Venus. Connectionist networks can be wired so as to display systematicity, but they can as easily be wired not to display systematicity. Change the representational conventions in a network and it no longer generates a systematic set of representations.[7] This is one of the things that Fodor, *et al.*, ought to say in response to Pollack (1990: 90–95) and to Niklasson and van Gelder (1994: 294–299).

2. Responses to the Systematicity Argument

The foregoing provides a brief explanation of the structure of the systematicity argument and how it bears on Connectionism, but this structure might be better appreciated by explaining how it deals with certain pro-Connectionist responses. For this purpose, not all objections are equally illuminating. For example, denying that thought is systematic is one way of rebutting the argument, but it is not the sort of objection to be pursued here, since, aside from the fact that it is not a very promising response, it does essentially nothing to help us understand the argument or its place in the Classicist/Connectionist debate.

2.1. *Matthews Response*

Defenders of Connectionism have sometimes taken to responding to the systematicity argument by challenging its methodology, charging that the argument is either a trick or that it is thoroughly confused. Quite recently, Robert Matthews has offered an entertaining, if misguided, analysis in this vein:

> New York street hustlers play a shell game called 'three-card monte.' The dealer shows the victim three cards and asks him to keep an eye on one card, usually the ace of spades. The dealer proceeds to shuffle the cards, moving them from one hand to another, all the while accompanying the card play with a rhyming patter. The play soon stops, the three cards are spread out face down, and the victim bets on which of these three cards is the designated card. The dealer accepts the bet, turns over the card indicated by the victim, turns over a second card (the correct card), collects the bet, and moves on to his next victim. A skilled dealer does not have to cheat to win; nor does he need be particularly quick in his shuffling. In fact, the shuffling is quite slow and deliberate. The crucial sleight of hand takes place right at the beginning of the game. Having succeeded in getting the victim fixated on the wrong card, the dealer does not want to confuse him, since he might inadvertently guess the correct card.
>
> I belabor this account of three-card monte, because it seems to me that the arguments mounted by Fodor and Pylyshyn (1988), Fodor and McLaughlin (1990) and McLaughlin (1993a, 1993b) against connectionism are a philosophical version of three-card monte. (It is probably no accident that Fodor was raised in New York City, McLaughlin across the river in Jersey City.) The game is played with three concepts, *systematicity, implementation*, and *explanation*. The strategy of the dealer is basically this: give the victim only a fleeting look at the concept that does all the work in the arguments, viz., that of explanation, while distracting him with talk of implementation and systematicity. (Matthews 1994: 347–348)

Without a doubt, there are many expository difficulties in the papers by Fodor *et al.*, but Matthews goes too far in claiming that the systematicity argument is a hustler's game. As was indicated in Section 1, the argument is in important respects just like one of many arguments in favor of Copernican astronomy over

Ptolemaic astronomy. It is also like certain arguments that Charles Darwin gave in favor of evolution with common descent against the theory of special creation. Thus, if Matthews is to reject the methodological legitimacy of the systematicity argument, he must either show why the foregoing analogies are not applicable or explain why so many apparently respectable scientific arguments are in fact mere hustler's games.

It is worth emphasizing the good methodological credentials of the explanatory standard invoked in the systematicity argument in the face of the Matthews claim that "connectionists are going to have to insist on their right to change what counts as an explanation of systematicity" (Matthews 1994: 348). Such a claim is likely to resonate to many different sensibilities. It is likely to appeal to those who are sympathetic to the idea of Connectionism as something revolutionary involving a Kuhnian "paradigm shift" with new explanatory standards that are incommensurable with the old standards. It is also likely to appeal to those who believe that Connectionism needs to define itself, rather than letting opponents define it. In any event, it is not clear what Connectionists can gain simply by asserting a right to their own standards of explanation. The history of science has shown that a certain standard of explanation, the one articulated by Fodor, *et al.*, has a track record in leading to the truth. The Copernican case has been given, as might others. Connectionists are free to count what they will as an explanation, but they are not free to make that new standard a standard that will lead to the truth. Whether following another explanatory standard will lead to the truth or not remains to be seen.

2.2. *Niklasson and van Gelder's Response*

Lars Niklasson and Tim van Gelder have recently offered another spirited, although less artful, response to Fodor and Pylyshyn's analysis (Niklasson and van Gelder 1994). Among other things, they charge that the systematicity argument is confused and confusing. They write:

> *Why* has it been so unclear whether these [connectionist] models actually show that connectionism meets the challenge? Our view (apparently shared by Matthews ...) is that the most important reason has been obscurity in the concept of systematicity itself. In their 1988 paper Fodor and Pylyshyn discussed systematicity at length, but provided no succinct and precise characterization of it; at best, they gestured at the phenomenon with hints, analogies, and anecdotal observations. ... Consequenty, despite the controversy created by the paper, there was simply no clear concept of system-

> aticity available, and it was entirely unclear what kind of modeling, if any, could demonstrate systematicity.
>
> In short, Fodor and Pylyshyn had set up a hurdle and challenged connectionists to jump it, but nobody knew quite where the top of the hurdle was. To make matters, worse, subsequent attempts by defenders of Fodor and Pylyshyn to clarify the concept of systematicity (e.g. McLaughlin, 1993a), not only failed to do so in any way that was of much practical help to connectionists, but to some extent shifted the ground. The hurdle was not only hard to see, it was moving as well. (Niklasson and van Gelder 1994: 288–289)

There is some truth in what Niklasson and van Gelder claim about a lack of clarity and the adequacy of definition in some of the papers by Fodor *et al.* One can, for example, detect terminological shifts from Fodor (1987) to Fodor and Pylyshyn (1988) to Fodor and McLaughlin (1990). There is also the sometimes elusive concept of "intrinsic connection" in passages such as the following (which is something like a definition of systematicity):

> *systematic* — by which I mean that the ability to produce/understand some of the sentences is *intrinsically* connected to the ability to produce/understand many of the others. (Fodor 1987: 159; cf. Fodor and Pylyshyn 1988: 39)

Finally, it is an open empirical question exactly how much systematicity there is to thought. *Exactly* what systematicities there are is an empirical fact that cognitive psychologists ought be in the process of discovering (cf., e.g., Fodor 1987: 153; Fodor and Pylyshyn 1988: 42–43).[8]

Be these points as they may, they do nothing to challenge the cogency of the systematicity argument. Certainly one will want a clearer specification of the actual systematic relations among human thoughts if one wants to develop a model of human psychology. Cognitive scientists in the cognitive modeling business will certainly feel some sympathy for this complaint. On the other hand, one does not really need much in the way of a clear specification of the actual systematicities in human thought in order to appreciate the force of the systematicity argument. If anything like the sort of thing Fodor *et al.*, are pointing to is correct, if there is virtually anything at all to the phenomenon of systematicity, then it will create problems for the Connectionist. If it is roughly true that the possible thoughts of a normal cognitive agent are contentfully related, then this is enough of an explanandum to cause problems for the Connectionist. The point is obvious if we review the Copernican-Ptolemaic example. Imagine a Ptolemaic astronomer responding to the Copernican argument about the motions of Mercury and Venus by complaining that the measurements of the positions of maximal deviation of Mercury and Venus have varied over time or that there remains considerable error in the measurements of the motions of Mercury and Venus. The Copernican can perfectly well concede that more precise and stable measurements of the motions of Mercury and Venus are desirable, and even necessary, for certain purposes. Nevertheless, the more precise measurements are not necessary in order to appreciate the problem that the motions of Mercury and Venus pose for Ptolemaic astronomy. If anything even remotely like these purported motions of Mercury and Venus are correct, then this has serious implications for Copernican and Ptolemaic astronomy. They give one a defeasible reason to prefer the Copernican theory over the Ptolemaic.

Having had their initial methodological say about the systematicity argument, Niklasson and van Gelder report on a network that is able to use back-propagation to learn a set of representations that is systematic in some of the senses that they take to be so crucial to the systematicity argument. This sort of network, however, does not engage the basic point made by Fodor and McLaughlin, namely, that the point of the systematicity argument is not just to get a network to produce a systematic set of representations. Rather, one must show how a systematic set of representations follows necessarily from Connectionist hypotheses. This they do not do, so they have no explanation of the systematicities they claim their network exhibits.

Having reported on their network simulations, in the concluding section of their paper, Niklasson and van Gelder have this to say:

> Over the years a variety of arguments have been advanced in support of the idea that the human cognitive architecture must be basically classical in form. It is interesting to ask why it is that in 1988, and in order to defend the classical conception against *connectionism*, Fodor and Pylyshyn came up with what, as far as we can tell, is an entirely novel argument. Presumably it is because they themselves felt that the traditional arguments were no longer effective; they could not be used to make a compelling case for the classical conception as against connectionism. … The systematicity argument can therefore be seen as a last ditch attempt to provide a decisive general argument in favor of the classical conception. Since the systematicity argument is undermined by the kind of results described here, it seems that there are no longer any powerful general arguments in favor of the classical view. (Niklasson and van Gelder 1994: 299–300)

This analysis is rather unconvincing, for reasons other than that the systematicity argument has not been undermined. Fodor and Pylyshyn 1988, discuss a productivity of thought argument for classicism (pp. 33–37) that is evidently meant to be a general argument against Connectionism, just like the systematicity argument. Fodor and Pylyshyn also indicate that they find nothing wrong with the productivity argument. They claim to introduce the systematicity argument alongside the productivity argument because they wish to provide an argument for the same conclusion as in the productivity argument, but that does not require the idealization to unbounded competencies (cf. pp. 36–37; Fodor 1987: 147–148). As they put it, with the systematicity argument "You get, in effect, the same conclusion, but from a weaker premise" (p. 37). Incidently, by tracing the systematicity argument back to the Chomskyan productivity argument(s), one can more easily see the sense in which Fodor and Pylyshyn were claiming that the arguments for classicism are traditional (cf., e.g., Fodor and Pylyshyn 1988: 33, 34, 49; Niklasson and van Gelder 1994: 289).

2.3. *Smolensky's Tensor Product Theory*

One of the most vigorous exchanges in the Classicism/Connectionism discussion of the systematicity argument has been between Paul Smolensky (1987, 1991) and Fodor *et al.* In this exchange, Smolensky suggests that Tensor Product Theory (TPT) might provide an explanation of systematicity, where Fodor and McLaughlin have offered numerous reasons why it does not. Although this exchange has been quite complicated, what has so far been said about the form of the systematicity argument will illuminate some important threads in the exchange.

The core idea underlying TPT is simple. Syntactically and semantically atomic representations are vectors. Thus, a syntactically and semantically atomic representation of John might be the tensor (12 33 42), a syntactically and semantically atomic representation of Mary might be the tensor (10 45 32), and a syntactically and semantically atomic representation of the two-place relation of loving might be (23 87 14). Complex mental representations are constructed from atomic mental representations by either tensor multiplication or addition. Consider the construction of the tensor representation of John's loving Mary. To represent this, one needs some indication of the fact that John is the subject of the proposition, Mary is the object, and loves is the relation in which John and Mary stand. This is achieved through tensor multiplication with row vectors.

Thus, the vector for John will be multiplied by a subject vector, say, (2 3), the vector for Mary will be multiplied by an object vector, say, (3 4), and the loves vector will be multiplied by a relation vector, say, (8 3). This yields the following:

John-subject	=	24	66	84
		36	99	126
loves-relation	=	184	696	112
		69	261	42
Mary-subject	=	30	135	96
		40	180	128.

These vectors may then be combined into a representation of John's loving Mary through vector addition of the John-subject, the loves-relation, and the Mary-subject vectors. Thus,

John loves Mary	=	238	897	292
		145	540	296.

In general, then, roles are bound to fillers (variables are bound to values) through tensor multiplication, where filled roles are combined through tensor addition. One can see that, given that John and Mary are tensors of the same rank with the same number of components (requirements that must be met in order for both of the tensor operations to be well defined), it follows that if a system can represent John loves Mary, then it can represent Mary loves John. So, TPT has at least taken one step in the direction of explaining the systematicity of thought.

Fodor and McLaughlin (1990), make a number of important criticisms of TPT but there remains one that seems deserving of special emphasis in the present context. Even if TPT is able to explain the systematicity of thought, this does not provide a vindication of fundamentals of *Connectionism*. It does not provide a vindication of the apparatus of nodes with weighted connections between them or the theory of learning/memory based on weight changes. A system that instantiates the TPT theory will explain systematicity in virtue of the fact that it is a TPT system, not in virtue of the fact that it is a network of nodes and connections. One can appreciate this by reflecting on the fact that TPT can be realized in networks, but it can be realized in non-network systems as well. A realization of TPT in a non-network system might necessarily produce a systematic set of representations, hence may explain the systematicity of thought. The point being made here is closely related to the observation that Connectionist networks *can* explain the systematicity of thought by implementing a Classical language of thought. This, however, would not vindicate *Connectionism*, since it is the Classical theory that does the explaining when the Connectionist system implements a Classical system. What is needed, then, is to have a Connectionist system explain the systematicity of thought, *in virtue of being a Connectionist system*, i.e., in virtue of being a system of nodes with weighted connections between them. As has been emphasized before, the problem for the Connectionist is that, given just Connectionism, rather than something like Tensor Product Theory, one can as easily come up with a network that generates a systematic set of representations as a network that does not generate a systematic set of representations.

3. Why Classicism fails to explain Systematicity

So far I have tried to explain the structure of the systematicity argument and why it proves to be such a challenge to Connectionist theories of cognition. Here, however, I want to argue that the explanatory standard set by the argument is so high that even Classicism fails to meet it. Recall that the crucial flaw in the Connectionist attempts to explain systematicity lies in the fact that one finds hypotheses that, from the Connectionist perspective itself, one could as easily adopt as not adopt. The Connectionist can hardwire or train a network to generate a systematic set of representations, but can also hardwire or train a network to generate a non-systematic set of representations. This is parallel to Ptolemy's problem with having to say that, from the perspective of the Ptolemaic theory, the deferents of Mercury, Venus, and the Sun are collinear, but could as easily have not been collinear. Unfortunately for Classical accounts of cognition, this problem also arises for Classicism. There are Classical systems with systematic sets of representations and Classical systems without systematic sets of representations.

Recall that Fodor's version of Classicism asserts that attitudes are computational relations to mental representations and that mental representations have compositional structure. From this alone, however, it does *not* follow that if a cognitive agent can think that John loves Mary, then it can *ipso facto* think that Mary loves John. Having compositional representations simply does not entail that a capacity to produce the mental sentence 'John loves Mary' brings with it automatically the capacity to produce the mental sentence 'Mary loves John'; a system can display a perfectly standard form of syntactic and semantic compositionality without having the capacity to write formulae of the form aRb if having the capacity to write formulae of the form bRa. Put simply, syntactic and semantic compositionality do not entail systematicity of representation.

Here is another way of making the point. Consider a Turing machine that computes over the simple alphabet {John, Mary, Jane, loves, hates, fears}. The Turing machine's program might allow it to write combinations such as:

'John loves Mary'	'John hates Mary'	'John fears Mary'
'John loves Jane'	'John hates Jane'	'John fears Jane'
'Mary loves Jane'	'Mary hates Jane'	'Mary fears Jane'

but not allow certain other combinations, such as:

'loves John loves,'
'loves hates loves,'
'is-taller-than,'

and, most notably:

'John loves John'	'John hates John'	'John fears John'
'Mary loves John'	'Mary hates John'	'Mary fears John'
'Mary loves Mary'	'Mary hates Mary'	'Mary fears Mary'
'Mary loves Jane'	'Mary hates Jane'	'Mary fears Jane'
'Jane loves John'	'Jane hates John'	'Jane fears John'
'Jane loves Jane'	'Jane hates Jane'	'Jane fears Jane.'

Such a Turing machine displays syntactic and semantic compositionality, since it allows the construction of various syntactically complex representations from the syntactic primitives {John, Mary, Jane, loves, hates, fears}, and allows that the meanings of these complex expressions is determined by the meanings of the parts and their mode of composition. Still, the Turing machine's ability to write a formula that means John loves Mary does not bring with it the ability to write a formula that means Mary loves John. More generally, a Turing machine's ability to write a formula that means one thing does not entail an ability to write any other formula whatsoever. The same point applies to all Turing-equivalent computational formalisms.

It is important to bear in mind the full force of the claim that compositionality does not entail systematicity. The point is that compositionality does not entail that there be any systematic relations between possible representations *at all.* The simplest way to see this is through the extreme case in which a compositional language allows for only one well-formed formula. A language with only one syntactically and semantically composite formula, say, 'John is tall,' cannot be systematic. In a slightly less extreme case, one might have a language with only two formulae that have syntactic ad semantic structure, but that represent only John's loving Mary and John's being tall. These might be the only two sentences the grammar of the language allows. It won't do, therefore, to try to defend the Fodorian systematicity argument by saying that compositionality does not entail any particular set of systematic relations among sentences, but that compositionality does entail that there be *some* systematic relations. This attempted defense relies on a mistake. Compositionality does not entail any systematic relations whatsoever.[9]

The hypothesis of compositional mental representations must obviously be complemented by an additional hypothesis, roughly to the effect that a Classical architecture can write a formula of the form aRb if it can write a formula of the form bRa. McLaughlin (1993a, 1993b), makes this point explicit. Thus, in order to explain systematicity, Classicism must claim that having an attitude toward a proposition P is a matter of (1) having a mental representation R that means that P, (2) standing in a computational relation to a mental representation R, (3) R having syntactic and semantic compositional structure, and (4) having a mental program that, for example, can write 'John loves Mary' if it can write 'Mary loves John,' that can write 'John loves Jane' if it can write 'Jane loves John,' and so on. Without (4), or some other comparable hypothesis, the systematicity of thought will simply not be a consequence of Classicism at all.

The problem for the Classicist is simply that condition (4) plays a role in the explanation of systematicity that the collinearity hypothesis plays in the Ptolemaic explanation of the motions of Mercury and Venus. It is an arbitrary hypothesis tacked on to the others. A computer with a compositional language of thought could be such that it writes 'John loves Mary' just in case it writes 'Mary loves John,' but it could just as easily be such that it writes 'John loves Mary' just in case it writes 'Peter despises Andrew,' or the Last of the Mohicans, for that matter. A geocentric system could be such that the deferents of Mercury, Venus, and the Sun are collinear, but it could just as easily be such that the deferents of Mercury, Venus, and Jupiter are collinear. So, Classicism does not explain systematicity.

The fact that mere compositionality does not entail systematicity, as described above, puts many earlier attempts to meet the challenge of explaining systematicity in a new light. As Fodor *et al.*, set up the explanatory challenge, they claimed that Classicism postulated structured representations, where Connectionism postulated unstructured representations. In response, some Connectionists tried to develop "non-Classical" structured representations (cf., e.g., Pollack 1990; Smolensky 1990; van Gelder 1990). But, the foregoing examination of the systematicity argument reveals that Fodor *et al.*, themselves failed to fully appreciate the explanatory standard at work in their argument. They failed to appreciate that postulating structured representations was, in itself, an insufficient basis with which to explain the systematicity of thought. An additional hypothesis is needed to explain the systematicity of thought, but it is this additional hypothesis (or some equipollent hypothesis) that is the undoing of the Classical explanation of systematicity. Fodor *et al.*, were to some extent misled themselves, hence misled those who tried to respond to them.

4. The Role of Systematicity in the Classicism/Connectionism Debate

The last two sections have argued that neither Connectionism nor Classicism, as they presently stand, explain the systematicity of thought. From this, one might naturally wish to infer that the explanatory standard that Fodor *et al.*, have been invoking is too demanding. Scientific theories, or at the least, cognitive science theories, cannot be made to meet this standard. As a counterweight to this inference, one should recall that explanatory arguments of the form sketched in this paper have been quite influential in fomenting scientific revolutions, such as the Copernican and Darwinian revolutions.[10] The moral of the story is that it is inadvisable simply to ignore the systematicity argument or claim that it is a mere trick. It is an argument that a theory of cognition must come to grips with.

Another moral of the foregoing analysis of the systematicity argument is that one cannot meet the explanatory challenge that has been set forth simply by producing a model that displays or exhibits systematic representations. Exhibiting systematicity is one thing, explaining it another. Explaining is more demanding. More important and subtle than this, however, is the observation that one cannot meet the explanatory challenge of the systematicity argument merely by tacking additional hypotheses on to either Connectionism or Classicism. To say that Classicism is the theory that there are compositional mental representations and that there is a computer program of the mind that allows one to produce say, aRb if one can produce bRa, will not provide an explanation of the systematicity of thought for exactly the reason that Ptolemy's theory of geocentrism plus the hypothesis of the collinearity of Mercurial, Venusian, and Solar deferents did not provide an explanation of the motions of Mercury and Venus relative to the Sun. To say that Connectionism is the theory that there are nodes with weighted connections between them and that these networks, say, implement Tensor Product Theory fails to explain the systematicity of thought in just the way the Ptolemaic theory of geocentrism plus the hypothesis of collinear Mercurial, Venusian, and Solar deferents failed to explain the relative motions of Mercury, Venus, and the Sun. What seems to be needed to explain the systematicity of thought, roughly speaking, is a single, new deeper hypothesis concerning mental representations that entails that thought is systematic. If this diagnosis is correct, then we can see that Classicism was, in a sense, at least on the right track. The Classical theory was at least a theory of the structure of mental representations. By contrast, Connectionism looks to be off on the wrong foot from the very beginning. Hypothesizing nodes with weighted connections and weight change procedures seems to be orthogonal to the explanation of the systematicity of thought. Nodes and weighted connections, it appears, can only be related to the systematicity of thought via some conjunctive strategy. Connectionist networks seem only to exhibit systematicity because they add some additional representational hypothesis.

It is sometimes suggested that what is important about Connectionism is not the nuts and bolts technical details of nodes, connections, and weight change procedures, but some rather more abstract possibilities that networks open up, such as the use of "soft-constraints" or "superpositional representations" (cf. Smolensky 1988; van Gelder 1991). Such analyses, however, obscure the significance of technical developments such as Hopfield nets (Hopfield 1982, 1984), the Boltzmann machine weight change procedure (Hinton and Sejnowski 1986), and back propagation (Rumelhart, Hinton and Williams 1986). Indeed, such analyses threaten to obscure the fact that, throughout the history and prehistory of connectionism, relatively technical developments in neurobiology, mathematics, and computation have sustained a core of connectionist research involving nodes, connections, and modifiable weights (cf., e.g., Aizawa 1992, 1995). These technical developments include everything from the study of spinal reflexes in the decerebrate cat and the squid giant axon preparation to the mathematical modeling of neurons with differential equations and the perceptron learning procedure. The research has included everything from purely neuroanatomical and neurophysiological studies of neurons, to computational modeling of neurons as interesting in their own right, to engineering studies of what can be done with mathematically specified neurons.

The systematicity argument suggests that Connectionism, with its commitment to nodes with modifiable weighted connections between them, is fundamentally ill-suited to be a theory of cognition. Connectionist cognitive psychology ought to be abandoned. Connectionism is committed to the wrong level of analysis for cognition. This dos not mean that all is lost for Connectionism. Quite the contrary. There are many brands of Connectionist research centered on the hardware of nodes, connections, and modifiable weights that are not cognitive psychology. The current breadth of Connectionist research in computer science, engineering, mathematics, and physics, as well as some one hundred years of research in psychology, neurobiology, neuropsychology, and mathematical biophysics bears this out. Connectionism may not give us good cognitive psychology, but it may have much to contribute in other areas of scientific inquiry.

Notes

1. For convenience, I will often refer to these three individuals a 'Fodor *et al.*,' even though there are probably differences of opinion among them concerning the systematicity argument.
2. The following example is borrowed from Glymour (1980), chapter 5, ad Kuhn (1957), chapter 5, who provide interesting discussions of this and related cases. In Aizawa, (in progress), I discuss examples of the explanatory strategy found in Charles Darwin's *Origin of Species*.
3. One might claim that the Copernican theory has an explanation of the motions of Mercury and Venus, where the Ptolemaic theory has none. Fodor *et al.*, appear to have a preference for saying this sort of thing. I see no reason for saying one thing rather than the other.
4. Cf. Fodor and Pylyshyn (1988: 7–11). There is, of course, much more to Classicism and Connectionism, even as pertains to the systematicity argument, but the status of mental representations is what is crucial for the present discussion.
5. Cf. Fodor and Pylyshyn (1988: 12ff).
6. Another sort of representational scheme would be to associate different regions of activation state space with distinct representations. Exactly how one produces systematic and non-systematic sets of representations is not important, only that one have available some Connectionist means of producing both.
7. It is perhaps worth emphasizing that networks might fail to generate a systematic set of representations for reasons other than mere changes in linguistic conventions. For example, take a network that generates a systematic set of representations on its output nodes. Such a network might be given a new set of weights, as through back-prop, so that the set of representations on the output nodes is not systematic. This, again, indicates how a connectionist network might be made to produce a systematic set of sentences or not. This is what undermines the Connectionist claim to having an explanation of systematicity.
8. Fodor *et al.*, are guilty of other types of infelicities as well. One of these will figure prominently in the subsequent discussion.
9. The foregoing is meant as a response to comments made independently by Donna Summerfield and Kevin Falvey. I don't, however, believe that they are satisfied with my response.

10. For a discussion of some Darwinian arguments apparently having this form, cf. Kitcher (1985) or Aizawa (forthcoming).

References

Aizawa, K. 1992. Connectionism and artificial intelligence: History and philosophical interpretation. *Journal for Experimental and Theoretical Artificial Intelligence* 4, 295–313.

Aizawa, K. 1995. Some neural network theorizing before McCulloch: Nicolas Rashevsky's mathematical biophysics. In *Proceedings of the International Conference on Brain Processes, Theories, and Models,* J. Mira-Mira (ed), Cambridge, MA: MIT Press.

Aizawa, K. Forthcoming. Explaining systematicity. *Mind and Language.*

Fodor, J. 1987. *Psychosemantics.* Cambridge, MA: MIT Press.

Fodor, J. and McLaughlin, B. 1990. Connectionism and the problem of systematicity: Why Smolensky's solution doesn't work. *Cognition* 35, 183–204.

Fodor, J., and Pylyshyn, Z. 1988. Connectionism and cognitive architecture: Why Smolensky's solution won't work. *Cognition* 35, 183–204.

Glymour, C. 1980. *Theory and Evidence.* Princeton, NJ: Princeton University Press.

Hinton, G. and Sejnowski, T. 1986. Learning and relearning in Boltzmann machines. In *Parallel Distributed Processing: Explorations in the Microstructure of Cognition,* vol. 1, D. Rumelhart, J. McClelland, and the PDP Research Group, 282–317. Cambridge, MA: MIT Press.

Hopfield, J.J. 1982. Neural networks and physical systems with emergence collective computational capabilities. *Proceedings of the National Academy of Sciences, USA,* 79, 2554–2558.

Hopfield, J.J. 1984. Neurons with graded response have collective computational properties like those of two-state neurons. *Proceedings of the National Academy of Sciences, USA,* 81, 3088–3092.

Kitcher, P. 1985. Darwin's achievement. In *Reason and Rationality in Natural Science,* N. Rescher (ed), 127–189. New York, NY: University Press of America.

Kuhn, T. 1957. *The Copernican Revolution.* Cambridge, MA: Harvard University Press.

Matthews, R.F. 1994. Three concept monte: Explanation, implementation, and systematicity. *Synthese* 101, 347–363.

McLaughlin, B. 1993a. The Classicism/Connectionism battle to win souls. *Philosophical Studies* 70, 45–72.

McLaughlin, B. 1993b. Systematicity, Conceptual Truth, and Evolution. In *Philosophy and Cognitive Science,* C. Hookway and D. Peterson (eds). Cambridge, UK, Cambridge University Press.

Niklasson, L.F., and van Gelder, T. 1994. On being systematically connectionist. *Mind and Language* 9, 288–302.
Pollack, J. 1990. Recursive distributed representations. *Artificial Intelligence* 46, 77–105.
Rumelhart, D., Hinton, G. and Williams, R. 1986. Learning internal representations by error propagation. In *Parallel Distributed Processing: Explorations in the Microstructure of Cognition*, vol. 1, D. Rumelhart, J. McClelland, and the PDP Research Group, 318–362. Cambridge, MA: MIT Press.
Smolensky, P. 1988. On the proper treatment of connectionism. *Behavioral and Brain Sciences* 11, 1–74.
Smolensky, P. 1990. Tensor product variable binding and the representation of symbolic structures in connectionist systems. *Artificial Intelligence* 46, 159–216.
van Gelder, T. 1990. Compositionality: A connectionist variation on a classical theme. *Cognitive Science* 14, 355–384.
van Gelder, T. 1991. What is the 'D' is PDP: A survey of the concept of distribution. In *Philosophy and Connectionist Theory*, W. Ramsey, S. Stich, and D. Rumelhart (eds), 33–59.

Connectionism, Tri-Level Functionalism and Causal Roles

István S.N. Berkeley
Department of Philosophy
University of Southwestern Louisiana

1. Introduction

As Block (1978: 270) notes, there are a "... bewildering variety of functionalist theories ...." However, according to Block (1980a: 173) all forms of philosophical functionalism have a common central thesis. This is that (to cite Sterelny's (1990: 2) formulation):

> the essential feature of any mental state is its causal role.

One of the crucial difficulties which advocates of functionalism have to face is to find an adequate account of the notion of a 'causal role.' One approach which has been adopted to deal with this difficulty is to consider causal roles within non-biological computational systems. In this paper I will illustrate this approach and argue that connectionist systems can (potentially) play a useful role in functionalist theorizing. This is not the first time that functionalism and connectionism have been conjoined (see for example, Clark 1989). However, the argument I offer is new and more detailed than previous discussions.

2. Causation

One of the difficulties which has to be faced when developing an account of the notion of a 'causal role' is the latitude which can be ascribed to the notion of causation. Although straightforward physical causation (governed, hopefully, by the laws of physics) is unproblematic, there are cases where the notion of cause is invoked and, at first glance at least, the causal relation appears to be extended or metaphorical at best. Unfortunately, mental states are one of the more common instances where an extended notion of cause seems to be used. Consider, for example, the claim that "My fear of all dentists caused me to hate him" or a statement like "I fell in love with her, because of her sense of humor." In both these examples, even if there are underlying physical process, they are not explicit.

This being the case, it would seem that a full account of the notion of a causal role must be able to handle not only what I will call 'brute physical causation,' but must also have some kind of story to tell about the extended notion of causation. This is especially important given that many of our folk psychological descriptions of mental states and processes seem to involve an extended notion of causation. Pretty clearly, the best way to go about providing an account of the extended notion of causation would be to show that it is, in fact, just a special kind of physical causation. Fortunately, Pylyshyn's (1984) three levels of description for cognitive systems offer a means by which such an account can be generated.

3. Levels

To see how exactly how Pylyshyn's three levels of description can be used to clarify the notion of a causal role, I will sketch the teleological relations between the three levels. The examples I use below are extremely simple and it is not my intention to give the impression that such simple mechanisms could have intentional states. Pylyshyn (1984) is quite clear that higher level descriptions should only be given if such descriptions offer generalizations which are not captured at the lower levels. In the examples below, this is not the case. My purpose here is merely to illustrate how the causal roles of mental states *could* be explained by reference to the physical causation.

3.1. *The Implementational Level*

When a system is described at Pylyshyn's physical or implementational level, the system is described in terms of its physical properties and in relation to the physical (or biological) laws which govern its operation. This being the case, the physical or implementational level is characterized by brute physical causation. To illustrate this point, it is worth considering the operation of a common component of computational devices, the AND gate.[1]

There are a number of different arrangements of physical components which can implement an AND gate. AND gates can be constructed using vacuum tubes, semiconductors and even electromagnetic relays (See Jacobwitz 1963: 139–142). I will focus on (one possible configuration of) a semiconductor AND gate here.

A simple AND gate has two inputs and a single output. The gate is constructed from a resistor and a diode with the appropriate interconnections between them. Figure 1 illustrates in schematic form the arrangement of these components.

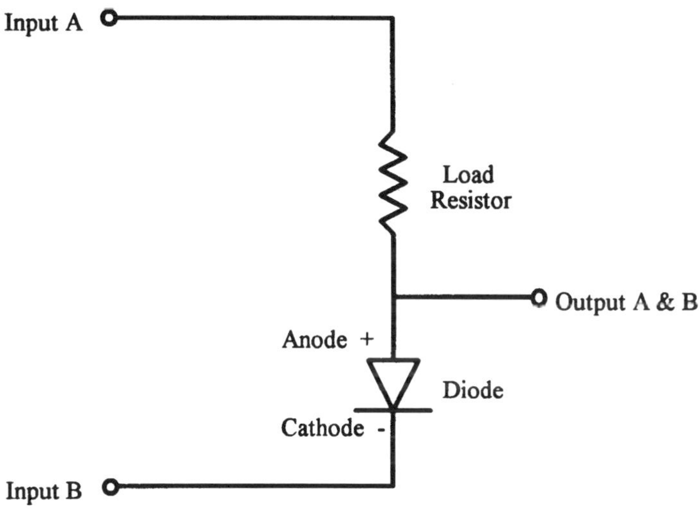

Figure 1. A Semiconductor AND Gate

The operation of the gate is the result of the interaction of the physical (in this case, electrical) properties of the components. A diode is basically an on-off switch which is voltage sensitive. A diode permits current (i.e. a stream of electrons) to pass through it if the potential (i.e. voltage) between its anode and cathode is such that the anode is positive with respect to the cathode. If the anode is negative with respect to the cathode, then no current will pass through the diode.[2] Now, consider the response of the circuit detailed in Figure 1 under various conditions (in doing this, I assume that the reader has some basic familiarity of with the basic electrical properties of simple components, such as resistors). For the purpose of this example, I will suppose the inputs can only be either +1 volts or −1 volts. As a matter of fact, these values are unimportant, provided that there is some reasonable difference between a high and low voltage input.

In the first instance, suppose that inputs A and B are both at –1 volts. Under these circumstances, there is no difference in potential between the anode and cathode of the diode, so it does not permit any current to flow through it. Consequently, there is no voltage drop across the load resistor and thus the output is just equal to the voltage at input A. That is to say, the output voltage in this case would be –1 volts. Next consider what happens if input A is at –1 volts and input B is at +1 volts. In this case, the cathode of the diode would be positive with respect to the anode and as in the previous case, no current would flow through the diode. The upshot of this is also similar to the previous case; there would be no voltage drop across the load resistor and consequently the output voltage will be equal to the voltage of input A at –1 volts.

The third combination of input voltages, when input A is at +1 volts and input B is at –1 volts, is a little more interesting. In this instance the anode of the diode is positive with respect to the cathode and consequently current will flow through the diode. As a result, input voltage B, in this case –1 volts, is directly connected to the output. Thus, the output voltage is the same as the voltage at input B, which is –1 volts. The final possible combination of inputs is when both inputs A and B are at +1 volts. In this case, due to the voltage drop which develops across the load the resistor, the anode of the diode is negative with respect to the cathode and so no current flows through the diode. As a result, the output voltage is going to be equal to the voltage of input A — that is to say, the output will be +1 volts in this case.

From the above it should be clear that, assuming that inputs A and B can only be at the two voltages –1 and +1, the output of the circuit in Figure 1 will always be equal to the lowest of the two input voltages or equal to both input voltages, if they happen to be the same. So, it is only when both input A and input B receive inputs of +1 volts that the output will be +1 volts. In all other instances, the output will be at –1 volts. More importantly though, this behavior is entirely the result of the electrical properties of the components of the circuit. Hence, this physical or implementational level description of the circuit is just governed by the familiar causal laws of physics.

3.2. *The Symbolic Level*

Next the symbolic level description of the AND gate needs to be considered. Of course, in the case of the AND gate, describing the mechanism at this level offers no additional generalizations. However, describing an AND gate at this level serves to illustrate the kind of teleological relations which will obtain between the levels in properly cognitive systems. Descriptions at the symbolic level must specify both formal rules and representations upon which those formal rules operate (exactly what the representations themselves stand for, of course, is specified at the semantic level). In the case of the AND gate, the basic representations at the symbolic level are just three variables (I will call them x, y and z). These variables correspond to the two differential voltages at the physical level which may be applied to the inputs A and B and the resultant voltage from the output of the AND gate. It is important to note that this relationship is only one of correspondence; the variables do not 'stand for' the voltages, as this would be to introduce semantic considerations. Rather, the particular voltages which are applied to inputs A and B and emerge from the output are just particular instantiations of the variables x, y and z.

There are a number of ways in which the rule(s) governing the operation of the AND gate upon the variables x, y and z can be specified at the symbolic level. Perhaps the simplest way is by saying that z is the result of the two-place function AND(x,y). Although such a description satisfies the general requirements of a symbolic level description, it might be objected that such a rule does not adequately specify the function.

In order to give a fuller description of the AND gate, it is important to note that the variables x, y and z can only take two possible values. For the purposes of this example, let us assume that the two values are **p** and **q** (exactly what **p** and **q** stand for will be specified at the semantic level). Now we can specify in more detail the rule, at the symbolic level, governing the operation of the AND gate. The rule may look something like this;

IF (x = y AND x is **p**), THEN z = x
ELSE z is **q**

The first thing to note about this rule is that, until such times as specific values are assigned to **p** and **q**, there is no way to distinguish between this rule and the rule for OR. The important point though is that the rule describes, in a systematic way, the operation of the physical components of the AND gate. A consequence of this is that this symbolic level description is simply the result of the physical causal laws which govern the operation of the AND gate at the implementational level. Consequently, any apparent causal relations at the symbolic level are just the result of causal processes at the implementational level. The other important point to note is that this rule will characterize at the symbolic level, the operation of *all* AND gates, regardless of whether they are constructed from solid state diodes, vacuum tubes, relay switches, or neurons. That is to say, there is a characteristic many-to-one relation between the implementational level causal laws and this description at the symbolic level.

3.3. *The Semantic Level*

Now that we have the AND gate described at the implementational and symbolic levels, it is time to turn attention to the semantic level description of the device. Pylyshyn (1984: 66) offers two criteria which a system must satisfy in order to have a semantic level description. Firstly, the system must be such that all it's behaviors can be described by formal rules which hold for all the states of the system. The AND gate satisfies this condition, as the formal rule and states of the system are all described at the symbolic level. The second criterion is that the regularities of the system must be such that they respect the semantic interpretation given to the system. The AND gate satisfies this criterion too, as will be demonstrated. Thus, the AND gate is a good candidate for a system with a semantic level description. Once again though, it is important to recall that as a semantic level description of a simple AND gate does not give rise to any generalizations which are not captured at the implementational or symbolic level, there are no grounds for claiming that an AND gate is cognitive in any way.

The semantic level description of the AND gate is given simply by the truth-table for the connective AND. That is to say, the semantic level description for the gate is just,

Table 1. Truth Table for AND

A	B	A AND B
True	True	True
True	False	False
False	True	False
False	False	False

Under this interpretation, the state **p** from the symbolic level description is assigned the value True and **q** is assigned the value False.

As a matter of fact, this is only one possible semantic level description of the device. If **p** is assigned the value False and **q** is assigned the value True, the result is the truth-table for the connective OR. This too provides an adequate semantic level description. That is to say,

Table 2. Truth Table for OR

A	B	A OR B
True	True	True
True	False	True
False	True	True
False	False	False

would be an alternative semantic level description of the gate, as it too satisfies both of Pylyshyn's criteria. The fact that there are two distinct semantic level descriptions of the device I have been calling an AND gate (of course, it could have just as well have been called a OR gate), serves to illustrate that there is a many-to-one relationship between the semantic and symbolic levels.

So far so good, but what of putatively causal talk at the semantic level? Someone might wish to say of the AND gate, for example, that "The gate gave the answer true, because both of the conjuncts were true." In a case such as this, although it is arguable that the notion of cause invoked here is the metaphorical one, the causation is, in fact, firmly rooted in physical causation, governed by physical laws at the implementational level. There is an isomorphism between the interpretation at the semantic level and the rules and representations at the symbolic level. These rules and representations are, in turn, a consequence of the physical laws which govern the interaction of the components of the gate at the implementational level. Consequently, although causal talk at the semantic level may appear to involve an extended notion of causation, it need not do so.

Of course, only a subset of all the implementational level properties of the gate are relevant to the higher levels. For example, the precise voltages applied to the inputs of the gate do not matter one iota, provided that the voltages are sufficiently distinct for the gate to operate. Thus, higher level causal talk about systems, such as the AND gate, which can be described at the semantic level should, in principle, be underpinned by causation at the brute physical level. Of course, whether or not this is the case in the instance of a particular system is an empirical matter, to be determined in the domain of Cognitive Science.

4. Causal Power

The importance of this extended discussion of the AND gate is to illustrate how adopting Pylyshyn's three levels of description offers a means of giving a substantive account of the notion of a 'causal role' in the functionalists central thesis. Naturally, it is important to ensure that there is some mechanism which can instantiate the necessary operations for any putative causal role. Fortunately, assuming that the Church/Turing thesis is true and that cognition is a computational process, guarantees that if the operation under consideration is computational, then a Universal Turing Machine (hereafter, UTM) will, in principle, be able to perform the relevant physical manipulations.[3] The program for such a UTM would then constitute the symbolic level description. It is then just a matter of supplying an appropriate semantic interpretation, and the 'causal roles' at each level can then be evaluated in order to individuate states (putatively mental or otherwise).

5. Connectionism and Functionalism

Hopefully a reasonably convincing case has been made that systems which can be described by Pylyshyn's three levels are at least consistent with the central thesis of functionalism. Furthermore, a case has also been made that such systems can also provide a means of specifying, in some detail, what is to count as a causal role. Assuming that these matters may now be accepted as argued, it seems to follow that any particular system which can be adequately described at Pylyshyn's three levels would also be consistent with a functionalist account of mental states, provided that such a system were truly cognitive. The question I now propose to address is whether or not connectionist systems can be described appropriately at all three levels. To anticipate a little, I will argue that connectionist systems can, indeed, be usefully described at all three levels and consequently that connectionist systems are entirely compatible with functionalism. In order to argue this case in detail, it will be helpful to have an example to work with. I shall begin by describing this example.

Consider a very simple network, which consists of three units arranged in two layers with connections between them. Such a network is depicted in Figure 2.

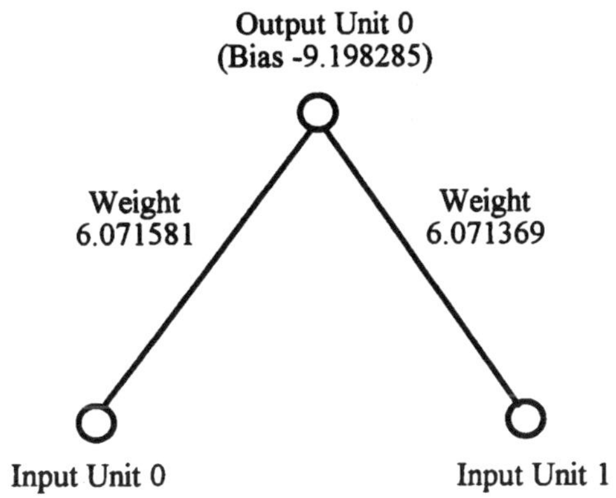

Figure 2. The And0 Netword after Training

This is an actual network which was trained using a backpropogation procedure (See McClelland and Rumelhart 1988: 121–137, for a technical discussion of this procedure). Needless to say, equivalent networks could have been generated using another training procedure, or by hand coding the weights and bias. I will call this network 'and0.' For completeness, I will briefly describe the technical details of the training. Those who are not interested in such details should skip the next paragraph.

The units in the and0 network use the common logistic activation function. The activation $\boldsymbol{a}$ of a unit i is determined by the following function,

$$\boldsymbol{a}_i = \frac{1}{1 + e^{-\text{netinput}}}$$

where *netinput* is determined in the following manner,

$$netinput_i = \sum_j \boldsymbol{w}_{ij} \boldsymbol{a}_j + \text{bias}_\text{i}$$

That is to say, $netinput_i$ is equal to the sum of the weights from i to j ($\boldsymbol{w}_{ij}$) times the activation of unit j ($\boldsymbol{a}_j$) plus the bias of unit i (bias_i), for all units j which are connected to unit i. The and0 network described in Figure 2 converged in 1663 sweeps, with a learning rate set at 0.1, a momentum set to 0.9 and the weight and bias start parameters set randomly between 0 and 0.3. The criterion for a hit was set at 2.5e-03 and the sum squared error of the output unit, after training was 1.509182e-03.

The and0 network has input and output behavior which is identical to that of the AND gate described above. (Those who are skeptical about this fact can verify this claim by doing the necessary calculations). This being the case, the descriptions of the and0 network at both the symbolic and semantic levels are (or could be, in the case of the semantic level) exactly the same as those of the AND gate. Of course, like the AND gate, the higher level descriptions of the and0 network offer no new generalizations and as a consequence, the and0 network is similarly non-cognitive. The AND gate and the and0 network do differ from one another at the implementational level however, as the and0 network was instantiated upon a SPARC station. The important point though is that if the tri-level description of the AND gate is compatible with functionalism, then given that the tri-level description of the and0 network is the same as that of the AND gate at the symbolic and semantic levels, it follows that the and0 network must also be compatible with functionalism. The differences in the physical causal processes which underlie the AND gate and the and0 network are unimportant. It is sufficient that there is for each some physical system upon which the device can be instantiated. Consequently, connectionism *is* compatible with functionalism. After all, what is sauce for the goose is sauce for the gander!

6. Networks and Causal Powers

Of course, just establishing that connectionism is merely *compatible* with functionalism is not to establish too much. A more significant result would be to show that networks have the required causal powers to instantiate the operations necessary for *any* putative causal role. Of prime importance in this context once again is the Church/Turing thesis. The reason this is important is because, given the assumption that cognition is a computational process, it is important to ensure that the systems under consideration, in this case networks, have the computational power to perform all the computations which might be involved in cognition. Of course, if the Church/Turing thesis is correct, then a UTM or any system which is computationally equivalent to a UTM (that is, a system which is 'Turing equivalent') *will* have the necessary computational power. Without Turing equivalence, there would be no guarantee that connectionist systems would have the computational power required to perform all the computations that might be involved in cognition.

Fortunately, under the appropriate circumstances, it has been shown that networks are Turing equivalent. In fact, the proof of this was one of the earliest results in work with networks. The Turing equivalence of networks, supplemented by indefinitely large memory stores (roughly equivalent to the infinite tape of the UTM) was shown by McCulloch and Pitts in 1943. This result serves to guarantee that networks will indeed (if they are permitted supplemental memory, as necessary) have the computational power required to perform all the computations that might be involved in cognition. This being the case, it would seem that networks are more than merely 'compatible' with functionalism. Instead, the combination of networks and functionalism would seem to offer as natural a means of investigating human cognition as is offered by more traditional computational systems.

7. An Objection

In a well known paper, Fodor and Pylyshyn (1988) argue that connectionist networks either lack the necessary causal properties to model certain crucial aspects of cognitive functioning, or they just amount to proposals at the implementational level and consequently offer nothing new to the study of cognition. Now, it might seem that the examples of the AND gate and the and0 network discussed here just serve to support the latter disjunct of this conclusion. Indeed, it might be objected that the conclusion argued for in this paper is no surprise at all, given Fodor and Pylyshyn's arguments. Space limitations do not permit a detailed response to this objection, so I will only briefly sketch what I believe to be the appropriate response.

Crucial to the argument made in this paper is the claim that the and0 network and the AND gate have identical descriptions at the symbolic level (and when appropriate, at the semantic level too). Thus, there seems to be little alternative but to accept that Fodor and Pylyshyn are correct that the and0 network is merely an implementational variant of the AND gate. However, it does not follow from this fact that *every* network is an implementational variant of some other system. Recent empirical work on interpreting trained network structure suggests that networks may be used to make novel *cognitive* proposals (cognitive proposals are, by definition, not at the implementational level).

Berkeley *et al.* (1995) describe a network called L10 which was trained on a variety of logic problems. When the network was interpreted, it had developed a number of rules with which to successfully handle these problems. Although some of the rules developed by L10 were the same as the rules of classical logic, such as Modus Ponens, the network had also developed a number of entirely novel rules. These novel rules were clearly cognitive, in so much as that they could be used to make predictions which could be empirically tested on human subjects. If the L10 network were to be describable at all three of Pylyshyn's levels (as it appears to be) and to offer useful generalizations at each of these levels, then it would seem that L10 and networks like it could, under the appropriate circumstances, play a useful role within functionalist theorizing. Indeed, as the results of interpreting L10 are novel, the network may serve to throw light upon hitherto unexplored aspects of cognition. This being the case, the objection raised above can be handled.

8. Conclusion

If the arguments offered in this paper are taken to be convincing, then there seem to be good grounds, at least *in principle*, for including the results from connectionist research in functionalist theorizing.

Notes

1. It is worth noting that the name 'AND gate' involves considerations which are beyond the merely implementational or physical level.
2. The precise physical reasons why a diode behaves this way need not concern us here. For more details, see your friendly local electrical engineer.
3. Of course, it is not possible to build an *actual* UTM, due to the requirement of a tape of infinite length. The Church/Turing thesis also guarantees that, provided the operation is computational, there would also be some finite Turing Machine which could be built and would be up to the job.

References

Berkeley, I., Dawson, M., Medler, D., Schopflocher, D. and Hornsby, L. 1995. Density plots of hidden value unit activations reveal interpretable bands. *Connection Science* 7(2), 167–186.

Block, N. 1978. Troubles with functionalism. In *Readings in Philosophy of Psychology*, N. Block (ed), 268–305. Cambridge, MA: Harvard University Press.

Block, N. 1980a. Introduction: What is functionalism? In *Readings in Philosophy of Psychology*, N. Block (ed.), 171–184. Cambridge, MA: Harvard University Press.

Block, N. (ed). 1980. *Readings in Philosophy of Psychology*, (2 Vols.). Cambridge, MA: Harvard University Press.

Boden, M. (ed). 1990. *The Philosophy of Artificial Intelligence*. Oxford: Oxford University Press.
Clark, A. 1989. *Microcognition*. Cambridge, MA: MIT Press.
Fodor, J. and Pylyshyn, Z. 1988. Connectionism and cognitive architecture: A critical analysis. *Cognition* 28, 3–71.
Jacobowitz, H. 1963. *Electronic Computers*. New York: Doubleday & Co.
McClelland, J. and Rumelhart, D. 1988. *Explorations in Parallel Distributed Processing: A Handbook of Models, Programs, and Exercises*. Cambridge, MA: MIT Press.
McCulloch, W. and Pitts, W. 1943. A logical calculus of the immanent in nervous activity. In *The Philosophy of Artificial Intelligence*, M. Boden (ed), 22–39. Oxford: Oxford University Press.
Pylyshyn, Z. 1984. *Computation and Cognition*. Cambridge, MA: MIT Press.
Sterelny, K. 1990. *The Representational Theory of Mind*. Oxford: Blackwell.

Emotion and the Computational Theory of Mind

Craig DeLancey
Department of Philosophy
Indiana University

1. Introduction

Cognitive scientists and philosophers often distinguish levels of description appropriate to minds and brains. One such distinction is drawn between levels of processing dependence, where a lower level process must be complete before the higher level one can be, because the higher level process is constituted by the lower level ones. This is a notion common to computer science, where a high level language is one where some commands are constructed out of several commands of a lower level language. Another distinction in levels is made between appropriate theoretical levels, such as Dennett's distinction between the design stance and intentional stance (1971), and Marr's distinction between computational, algorithmic, and hardware levels (1982) (see also Millikan 1990; Newell 1981). These levels are distinguished on the supposition that there are different theoretical vocabularies appropriate for each; for example, each level might have a corresponding science with its own taxonomy, such as computation with its symbols and states, and neural implementation with its neurons, neurotransmitters, and so on. These are called "levels" because there is often, though not always, a commitment that one is reducible to or supervenes upon another (and if the commitment for reduction is strong there may be no distinction in kind between processing dependence levels and theoretical levels). Cognitive scientists typically circumvent coming down explicitly for one kind of reductive commitment or another, and simultaneously roughly respect these theoretical levels while also giving explanations which mix them. A model for a word recognition processes, for example, might make reference both to semantic priming and the plausible neural structure and limitations of the sensory system(s) involved. However, other explanatory frameworks require that we carefully distinguish different theoretical levels, separating out the implementational from the semantic, the biological limitations from the logical ones. One such approach, the computational theory of mind, has the advantage of being pitched at the level of symbol processing, and as a result is implementation-independent. This has been one of the advantages of the computational theories, since there is broad intuitive appeal to the notion that different kinds of systems, such as robots or alien organisms, could be intelligent even if their physical structure were not much like our own (e.g. Putnam 1964, 1975).

This feature of the computational theory of mind has not gone unchallenged. In a different context, Churchland and Sejnowski have argued that we might waste time and energy by ignoring lower theoretical levels:

> Computational space is undoubtedly vast, and the possible ways to perform any task are probably legion. Theorizing at a high level without benefit of lower-level constraints runs the risk of exploring a part of that space that may be interesting in its own right but remote from where the brain's solution resides. (1992: 11–12)

Others, taking a more radical tack, have argued against purely representational theories of mind, one of which would be computationalism, on the grounds that representations do not earn us anything more in our best theories of mind than is already had in non-representational descriptions (Ramsey, Ms; see also van Gelder 1995). Representations might even, as Brooks holds, confuse things:

> When we examine very simple level intelligence we find that explicit representations and models of the world simply get in the way. It turns out to be better to use the world as its own model … Representation is the wrong unit of abstraction in building the bulkiest parts of intelligent systems. (1991: 140)

To these objections the computationalist has responses. To the former, she might likely say that she is interested not in making a physiologically accurate model of the human mind, but rather in finding the essential features of any intelligent agent. Or, in more traditional AI, one might go so far as to say she wants only to solve a problem, such as making a computer that can talk or recognize faces. In these cases, exploring the computational space need not be beholden to human physiological features. And to the latter and more radical objection, the computationalist is certainly eager to respond with lengthy debate (e.g. Clark and Toribio 1994). Much of the work that has gone into representational theories of mind has been dedicated precisely to delineating when something is functioning as a representation, how such representations are fixed to their objects, how they are used in inference, and how they might be reduced to existing physical structures. At this early stage in the sciences of mind, there seems to be no known way to settle the debate between representationalists and non-representationalists (or *anti-representationalists*, as they are often called) without begging the question.

However, the case for computationalism is more dubious when we acknowledge that there are mental phenomena that require, for a proper accounting, that we get below the level of symbol processing. Such phenomena show us that a computational theory of mind which hopes to have a sufficiently complete description of the mind is hoping in vain. Chief among these phenomena I take to be emotion. Emotions pose a problem to computationalism because by their very nature they act across the traditional levels of description that computationalists necessarily respect. Emotions are intentional, but can be disassociated from their objects; they are cognitive, but also have essential somatic elements; they can be associated with judgments and perceptions, but actually they can be shown to influence perception and judgment, and therefore perhaps to be in some sense prior to perception and judgment. It is these kinds of features of emotion that have made it the most neglected feature of mind in contemporary philosophy of mind: emotions cast us back again and again into the problem of an integrated, embodied agency, something for which we do not yet have a good account.[1]

In this paper I shall review contemporary evidence (1) that emotions influence perception, and do so at a pre-cognitive level; (2) that although related to judgments and other cognitive acts in significant and important ways, emotions are separable from judgments and other cognitive acts which some theories identify with specific emotions, thus belying any prospect of fully accounting for emotions as judgments or similar cognitive states; and (3) that emotions that are essentially linked to the extended body, and that are thus not reducible to cognitive states, are necessary for rationality and hence cannot be ignored by computationalism. In conclusion, I shall detail how these features pose insurmountable problems for computationalism, and I shall suggest some directions in the philosophy of mind that accounting for embodied emotion might require.

Focusing on perception and rationality is advantageous because it seems unlikely that any computationalist can escape these issues by reducing the size of her domain. The evidence that I will review is largely from cognitive psychology and neuroscience, since such evidence is, in this case, perhaps the most substantial. However, I urge those readers who are wary of possible scientism, and therefore likely to be put off by this method, to read on and to consider my appeals to contemporary science as embellishments. The comments I make here could have been made by any philosopher who was comfortable with a few plausible claims about emotion (namely, emotions influence perception, emotions are bodily, emotions are integral to rationality), based, perhaps, on one's own phenomenological evidence.

Three notes about terminology. First, throughout I shall understand computationalism to be a representationalist functionalism: a view that the mind is sufficiently characterized as a processor of representations and propositional contents, that these representations and propositional contents are part of a structured and internal system of representations, and that each functional unit can be, in principle, analyzed into other functional units, until ultimately the bottom or implementational functional units can be fully described in purely physicalist (i.e. non-intentional or non-mental) discourse. I will call this analysis *the analytic discharge of the homunculus*. This in-principle analysis is not relevant to the functional description, and is required only that the theory not be dualist; the computationalist supposes that the details of the functional analysis are not relevant to the functional level of computation itself. In general, the computer is taken to be the prime exemplar of how such a thing is possible, and it is supposed that the functional systems in question can be instantiated in a computational architecture — hence the term computationalism. But the conclusions drawn in this paper will apply more broadly to similar representational theories of mind which rely upon the manipulation of representations and propositional contents to account for perception, rationality, and the other relevant features of mind discussed in this paper. Second, I shall use the terms "representation" and "symbol" interchangeably. There is much laxity in how the term "representation" is used, and cognitive scientists, I fear, apply it far too liberally. However, here I take it to mean something like Haugeland does (1991): a representation "stands in" for something in the environment of the agent which may not be present, and it does so as part of a representational system, and as such the representation can coordinate behavior with the environment. For the computationalist, the representational system must be inside the organism (where we can understand "inside" to mean at least within the boundary of its skin). I also add the proviso that the intentional may be a superset of the representational, leaving open the possibility that there are states which can be said to be "about" something else, but which also do not meet all the criteria of being representations.[2] Third, I will use "cognitive" to identify processes which are conscious (open to introspection) and volitional, or which although unconscious either could be conscious or are of the same kind as conscious processes typically proposed by the computational view of mind, such as rule following or searching of lists.

2. Precognitive Emotions and Emotional Congruence in Perception

We might begin a consideration of the relation between emotion and cognition by wondering whether emotion follows cognition, cognition follows emotion, or the two can occur independently (we shall consider the possibility that emotions are cognitions in the next section). Although models that presume emotion follows cognition have been predominate for several decades, the answer seems to be that all three relations are possible in different instances.

It is commonly accepted today that emotions can be generated or caused by cognitive states: even without reviewing evidence it seems common sense that we often have emotional reactions because we make a complex judgment about a situation and draw consequent inferences. But that emotions can follow cognition does not entail that all emotional reactions follow cognition. In fact, there is substantial evidence that emotional reactions can occur without cognition, based both on the speed of some emotional responses and the fact that they occur sometimes without the recognition that cognition would seem to require. Zajonc, for example, has observed that subjects show a liking for stimuli that they were visually shown for such brief times that they later cannot say whether or not they recognize the stimulus, but they still reliably show preferences for it over new stimuli (1968). This mere exposure effect has been widely studied and is well documented. Zajonc found even that the exposure could be so brief that the effect could be shown when stimuli were displayed to subjects for only a millisecond (Kunst-Wilson and Zajonc 1980). Such fast exposure times — much faster than is normally required for conscious recognition of a stimulus — accompanied as they are by no report of recognition and no indication of recognition in forced choices, suggest that emotional evaluation occurs along a separate track from cognitive judgments. This separate track means that cognition can cause emotions, as traditional cognitive models of emotion suppose, but that also emotions can occur on their own, and do so more commonly and quickly than cognitive evaluation of the environment (Zajonc 1980).

In addition to occurring after cognition or independently of it, emotion might in turn influence such primary cognitive processes as perception. Again, this too seems common sense. If someone is home alone, frightened, she may mistake her running cat as a danger and leap back in shock. If one is in love, her friends might accuse her of seeing the world through rose colored glasses. If someone is depressed, she may continually interpret events in a negative way. And so on. This phenomenon, if it does occur, is called emotional congruence in perception. "Emotional congruence" is the term for the tendency of cognitive processes to agree along some dimension with emotional state. Emotional congruence has been studied at higher level cognitive processes than perception, such as in the formation and recollection of memories (for review see Blaney 1986; Singer and Salovey 1988). Unfortunately, the intuitively appealing notion of emotional congruence in perception has actually been quite difficult to demonstrate in the laboratory. However, some recent experiments by Niedenthal *et al.* (1994) have provided empirical evidence of emotional congruence in perception.

Niedenthal *et al.* reconsidered some theories of the emotional influence on perception of words (and a few other things). These theories modeled emotions as nodes in a semantic network. What they found was that although research of this kind had produced little results, this was to be expected when the prevailing models were analyzed. The experimenters under review (e.g. Clark *et al.* 1983) treated categorizing by affect as a one dimensional feature, with negative or positive poles, and grouped words to be tested for congruence effect into positive, negative, and neutral groups. But this approach should be expected to lead to little or no significant results. If we adhere to the spreading activation associationist model that has been tested in these experiments, either of the valences that was activated in a network could be expected to increase activation to a large number, roughly a third to a half, of the other nodes in the network. The result would be to produce little differentiable effect. If emotions were instead complex multidimensional categories, their influence could be expected to be lost in the roughshod simplicity of the bivalent models.

In response, Niedenthal *et al.* created a group of experiments where emotional influences were studied on a multidimensional model and on a bivalent scale. They also studied emotional influences cross-modally, using music to evoke the emotional state and testing for word recognition or gender facial recognition (where the pictures of faces were expressing affect), thereby avoiding problems of other semantic interference that could result from using words to both evoke and test an emotional influence. As expected, these experiments showed no significant results for the bivalent categorizations, but significant affective influence of emotion on the more complex emotional categories. Subjects that listened to sad music had a quicker response in identifying sad words than happy ones, and vice versa. Similarly, subjects who listened to happy music were faster at recognizing the gender of a happy facial expressions than were subjects who listened to sad music, and vice versa. Niedenthal *et al.* therefore produced a significant observation of emotional congruence in perception. This result is supported by on-going research in the same lab (Niedenthal *et al.* Ms; Halberstadt *et al.* 1995). Combining the results of Zajonc and Niedenthal *et al.* gives a picture like that in figure 1; emotions operate on a separate track from cognition, reacting in their own ways to the environment, but each can influence the other either by emotions causing emotional congruence or by cognitive states causing emotions to occur.

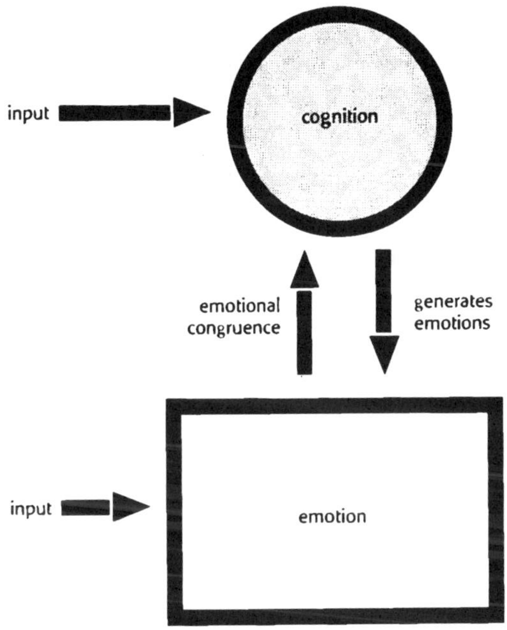

Figure 1. Emotion and Cognition as Independent

One could plausibly argue that if emotions influence perception then they are in some sense, at least in these cases, prior to perception — a conclusion that coheres well with Zajonc's suggestion of a separate but related emotional track. The computationalist would seem therefore to be left in need of an account for influences which fall outside her purview. But this need not be the case, since models which have been used in this research include semantic activation networks (e.g. Bower 1981). These models treat the effect of emotion on perception as activation spreading from an emotional node to related semantic nodes and increasing their own activation. Prima facie, such models place emotion and the other nodes on the same level. Such approaches are, I think, flawed, as I outline below. Furthermore, the same result could be achieved in other ways: one neglected possibility is to treat an emotion as a specific set of changed parameters of the model, in this case, a dropping of the threshold of related nodes, and perhaps a rising of the threshold for others. But given some of the existing models, the computationalist could accommodate both Zajonc and Niedenthal by having computational models of the effects of emotion that run in parallel to and are linked with cognitive models. And in fact, although modeling activity at the level of perception or below, and hence below higher cognitive processes like inference, the models considered by both Zajonc (1980) and Niedenthal (1994) are amenable to a computational theory of mind. Figure 2 shows the revised scheme.

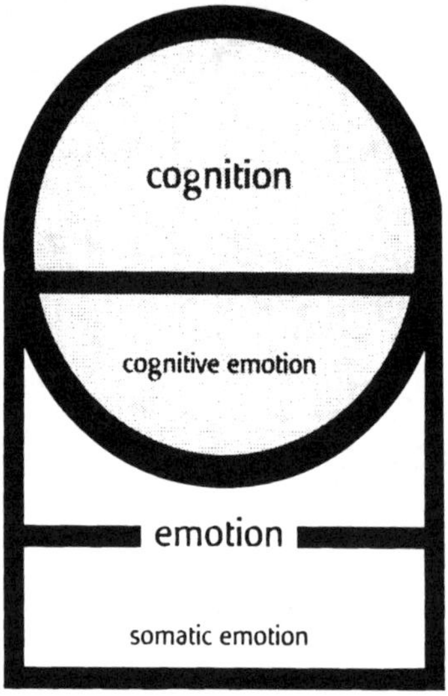

Figure 2. Cognition Encompassing Cognitive Emotional Features

However, even if we grant the computationalist a semantic model of the influence of emotion on perception, there is a problem. To treat emotions as categories in a semantic activation model means that they are, from a purely cognitive position, structurally indistinct from any other broad cognitive category, such as "game" or "uninteresting." Yet surely not all such categories are properly called emotions. Presumably what separates emotion from these other categories is the host of other phenomena that accompany an emotion, such as the somatic body responses. What then becomes the pressing issue is the relation between these two things: the cognitive, semantic emotion and the bodily, physiological passion. The model of Niedenthal *et al.* is meant only to describe the possible semantic effects of emotion upon perception, if such effects are properly described as being categorizing; they are willing to mix the theoretical levels and consider that the emotional nodes in the connectionist model will be part of other networks, such as networks that model or represent somatic activities of emotion (1994: 105–08). The computationalist, with her commitment to purity of levels, is forced into maintaining a distinction between the fully cognitive, mental emotion running in parallel to a bodily — and therefore external to the mind and the computational theory itself — reaction or correlate to the emotion. But can a merely semantic model, which leaves open the possibility for correlated somatic passions but does not incorporate these into the model of mind itself, properly account for emotion? Or, to put it another way, can we expand the realm of the cognitive to include some of the emotional processes, leaving out the somatic ones, and thereby primarily identifying emotions with cognitive states? To see why not, we need to turn to neuroscience.

3. The Neuropsychology of Cognitive and Embodied Emotions

We will take from philosophy a theory which explicitly identifies emotions with cognitions and separates (a physicalist account of) somatic emotion from the cognitive function of emotion. One such group of theories of emotion is found in the cognitive theories of emotion that identify emotions with judgments, such as the theory that emotions are judgments of value (hereafter *EJV*). In philosophy, Robert Solomon has been a leading contemporary advocate of this theory, and, as I understand her work, a strong new champion of *EJV* is Martha Nussbaum.[3] Both Nussbaum and Solomon hold that emotions are cognitive.[4] An agent has an emotional experience when she makes a judgment, based upon certain beliefs, and that judgment is a certain kind of judgment of value. Since *EJV* is an identity theory, it follows that emotions cannot possibly occur without their corresponding judgments of value, and that the emotional judgments of value cannot possibly occur without their corresponding emotions.[5] Emotions are not parts or constituents of judgments of value, and the judgments of value are not parts of emotions; each emotion *is* some judgment of value. *EJV* is a theory that an emotion is an intentional state — specifically a judgment, but perhaps a judgment of a complex form — and the object of the judgment has a value to the subject given (at least in part) by the subject's beliefs about that object.[6,7] Nussbaum's theory of value is based upon the Greek notion of *eudaimonia*, sometimes translated as flourishing. This notion is not incompatible with a biological interpretation of judgments of value, and hence can in principle be extended to non-human animals. For his part, Solomon holds that "the ultimate object of our emotional judgments is always our own sense of personal dignity and self-esteem" (1976: 190). For our purposes here, these notions of value are not significantly different.

As an example illustrating *EJV*,[8] consider the death of a loved one. Confronted with this agonizing event, a person experiences grief because she makes the painful judgment that this loved one is no longer going to be part of her life and her life goals. This judgment is heavily laden with value because she herself believes that the loved one played a significant part in her life goals; her loved one was part of her own flourishing — her eudaimonia — and the painful emotional experience of grief is the recognition of this breach in her eudaimonia. Thus, her beliefs about her loved one are what make her judgment of her loved one's death a strong value judgment, and this evaluative judgment is the experience of grief.

Yet it would seem intuitive that emotions can be dissociated from their corresponding judgments. If one becomes happy because she concludes, after reviewing her bank account, that she has more money than she suspected, it is plausible that this happiness can continue after the judgment has been forgotten. But an identity theory requires that the judgment somehow still be present and occurring. Free floating moods, also, are ruled out on this theory; a mood like anxiety must be a judgment, even if that judgment is for some reason suppressed from introspection. But whereas these different intuitions are inconclusive, it is not difficult on strictly neuropsychological grounds to show that the identity claim of *EJV* is very likely wrong. In abnormal subjects, at least, we can show both that the emotion occurs without the supposedly identical kind of judgment, and, vice versa, that the kinds of judgments thought to be identical with some emotions can occur without the corresponding emotion. This double dissociation will apply to most theories that identify an emotion with a cognitive process, since the results reviewed here suggest that there are dedicated,[9] and plausibly pre-cognitive, emotional systems, and furthermore that cognitions linked to emotion require somatic emotional states.

To establish that an emotion can occur without the corresponding kind of judgment of value, we need to find a case where an emotion is not accompanied by the type of judgment of value that is supposedly identical to that emotion. One now classic example is available in the literature of decerebrate animals (live animals where the cerebral hemispheres are surgically disconnected from the brainstem) and decorticate animals (live animals which have had all or significant portions of the cortex removed). It has long been known that partially decerebrate animals can show some strange "quasi-affective" reactions to trivial stimuli. Furthermore, animals with intact diencephalons but lacking some portion of the frontal lobe or other prosencephalon structures (generally, animals where much of the cortex and limbic system are ablated, but deeper structures remain), can show affective behavior that is excessive for a given stimulus. Research in frontal lobe lesioned cats by Cannon and Britton (1925) showed that this phenomenon, which they called "sham rage,"[10] could last for long periods of time, and was not directed. These emotional attacks could be brought on sometimes by merely touching the animal, and included all the physiological signs of energetic rage. Bard (1928) found that this rage reaction required that the posterior portion of the hypothalamus be intact.

These results suggest that diencephalon structures (such as the thalamus and hypothalamus) have a specialized role in energetic emotion. Without the diencephalon, animals show reactions that mimic affect but do not have any "energy."[11] With the diencephalon — with at least the posterior portion of the hypothalamus — intact, but with the frontal lobe lesioned, animals show wild, exaggerated affect. This suggests that the frontal lobe or other higher prosencephalon structures inhibit diencephalon structures and keep emotions in check, and that diencephalon structures are, at least in part, responsible for producing this energetic affect. But this would be contrary to *EJV*. Emotions occurring exaggeratedly as a result of over-excitement of an emotion producing structure suggests not an identity between emotions and judgments of value, but rather that, at best, emotions are triggered by judgments of value, and accompany such judgments.

The sham rage behavior suggests another problem for *EJV*. These cats lack intact frontal lobes; it is reasonable to conclude that they may lack some of the cognitive skills that would allow for the kind of judgments that *EJV* requires. For example, they lose the ability to sequence their actions, such as performing grooming behavior in the correct order. But the cats can go into rage reactions, on and off, for hours in reaction to trivial stimuli. And these rage reactions are typically labelled as "misdirected" — for example, a cat pinched on the tail attacks not the wrong-doer, but whatever is directly in front of itself. And once initiated, their rage reactions seem to continue without any corresponding belief or judgment being requisite; in other words, it would seem that their rage lacks the kind of intentional object *EJV* requires, or at least their rage fails to discriminate properly between possible intentional objects. Instead of directed behavior, the excessive affect seems to result from the system for rage being hyper-activated and running on unchecked.

Contemporary neuropsychological research in other areas supports this view. In an article summarizing some recent research on emotion and memory,

Joseph LeDoux reports that he and other researchers found that aural fear potentiation could occur even when portions of the auditory cortex were lesioned. They uncovered sub-cortical thalamic pathways connecting the auditory sensory region with the amygdala (a group of nuclei that is part of the limbic lobe and is at the base of the temporal lobe). What was lost in rabbits whose auditory cortexes were lesioned was tone discrimination abilities. Rabbits would show fear to tones that were different in pitch to the conditioned tone, whereas with the auditory cortex intact they would not (1994: 54). Again, the portrait that emerges is that emotions regularly include high level processing, but can in abnormal subjects still occur with such processing circumvented.

We can also dissociate in the other direction: there are cases where the judgment of value one would expect to be identical with an emotion for the *EJV* theorist occurs without the corresponding emotion. One such group of examples might be found in patients with frontal lobe damage who show defects of affect. One such patient has been studied by Damasio, Tranel, and Damasio (1990; see also Damasio 1994). This patient, EVR, underwent a bilateral excision of the orbital and lower mesial cortices (roughly, the frontal lobe cortex behind the forehead and just above the eyes) and thereafter developed an "acquired sociopathy," showing subtle but very grave defects of affect, even though EVR showed no significant cognitive deficits, such as loss of the conceptual categories we might associate with affect. EVR and patients with similar lesions were studied in an experiment where they were shown pictures meant to illicit an emotional response, while their electrodermal skin conductance responses were monitored. These pictures were mixed with pictures that were emotionally neutral. As controls, normals and subjects with brain damage other than in the frontal cortex were used. EVR and some other frontal lobe damage patients show very striking differences in their electrodermal skin conductance responses. Whereas the control subjects showed significant electrodermal skin conductance responses to the stimuli meant to evoke emotions, EVR and the other similarly damaged patients showed *virtually no* such responses. This indicates significant deficits in their somatic emotional reaction to pictures meant to illicit emotional reactions. Most interestingly, in an exit interview, EVR said that he had not experienced the kind of "feeling" that he thought he ought to have in relation to some of the stimuli.

It would seem that EVR holds the proper beliefs and makes the correct judgments requisite for the missing emotional experiences, and still lacks certain emotional responses. A defender of *EJV* could respond that EVR's deficit is in the somatic emotion alone: he is able to discuss situations and describe the results that certain actions would entail, and he has the cognitive emotion, but oddly he does not have the full emotion in as much as he lacks the somatic reaction that occurs in normals. And, in fact, we might expect a computationalist committed to a theory like *EJV* to consider this result almost unsurprising, since presumably the computationalist would certainly not deny that there are regular somatic reactions that accompany an emotion, but would seek instead to separate them from the cognitive element or vector of the emotion.

Many of the theories of emotion in philosophy, early natural science, and folk psychology share an association of emotions with the extended body, in distinction from the mind, or even the brain, alone. The reason for this might plausibly be said to be the phenomenological evidence. Whereas language use, mathematical reasoning, and planning about the future seem to be skills which require little contribution from the somatic body, the heat of anger or the ache of despair seem to be centered right in the pit of one's chest. Williams James captured this sense acutely when he wrote:

> *If we fancy some strong emotion, and then try to abstract from our consciousness of it all the feelings of its bodily symptoms, we find we have nothing left behind,* no 'mind-stuff' out of which the emotion can be constituted, and that a cold and neutral state of intellectual perception is all that remains. (1950: 451; James's emphasis.)

However, many have argued that there is a separate mental or cognitive aspect of the emotion, distinct from the bodily passions. Descartes, in his theory of passions, distinguished between passions and emotions. Emotions can be "internal" to the soul, but are usually accompanied by correlated passion in the body. For these passions, he readily gave explanations of their corresponding bodily states, which are caused when the brain sends different amounts and kinds of animal spirits to different parts of the body. In hatred, for example:

> the pulse is irregular, weaker and often quicker; we feel chills mingled with a sort of sharp, piercing heat in the chest; and the stomach ceases to perform its function, being inclined to regurgitate and reject the food we have eaten or at any rate to spoil it and turn it into bad humors. (1985. 363)

In Descartes's own theory we see tensions which still remain between the embodied states of passion and the mental or cognitive features of the emotion. We still are often saddled with a dual-explanation theory of this kind. The computationalist could argue that what is impaired in EVR is the part of the brain that causes the largely bodily somatic reactions to occur; he lacks part of the emotion, but the "high," cognitive part of the emotion could still be nothing more than a kind of judgment of value which EVR in fact does have.

Descartes's path poses serious problems for verification: what is an emotion without physiological responses? It would seem that, if defended in this way, *EJV* can amount to little more than a brute redefinition of emotion: emotions are just judgments, regardless of whether accompanied by anything like the states that are the usual referent of the term. Besides being methodologically dubious, this approach is clearly inappropriate when we take a closer look at Damasio's patients and discover some of the positive features of their emotional deficits. Following the surgery to remove his tumor and therefore also the significant portions of his frontal lobe, EVR made a series of disastrous judgments, went bankrupt, and was ultimately unable to care for himself. He also showed signs of sociopathic disorder, such as inability to maintain enduring attachment to a sexual partner. However, in standard tests he continued to show above average intelligence and no typical cognitive deficits. Investigating these peculiar failures in a previously successful and admired man, Damasio has been able to show consistent deficits in rationality in a group of patients with ventro-medial frontal lobe damage like EVR's.[12] In one class of experiments where a gambling task was used, these subjects did significantly more poorly than normals and subjects with other kinds of brain damage. Furthermore, control subjects showed skin conductance responses that increased with time during the gambling task, suggesting that different emotional associations are being learned and enhanced as the task continues and as the subjects gain more experience and skill. The ventro-medial frontal damage patients, however, did not show this increasing response, and they also did consistently poorly, choosing immediate reward over long term gains (1994: 212–22).

Damasio has proposed an explanatory theory of these phenomena which he calls the *somatic marker hypothesis*. His hypothesis is that the body acts as a theater for the emotions, as James suggested. Some of our emotions are innate; these primary emotions are still intact in EVR and similar subjects, so that they do show affective reactions to startling stimuli, for example. But some of our emotions can be trained, and thus associated with various stimuli. Patients like EVR have sustained damage to a neural system in the frontal lobe cortex which is necessary for these secondary emotions. As a result they have an impairment in these emotions, emotions which are cognitive in that they can be associated with contents through learning, but which are necessarily somatic since it is the body which in principle is the first target for the pain, pleasure, or more complex associations that a stimulus may have. The frontal lobe damage interrupts the somatic marker on different alternatives, so that when patients like EVR are confronted with alternatives in some situations, they lack an emotional element which in normals reduces the set of alternatives by marking some as painful, others as pleasurable.

Emotion, if Damasio is correct, is *necessary* to rationality. And not just the idealized cognitive emotions of *EJV*-like theories, but emotions that are visceral and embodied. Indeed, here Damasio's work has common ground with work in social psychology on the relation between introspection and decision. In a series of related articles Wilson *et al.* (1991, 1993, 1995) have uncovered evidence that when subjects are asked to articulate their reasons for their preferences,[13] they can change their preferences in ways that they later regret. There is a significant body of research on the possible disadvantages of introspection; for example, one study (Schooler and Engstler-Schooler 1990) showed that when subjects verbalized their memory of faces or other stimuli, they were less likely than controls to recognize the faces later. Wilson *et al.* extended this area of research by actually seeking to demonstrate that the kinds of decisions that one makes by introspecting at length, such as making lists of pro and con reasons for a choice, can have as a result that subjects change their decisions and choose things that later they would rather they had not chosen over other options. These findings suggest that choices are often directed by factors which cannot be articulated. It is tempting to suppose that Damasio's own hypothesis can be used here to develop a coherent picture that what usually influences decisions and is inarticulable is not just symbol processing that is unavailable to introspection, but is rather something wholly different in kind from logical or language-based analysis of propositions meant to assess the positive and negative features of each option. These influences are the emotional states themselves, which are primarily felt, visceral experience. These emotions give value to alternatives, and give reason a scale to measure against. The somatic marker hypothesis is therefore potentially strong evidence that Hume was very nearly right when he wrote that:

> we speak not strictly and philosophically when we talk of the combat of passion and of reason. Reason is, and ought only to be the slave of the passions, and can never pretend to any other office than to serve and obey them. (1951: 415)

4. Conclusion

We have reviewed evidence from cognitive psychology that emotions can operate independently of cognition, and that they influence perception. Next, we saw neuroscientific evidence that emotions, although normally cognitive, are separable from the judgments that the theory *EJV* would associate with them, and very likely this result generalizes to other identifications of emotions with high cognitive processes. This same body of evidence shows that emotions are essentially linked to the extended body, and is consistent with some of the interesting results from social psychology regarding introspection and decision. Finally, this necessary somatic element of emotions seems to be an essential part of functioning rationality.

Computationalism is therefore confronted with a series of problems:

1. If emotion operates sometimes independently of cognition, computationalism can at best model it as a separate and parallel computational process which can influence and be influenced by cognition.
2. Because of the commitment to implementation independence, the kind of bodily features accompanying emotions must be treated as contingent and unnecessary by a computational theory. Computational theories of mind are as a result naturally compelled to a neo-Cartesian theory of emotion: there will be emotions of the mind and also the contingent, and usually parallel, passions of the body. Integrating these two is a serious problem.
3. Emotion seems to influence perception. But computationalism cannot well account for the effects of emotion on perception. We have noted that treating emotions as categories makes them indistinct from other cognitive categories of similar generality, such as "game" and "uninteresting." Again, the result is a neo-Cartesian dualism between mental emotions and bodily passions: some categories, like the fearful, will have the merely incidental feature that they are oft accompanied by bodily passions.
4. But neo-Cartesianism must ultimately fail if Hume and Damasio are correct that the somatic dimensions of emotions are necessary for full rationality. Whereas rationality would certainly be a feature of mind for which the computational theory of mind would have to account, it is unclear how if at all it can account for this role of emotion. Again, computationalists can allow for emotions to operate as categories, sorting out depressing contents from happy ones, and so on. But in patients like EVR the ability to articulate such categorizations is present, but seems impotent to influence their reasoning in appropriate ways. Furthermore, what is needed here is not an assertion that there are orderings of the stimuli, but rather an account of what makes one stimulus painful, another pleasant, what makes a person cringe from one possible future and pursue eagerly another. Computationalism here comes along after the fact.

There is one escape that the computationalist might seek at this point: expanding the realm of the cognitive or computational to include the somatic emotional states (figure 3).

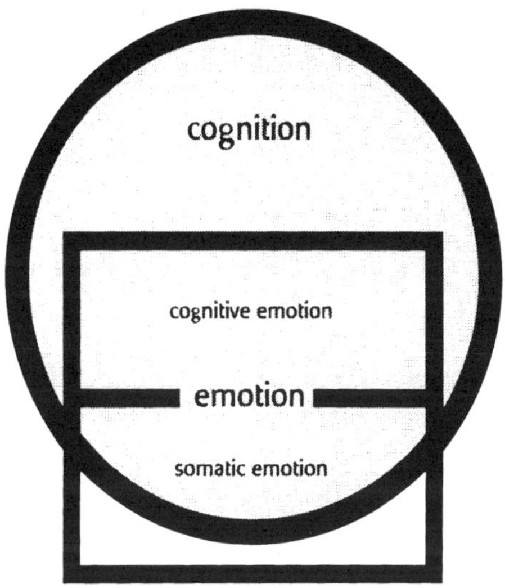

Figure 3. Cognition Encompassing all Emotional Features

It is not uncommon for cognitive scientists to talk about almost every level of description of mental activity as symbol, or at least information, processing. Thus, the neural mechanisms which underlie somatic marking in Damasio's theory, for example, would be wholly computational; the body itself (or rather, a mental model of it), its states, and the awareness of them would be computational; and the changes in perception, even though admittedly of a different level from the symbol processing which occurs in recognition, and even though perhaps continuous, would also be computational. This is perhaps forgivable in cognitive scientists, both because they may mean something different by "computation" and "representation" in these different cases, and also because they are not enmeshed directly in the metaphysical problems of naturalizing mind and therefore may be allowed to use representational or intentional description as a convenient shorthand in these situations. But this laxity in identifying the symbolic must lead right into the hands of the more radical anti-representational positions. For it is a consequence of this pan-representationalism that every bodily activity, from mathematical cognition to a reflexive eye blink, is a computational process. But the neural mechanisms which underlie an eye blink reflex are well understood, and surely here the representational account adds nothing to a purely physical explanation which is given in terms of neural firings and related chemical and electrical events. A computational description at this level undermines the non-pernicious homun-cular thesis of functionalism which requires that the bottom level be given a non-intentional or non-representational description. For computationalism not to be vulnerable to eliminativism, there must be levels of description where symbolic accounts either are necessary, or are at least appropriate because of a convergence with folk psychological concepts, and also levels where the functional analysis leaves off representational description.

Still, this leaves open the possibility that we revise computationalism and allow for it to be representational all the way down to the somatic, extended body, relinquishing the analytic discharge of the homunculus. This raises metaphysical concerns about where naturalism now fits. And since the relevant features of the body are contingent and specific to our environments, the loss of the implementation independence would change the theory significantly. Some of these problems may be surmountable: we could still lay claim to naturalism by being ascriptivists about intention (as per Dennett 1971), for example, allowing us to describe the body as intentional without committing to such description being reducible to the physical states. And the loss of implementation independence changes the sales pitch for computationalism but leaves much of it intact. But a more grave problem awaits.

There is some precedent for thinking of the body as intentional or representational. Zajonc and Markus (1984) have argued that we should think of the bodily states of emotion, particularly motor states, as the "hard representation" of an emotion: "The motor movement can *in itself — without kinesthetic feedback and without a transformation of that feedback into cognition* — serve representational functions" (76; their emphasis). This suggestion is provocative and, I think, largely correct: if Damasio is right that a bodily emotional state is something that we learn to associate with a certain stimulus, giving that stimulus a particular and perhaps complex valence, then we should be able to say that that emotional state is *about* that stimulus. Similar intuitions inspired the cognitivists in emotion, who begin with the observation that generally one fears something, is angry at someone, hopes that something, and so on. One reservation we might have is that the somatic emotional state has features which do not seem properly described as representational (as we have been using the term here). The somatic fear reaction will include an increase in heart rate and adrenaline levels, resulting in a preparedness to flee. This is a direct response to the environment, and not plausibly a part of a representational system where it can stand in for something else. It may be more precise, or at least more cautious, to say that the somatic body state is intentional, but not necessarily or wholly representational. Thus we might revise Zajonc and Markus's phrase, and say that the body provides the *hard intentionality* of emotions.

But even if we also overstep this concern and treat the relevant features of the intentionality of the somatic emotional state as being tractable for the computational representational system, the model will require that we distinguish the various roles of somatic emotional reactions in the representational system. What is needed for an account of the intentionality of the body is an account of the functions of relevant bodily states in their environment. Such a theory must refer to the biological nature and perhaps evolutionary history of the organism and its relation to its environment: all of our somatic reactions, such as our fear of great heights, disgust at the smell of rotting meat, empathy for close relatives, and so on, will only be properly explained in a theory of the role of these states, and this can only be done by referring outside the body to the environment in which the organism evolved, the environments in which it was raised, and the environment in which it now acts.

Ruth Millikan's theory of evolutionary determination of function is one such approach, since it would give a method for positing the intentional object of bodily intentional states using evolutionary history (1984, 1993). Beth Preston's work on interactive understandings of embodied agency is similarly minded; she illustrates her approach with an explanation of cat purring, itself part of a bodily affective state of the kind we need to explain:

> The reason cats purr when patted or fed, is to indicate that they are not about to bite the hand that feeds or pats … So purring has a very important and diversified social function, which is ultimately … understandable only in terms of the structure of feline social relations as a whole … Consequently, this aspect of the explanation of behavior does not recognize an inside-outside split, since there is no obvious sense in which the social structure is either inside or outside the individual member thereof. (1994: 184)

What Millikan and Preston share is a divergence from what Millikan calls "meaning rationalism."[14] This is the idea that psychology or any other science of mind can proceed by studying only the agent itself. Instead, Millikan argues that meaning cannot be fixed in one agent, but requires reference to the environment of the agent: "I no more carry my complete cognitive systems around with me as I walk from place to place than I carry the U.S. currency system about with me when I walk around with a dime in my pocket" (1993: 170). For Preston, this common commitment requires that we expand our plausible accounts of the functions of behaviors by referring to their roles in a society or in relation to their natural history. The problem of accounting for somatic emotional intentionality shows that these externalists are right: to understand the intentionality of the somatic emotions it will not always be sufficient to take recourse in the posited contents of the cognitive system; we will also need to understand the body's individual interaction with its environment sufficiently well to posit the function of different somatic emotional states in relation to that environment. And this, ultimately, will reach well beyond computationalism: if we relinquish both the analytic discharge of the homunculus and computationalism's commitment to a fully internal representational system, we will have moved into a different kind of theory altogether.

So this last resort, expanding the realm of computational cognition to include the somatic emotions, will not succeed in saving computationalism. I conclude that the computational theory of mind is inadequate. This result is evidence against the purest form of representational theories of mind which include a high degree of implementation independence. Of such theories only what we might call problem-solving AI — research which aims to solve technical problems like getting a radar system to distinguish planes from clouds, and so on — escapes this criticism.

Notes

1. This is not to suggest that the study of emotion is a body of unified knowledge that successfully integrates across these traditional divides. In fact, the study of emotion in philosophy and in psychology generally recapitulates the divide.
2. This distinction might be drawn more elegantly in a different terminology. What is often called a "representation" is the content of an (presumably, but not necessarily, common or reoccurring) instance of representation. When "representation" is used thus, concepts tend to be taken as typical representations. Changing our terminology, we could distinguish between representational content (what I have above called "representation") and any instance of representation (which I am calling an "intentional" state). Some instances of representation in this second sense may not be correlated with a common or reoccurring content, although in principle each has a correlated content. (I am indebted to Nino Cocchiarella for clarifying this distinction.)
3. Nussbaum delivered a paper at Indiana University, Bloomington, in 1994 arguing for a strong form of *EJV*, and it is this paper that led me to choose her as an example. Nonetheless, much of what I say here about the theory can be reconstructed from Love's Knowledge (1990) (see 291 ff). For Robert Solomon's position, see The Passions (1976). Solomon also holds an explicit identity theory: "The relationship between beliefs and opinions on the one hand and emotions on the other is not a matter

of causation or coincidence but a matter of logic" (179), and "An emotion is a *judgment*" (185; his emphasis).

Most other contemporary theories of emotion in philosophy neglect to study or incorporate the somatic or lower level mental processes, and so are equally inadequate when judged as theories of emotion. Most such scholars, however, lay no claim to anything more than description of certain cognitive or mental features of emotions.

4. Neither Solomon nor Nussbaum is likely to have any sympathy for computationalism; nonetheless, their theories are identity theories of the kind necessary for a computational theory that claims to give a complete, or at least sufficient, account of emotion.
5. By emotional judgments of value I merely mean to say: those kinds of judgments of value that are emotions according to *EJV*. There could be, under *EJV*, judgments of value that are not emotions; this is not important to the discussion that occurs here. From this point on in the paper I will be concerned only with those judgments of value that are supposedly identical to an emotion.
6. In fact, the value of the intentional object is for the subject essentially part of the intentional object (Solomon 1976: 178).
7. One could say that these states are perceptual — they amount to seeing things as something. For those who separate judgment from perception, wanting to reserve the term "judgment" for special kinds of conscious decisions, we may still be able to call these emotions "perceptions," while still retaining the principal features of *EJV*.
8. I borrow this example from Nussbaum.
9. For decades, the working hypothesis in this area of neural science has been that there are dedicated neural systems of emotion. One candidate has long been the limbic lobe (which lines the inside of the neocortex); recently attention has focused upon the amygdala, portions of the frontal lobe, and some other structures (for a review see Steinmetz 1994).
10. It could be objected that this behavior was not really rage; lacking judgment, it really was "sham" rage. Here, I will suppose that the rage behavior of the frontal lobe lesioned cats is indeed rage. Since the behavior of the cats reflects that of rage, I feel that the burden of proof would be to show that it is not rage; anything else would be begging the question in favor of *EJV*.
11. Bazett and Penfield, in their paper on decerebrate cats, reported that "Although the reflexes so far described seem superficially to have some relationship to emotion, they are clearly, even to a casual observer, merely pseudoaffective, since they differ from many spinal reflexes only by being rather more elaborate, and since in such a decerebrate animal they appear with a mechanical precision whenever the animal is put under the same conditions" (1922: 217).
12. I use the term "rationality" to identify reasoning and decision making which (at least) requires evaluation of responses and the results of which are open to evaluation by the society of the subject and the subject himself (e.g., decisions that EVR might later say were wrong, even though there was no mistake in the logic of the reasoning). Note that

where Hume uses the term "reason," in the quote below, he probably means something like I use "rationality" here.

13. Note that such articulations of reasons have in the past been used by cognitive scientists to provide algorithms that they supposed the subjects were using to make their decisions.
14. They do have differences: Preston is careful to distinguish herself from some of Millikan's positions (Preston 1994: 188). However, for our purposes here, they are relevantly similar.

References

Bard, P. 1928. A diencephalic mechanism for the expression of rage. *American Journal of Physiology* 84, 490–515.

Bazzett, H.C. and Penfield, W.G. 1922. A study of the Sherrington decerebrate animal in the chronic as well as the acute condition. *Brain* 45, 185–265.

Blaney, P. 1986. Affect and memory: A review. *Psychological Bulletin* 99, 229–246.

Bower, Gordon H. 1981. Mood and memory. *American Psychologist* 36, 129–148.

Brooks, Rodney. 1991. Intelligence without representation. *Artificial Intelligence* 47, 139–159.

Cannon, W.B. and Britton, S.W. 1924. Pseudoaffective medulliadrenal secretion. *American Journal of Physiology* 72, 283–294.

Churchland, Patricia S. and Sejnowski, Terrence J. 1992. *The Computational Brain*. Cambridge, MA: MIT Press.

Clark, Andy and Toribio, Josefa. 1994. Doing without representations? *Synthese* 101(3), 401–431.

Clark, D.M., Teasdale, J.D., Broadbent, D.E. and Martin, M. 1983. Effect of mood on lexical decisions. *Bulletin of the Psychonomics Society* 21, 175–178.

Damasio, Antonio R., Tranel, Daniel, and Damasio, Hanna. 1990. Individuals with sociopathic behavior caused by frontal damage fail to respond autonomically to social stimuli. *Behavioral Brain Research* 41, 81–94.

Damasio, Antonio, R. 1994. *Descartes' Error: Emotion, Reason, and the Human Brain*. New York: G.P. Putnam.

Dennett, D.C. 1971. Intentional systems. *Journal of Philosophy* 68(4), 87–106.

Descartes, Rene. 1985. *The Philosophical Writings of Descartes, volume I*. John Cottingham, Robert Stoothoff, and Dugald Murdoch (translators). New York: Cambridge University Press.

Halberstadt, Jamin B., Niedenthal, Paula M. and Kushner, Julia. 1995. Resolution of Lexical ambiguity by emotional state. *Psychological Science* 6 (5), 287–282.

Haugeland, J. 1991. Representational genera. In *Philosophy and Connectionist Theory*, W. Ramsey, S. Stich, and D. Rumelhart (eds), 61–90. New Jersey: Erlbaum.

Hume, David. 1951. *A Treatise on Human Nature,* L.A. Selby-Bigge (ed). Oxford: Oxford University Press.

James, William. 1950. *The Principles of Psychology, volume II.* New York: Dover Publications.

Kunst-Wilson, W.R. and Zajonc, R.B. 1980. Affective discrimination of stimuli that cannot be recognized. *Science* 207, 557–558.

Le Doux, Joseph E. 1994. Emotion, memory and the brain. *Scientific American* June, 50–57.

Marr, D. 1982. *Vision.* New York: W.H. Freeman.

Millikan, R.G. 1984. *Language, Thought, and Other Biological Categories.* Cambridge, MA: MIT Press.

Millikan, R.G. 1990. Truth rules, hoverflies, and the Kripke-Wittgenstein paradox. *Philsophical Review* 99(3), 323–53.

Millikan, R.G. 1993. *White Queen Psychology and Other Essays for Alice.* Cambridge, MA: MIT Press.

Newell, Allen. 1981. The knowledge level. *AI Magazine* Summer, 1–20.

Niedenthal, Paula M., Setterlund, Marc B. and Jones, Douglas E. 1994. Emotional organization of perceptual memory. In *The Heart's Eye: Emotional Influences in Perception and Attention*, Paula Niedenthal and Shinobu Kitayama (eds). San Diego: Academic Press.

Niedenthal, Paula M. and Halberstadt, James B. Emotional State and Emotional Connotation in Word Perception. Manuscript, Indiana University Cognitive Science Research Report 119.

Nussbaum, Martha. 1990. *Love's Knowledge.* Oxford: Oxford University Press.

Preston, Beth (1994). Behaviorism and mentalism: Is there a third alternative? *Synthese* 100, 167–196.

Putnam, Hilary. 1964. Robots: Machines or artificially created life? *Journal of Philosophy* 61, 668–691.

Putnam, Hilary. 1975. The Nature of Mental States. In *Philosophical Papers*, volume 2, *Mind, Language, and Reality.* Cambridge, MA: MIT Press.

Ramsey, William. Do connectionist representations earn their explanatory keep? Manuscript.

Schooler, J.W. and Engstler-Schooler, T.Y. 1990. Verbal overshadowing of visual memories: Some thing are better left unsaid. *Cognitive Psychology* 22, 26–71.

Singer, J.A. and Salovey P. 1988. Mood and memory. *Clinical Psychology Review* 8, 211–251.

Solomon, Robert. 1976. *The Passions.* Garden City, NY: Anchor/Doubleday.

Steinmetz, Joseph E. 1994. Brain Substrates of Emotion and Temperament. In *Temperament.* John E. Bates and Theodore D. Wachs (eds). Washington, DC: The American Psychological Association.

van Gelder, Tim. 1995. What might cognition be if not computation? *Journal of Philosophy* 7, 345–381.

Wilson, Timothy and Schooler, Jonathan W. 1991. Thinking too much: Introspection can reduce the quality of preferences and decisions. *Journal of Personality and Social Psychology* 60(2), 181–192.

Wilson, Timothy, Lisle, Douglas J., Schooler, Jonathan W., Hodges, Sara D., Klaaren, Kristen J. and LaFleur, Suzanne J. 1993. Introspecting about reasons can reduce post-choice satisfaction. *Personality and Social Psychology Bulletin* 19(3), 331–339.

Wilson, Timothy D., Hodges, Sara D. and LaFleur, Suzanne J. 1995. Effects of introspecting about reasons: Inferring attitudes from accessible thoughts. *Journal of Personality and Social Psychology* 69(1), 16–28.

Zajonc, R.B. 1968. Attitudinal effects of mere exposure. *Journal of Personality and Social Psychology Monograph* 9(2), 1–28.

Zajonc, R.B. 1980. Feeling and thinking: Preferences need no inferences. *American Psychologist* 35(2), 151–175.

Zajonc, R.B. and Markus, Hazel. 1984. Affect and cognition: The hard interface. In *Emotions, Cognition, and Behavior*, Carroll E. Izard, Jerome Kagan and Robert B. Zajonc (eds). Cambridge, UK: Cambridge University Press.

Remembering, Rehearsal and Empathy

Towards a Social and Embodied Cognitive Psychology for Artifacts

Kerstin Dautenhahn & Thomas Christaller
VUB-AI LAB
Brussels
German National Research Center for Information Technology (GMD)

1. Introduction

Our professional background are biology and artificial intelligence and we are mainly interested in the construction of intelligent autonomous agents based on biological and psychological findings and models. This paper outlines our research framework which grew out of considerations on cognition for artifacts, although we are aware that the successful implementation of these ideas is still a future goal. Instead of reviewing intensively literature of cognitive science and especially of cognitive psychology we focus on some points which are most relevant to our view as artificial life and artificial intelligence researchers. The ideas result from our search for a common framework for natural and artificial cognitive systems. This framework should account for embodiment and should be a 'holistic' approach. First, the body of a cognitive agent is not only the physical basis of cognition but a necessary condition and point of reference for perceptions and memory. Second, we do not pursue a 'modular' approach, i.e. we do not assume distinct cognitive modules (e.g. memory, learning, planning) with well-defined interaction protocols. Moreover, instead of modeling specific aspects (behaviors, functions) of a cognitive system, the general framework should be able to provide explanation on different scales and levels of description. This includes remembering, rehearsal, symbols and empathy.

In our concrete work on artifacts so far we have used a behavior-oriented programming approach. But we agree with Tsotsos's criticism of 'behaviorist intelligence' (Tsotsos 1995). We believe in the general idea of a bottom-up approach for achieving intelligence but we disagree with falling back into the 'dark ages' of behaviorism. Going beyond a criticism of state-of-the-art 'robot behaviorism' and on the basis of considerations about human cognition this paper discusses aspects of 'artificial cognition' which might contribute to a 'cognitive psychology for artifacts.'

The first two sections (2, 3) describe our conception of embodiment and social intelligence which are the basis of a physical and social embeddedness of cognition. Sections 4 and 5 outline our conceptions of remembering and rehearsal. The following section (6) proposes a framework to integrate these phenomena. Sections 7 and 8 discuss how symbols, language and empathy might be interpreted in the context of such a framework. Very first steps towards an experimental investigation along theses lines of argumentation are sketched in section 9.

2. The Bodily Basis

It is normal practice in AI to use as examples tasks for which there appears to be very little requirement for physical activity. However, there is much evidence to support the fact that cognitive capabilities are only possible through the interaction of body and mind. In the following we describe our conception of 'bodily embeddedness.'

2.1. *In Search for 'Embodiment': An Artificial Life Approach*

Our main focus of interest is the construction of artifacts which behave in a way similar to living organisms concerning the complexity of the interactions with the environment. This research issue is intensively studied in the relatively new research field 'artificial life.'

> Artificial Life is the study of man-made systems that exhibit behaviors characteristic of natural living systems. It complements the traditional biological sciences concerned with the analysis of living organisms by attempting to synthesize life-like behaviors within computers and other artificial media. By extending the empirical foundation upon which biology is based beyond the carbon-chain life that has evolved on Earth, Artificial Life can contribute to theoretical biology by locating life-as-we-know-it within the larger picture of life-as-it-could-be. (Langton 1989)

Artificial life research can be interpreted as an attempt to construct artifacts that behave like living organisms (in their own, 'characteristic' ways) instead of re-building or copying construction principles or behavior of natural, biological systems. A main research issue in Alife concerns the question of how 'intelligence' and 'cognition' in artifacts can be defined and achieved. A part of the artificial life community argues that the success of the 'constructivist approach to cognition' (see Stewart 1994) depends on the success of building physical agents, namely robots. A strong argumentation for these 'artificial life roots of artificial intelligence' is given in Steels (1994a). Artifacts which have to preserve their autonomy in a complex and unpredictable environment should develop a kind of 'intelligence' similar to that of living organisms. In order to identify fundamental principles of the development of intelligence in general we confined us to the study of 'social intelligence' (see section 3 and Dautenhahn 1995) and 'embodiment.' Natural living systems (animals, plants) do not simply 'possess' a body as one among several individual features. Instead cognitive abilities are realized and expressed through the body. We believe that cognition cannot be studied in isolation from the body, because cognitive abilities result out of the morphogenesis of the body and its interactions with the environment (see structural coupling, Maturana and Varela 1987). The body is not a fixed and pregiven 'actuator device,' but it is a dynamic and ontogenetically evolving entity.

2.2. *Bodies for Artifacts*

In the behavior-oriented research direction of robotics the need for an 'embodiment' of intelligent artifacts has been under discussion for years, stated, for instance, in 'intelligence without reason' (see Brooks 1991). In contrast to these approaches we do not agree that it suffices to run a control program on a physical body (with real sensors, real actuators, and a real environment) in order to fulfill the embodiment criterion. We doubt that such a robot will behave significantly different from a robot with simulated sensors and actuators. Consequently the robot is not able to distinguish whether it has a simulated body in a simulated world or whether it possesses a real body. This distinction can be easily performed by humans (see Stadler and Kruse 1986) using different mechanisms related to perception and sensation of the world. Instead we claim that the development of a 'conception' of the body, which is generally discussed as the acquisition of a 'body image' or 'body schema,' is necessary for embodied action and cognition. The conception of 'body sensation' and awareness can be used to distinguish simulated and real stimuli, e.g. observing the consequences of one's actions and interactions with the environment on the body. Additionally the morphology of a natural body has evolved in phylogeny in close interdependency with and mutual adaptation to the animate and inanimate environment. In the same way in ontogeny the body is shaped by the environment on the basis of given dispositions (see Maturana's ideas on structural coupling, Maturana and Varela 1987). Transferring this idea to artifacts implies the careful design and adaptation of a robot's body to the environment. In principle the design itself has to be evolvable and self-redesignable. An ideal solution would be bridging of the gap between (robot) hardware and software level. Although the ideas of realizing evolvable robots with an adaptation of both body and control mechanisms in a close interdependency is still a future goal, one should have this goal in mind and possibly think about short- or middle- term 'interim solutions' in order to investigate the mechanisms.

2.3. *Individual bodies*

Every natural system has a unique body and a unique cognition. Even twins with identical genome equipment are not 'equivalent' (in a mathematical sense). Embodied cognition depends on the experiences the individual collects during his/her lifetime. For two individuals there cannot be identical experiences. Even if they are physically close together their viewpoint cannot be identical. This leads us to one of several important consequences of being an embodied system:

- Bodies are physical entities and occupy a distinct volume in space at a specific time. No other individual can be at the same point in space and time.
- In our perception of human 'reality' we use the metaphor of a 3–dimensional space for locating and interacting with objects. Our body, too, the movements of its limbs and interactions with other objects are located in this geometrical space. The shape of the body changes during active movements or contacts with other objects.
- The body shows characteristic symmetrical and asymmetrical features concerning movement and orientation of the sensors. Therefore individuals take a certain perspective according to the orientation of the body.
- In order to preserve the integrity of the body important mechanisms for organisms are to detect contact with or distance to other objects.

- The active exploration of the environment through body movement is highly important for learning about the environment and the development of cognition. The second-hand knowledge of watching the world from a distance can only complement the first-person experiences of actively interacting with the environment.

2.4. *Body as a Social Tool*

The human species is the species where the most elaborated form of 'self-manipulation' of the body can be found. This includes decorating the body, actively manipulating its shape (e.g. through increase or decrease of weight) or using it as a 'device' for social communication: using markers on the body in order to indicate the position in a social hierarchy or using the body as a 'social tool' for threatening or as a 'social stage' to present a certain role or attitude. In Synnott (1993) Anthony Synnott argues that the human body, its attributes, functions, organs and the senses are not 'given,' but socially constructed. The body should be seen as a "social category with different meanings imposed and developed by every age and by different sectors of the population."

The human species is also the species where the most complex social interactions can be found. Group living is an advantageous strategy since it provides means of division of labor, cooperation etc. and gives rise to the development of tradition and culture. As a consequence members of a social group must get used to having conspecifics more or less close to their own body. This can become dangerous in group structures where the single members pursue their own interests. Only in groups with a strict and fixed hierarchy the individuals need not care much about their safety. This is the case either in dictatorship-societies or 'superorganism' structures (social insects) with anonymous contacts between its group members. But in most social groups the structure is more or less flexible. The social status has to be confirmed or updated regularly. Therefore certain 'regions' (distances around the body) developed in association with the behavior repertoire which can be executed within these regions. The closer conspecifics can approach one's own body, the stronger is usually the degree of familiarization (social bonding) with this person. In non-human primate societies 'social grooming' is done to a large extent via physical contact, using the body and its behavior as a means of communication. Humans later on developed the more effective means of highly elaborated language (see Dunbar 1993). Therefore social spaces around our body do not only represent the physical distance of an object within. But, in the context of conspecifics, the space is associated with an emotionally colored behavior repertoire. In social interactions the distance to a group member has to be judged according to the social status and the goals and interests of both individuals.

A huge amount of literature about body images, its definition and role in child development, its relationship to imagery, and the clinical consequences of distortions have been published (e.g. Van der Velde 1985; Mertens 1987; Auchus, Kose and Allen 1993). The definition of 'body image' varies widely in literature. According to Slade (1994) 'body image' should be viewed as a "loose mental representation of the body's shape, form and size." He identified at least seven sets of factors which affect its development and manifestation, including historical, cultural and social, individual and biological factors. Most of these factors are highly dynamic, i.e., varying over time on different time scales. Slade regards the history of sensory input to body experience as the basis for the general 'mental representation of the body.' A combination with at least another kind of variable, the biological variables which might change from day-to-day, indicates the time-changing manifestation of 'body image.' Therefore we suggest a 'dynamical system' point of view of 'body image.' Instead of viewing a 'body image' as a static, structural entity within the brain (a typical 'model' which can, for instance, be implemented in schemata-like representations which are traditionally used in many artificial intelligence approaches) it might be useful to treat 'body image' as a specific manifestation of the dynamics of a state-space built up by cognitive and bodily variables.

3. Social Intelligence

During the last decades, the social factor has come under close study in scientific discussions about the origin of human intelligence. The *social intelligence hypothesis*, which is also called the *Machiavellian intelligence hypothesis*, goes back to the ideas of several researchers, namely Chance, Humphrey, Jolly, Kummer and Mead (Chance and Mead 1953; Jolly 1966; Humphrey 1976; Kummer and Goodall 1985). According to the social intelligence hypothesis, (see Byrne and Whiten 1988 or Byrne 1995, chapter 13 for an overview) primate intelligence "originally evolved to solve social problems and was only later extended to problems outside the social domain" (Cheney and Seyfarth 1992). In contrast to non-human primates, which are able to handle complex social problems in a kind of 'laser beam' (domain-specific) intelligence, humans are able to transfer and adapt knowledge from one domain to the other. Cheney and Seyfarth (1992) state that "Monkeys ... do not know what they know and cannot apply their knowledge to problems in other domains." In this way they suggest to view non-human primates as good primatologists while humans can become good natural scientists. Figure 1 sketches the social intelligence hypothesis.

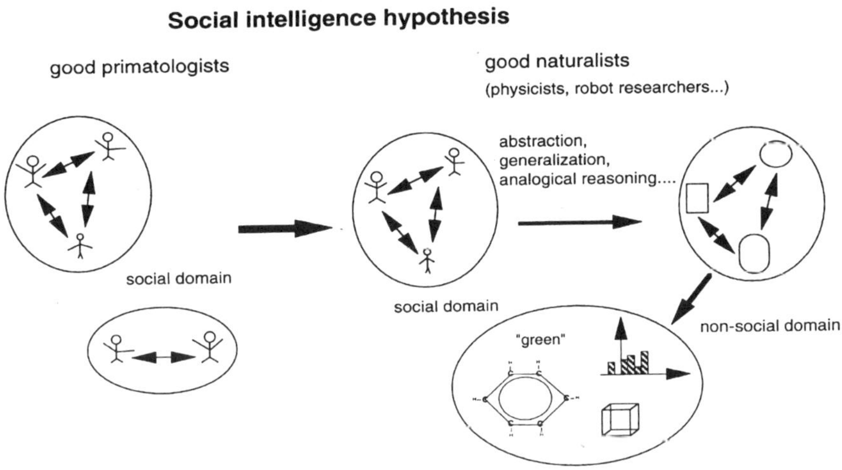

Figure 1. The social intelligence hypothesis. From social interactions to abstract problem solving.

Humans do not only have mental states and use them in a social context, but they are aware of their mental states (motives, beliefs, desires, intentions etc.) and can attribute mental states to others. With such a 'theory of mind' (Premack and Woodruff 1978) one can predict and analyze the behavior of both oneself and others. This enables one to establish and effectively handle highly complex social relationships and, at the same time, this kind of 'inner eye' (Humphrey 1986) allows a cognitive feedback, which is necessary for all sorts of abstract problem solving, e.g. mentally testing hypothesis, planning in advance, creative thinking or cognitive empathy. The social intelligence hypothesis claims in its strictest formulation that all these intellectual capacities evolved out of a social domain, i.e. out of interactions of embodied individuals. This would mean that even those human behaviors which are attributed to 'rationality' have a social and embodied basis (e.g. mathematical thinking). This is supported by the work of Mark Johnson (Johnson 1987) on metaphors:

> In considering abstract mathematical properties (such as 'equality of magnitudes') we sometimes forget the mundane bases in experience which

> are both necessary for comprehending those abstractions and from which the abstractions have developed. … Balance, therefore, appears to be the bodily basis of the mathematical notion of equivalence. (Johnson 1987: 98)

In a similar way, Daniel Bullock (Bullock 1983) suggests that the highly laborated human ability of symbolization, which seems to be an important characteristic of human cognition, stems from social interactions. "Symbolization is not the isolated thinker's act of representing a symbol to him- or herself." Instead, symbolization is a social act of agreeing, "of establishing a social convention — something that logically requires a minimum of two agents."

Intelligent agents are embedded in time and space. Complex social interactions evolve over time and the social relationships of an individual at a given point in time are the results of all social interactions in which the agent has been involved up to this point. We do not regard 'social expertise' as a set of special social rules (propositions), which are stored, retrieved and applied to the social world. Instead, social knowledge is dynamically reconstructed while remembering past events and adapting knowledge about old situations to new ones (and vice versa). In our work on physical artifacts, namely robots, we hypothesize that social intelligence might also be a general principle in the evolution of *artificial intelligence*, not necessarily restricted to a biological substrate. The next sections are based on this assumption of a physical and social embeddedness of cognition as the basis for the development of artificial intelligence for artifacts.

4. Remembering

Rosenfield (1993) presented an approach to memory which contradicts many approaches in computer science and cognitive science, namely the assumption that memory has to be regarded as a module which contains representations of concepts, words etc. Similar ideas can also be found in Bartlett (1932) who favors using the term *remembering* instead of *memory*. Rosenfield proposes a reinterpretation of many case studies in clinical psychiatry. His main statements which are also relevant for this paper are: (1) There is no memory but the process of remembering. (2) Memories do not consist of static items which are stored and retrieved but they result out of a construction process. (3) The body is the point of reference for all remembering events. (4) Body, time and the concept of 'self' are strongly interrelated.

In AI and more generally in computer science a concept of memory was developed by technical means. Of course this concept was shaped very much by the characteristics of digital computers.

If we want to organize a memory schema where we somehow trace the changes of a system over time we have two different approaches at our disposal. The first one is storing the whole status of the system before it changes. Remembering or forcing the system back into an earlier state of computation could then be achieved by just re-installing this status. The alternative is to store the inverse of those operations which cause the change. Now an earlier stage could be achieved by applying the stored sequence of inverse operators to the actual state of the system. The difference between both approaches is that the first one is cheap in computational terms but expensive in storage capacity while the opposite is true for the other alternative.

For knowledge-based systems a lot of different memory schemes were developed. One of the most influential one is the semantic network kind of memory. It originated from psychological work on semantic memories and one of its first implementations in AI was done by Quillian (1968). Technically speaking this is a labeled directed graph where the nodes are also labeled. When constructing a knowledge base it has turned out that it is impossible to come up with a (formally) complete and consistent version of it just from the beginning. To the contrary it is a laborious task to make and keep a knowledge base at least consistent. The whole knowledge base reflects exactly everything entered into it.

A different track was chosen by the so-called neural network approach which was inspired by the neurosciences. Here again it is basically a directed graph where only the nodes in the input and output layers are labeled. The labels on the links are weights which reflect the strength of the connection between two nodes. With this schema it is possible to account for statistical phenomena, e.g. how often a link was traversed. Learning can be done mainly by adding new nodes and changing the weights. Depending on the kind of network organization there are different trade-offs if and when such a network converges into a stable state. The performance of such networks when used as classifiers depends very much on the chosen learning sample and sequence and when to stop with the training of the network. While artificial neural nets allow to take time and usage into account they must still be carefully crafted and they must be prevented from changing their performance explicitly.

There are different approaches both in machine learning and neural nets which try to overcome the restriction that most of the techniques require a distinction between learning and performance phase, e.g. reinforcement learning (see Kaelbling, Littman and Moore 1995) and so-called life-long learning (see Thrun and Mitchell 1995). However, the retrieval is still not taken into account as a process which changes the memory. The question of course is why should it do that? Normally we think of a memory as a reliable storage of items which can be recalled as they were stored. The above mentioned work of Rosenfield and Bartlett suggest instead that the retrieval process is changing actively the content of what it retrieves.

In the following we discuss this phenomenon and give some hints on how to construct such memory schemes where the retrieval is as important as the learning process in shaping the memory. People do not care about a consistent memory because they have no means to compare different states of their memory. Their starting point is the present and their impression that at the very moment they are consistent. When asked to remember some episode the retrieval process takes care that the recalled episode is reconstructed from memory in a way which is consistent with the present state and self image of the person. This is a plausible strategy when we assume that we change ourselves from moment to moment and that the number of episodes to remember is so high that it is impossible to store them simply away as they are or to reinstantiate them in an objective manner. The environment of the remembered episode might have changed dramatically over time so that it will not integrate into the actual state of a person without any kind of adaptation. Then it is easier in terms of computation to make the episode consistent with the actual state than vice versa.

The metaphor we would like to explore is to look at the information processing system using a brain as a dynamic system where each trajectory representing its development over time will never enter the same point in its state space. This is inspired by Heraklits saying that we will never enter the same river twice because after the first time we and the river have changed through this act of entering. But what is achievable is that we can push ourselves as near as possible to some state we passed earlier. This might explain why remembering a fact is thought of as 'putting ourselves back in time.' The more cues we can associate the more reliably could this 'putting back' be done. But this gives us no hint about the distance to our former state because we do not remember single states and are therefore unable to compare the actual state with some former one. Remembering should be seen as a modulation process of the actual state of a cognitive system which tries to push the system partially as much as possible to a former state. This process changes both the remembered episode and the actual state of the system.

The major reason why this strategy is the favorite one for biological cognitive systems may be seen in the fact that the bodies of these systems are changing all over time. This is most obvious in their youth and in their old age. But all over lifetime every part of the body including the senses and the muscles change. The same perception of the world or of some inner state might be related to very different circumstances. E.g. the relative size of objects and conspecifics changes for an individual when he/she grows up. If there is no feedback over time in memory the individual might still hold the perception of 'something large' (e.g. referring to the body size of a relative) which turns out to be relatively small from the adult's point of view.

One last important aspect of such a memory should be mentioned here. Remembering is a very important process for building up expectations of what the system will experience in the present. This allows the system to use its perceptions as a kind of confirmation of what it already knows.

5. Rehearsal

Mental images are not merely individual images but rather whole sequences of images, a form of scenario, linked to other inner sensory impressions such as sounds and smells and emotional experiences. In a way we could use the metaphor of a mental film studio at our disposal,[1] in which we select and shape not only the props and scenery but also get to define the actors, their roles and how they relate to each other and one's own person. We are spectator, director and actor in one. Mental practice is a commonly accepted method of preparation for a sports event. Mental imagery and rehearsal is strongly related to dreaming and day-dreaming and can be assumed to be an important source for creativity. Mental images are not merely a form of photography or snapshot of what is represented upon our retina at any given moment. They are structured, linked to mental 'pictures' of other sensory organs and emotionally colored.

Mental images, their possible representation and function in humans, and mental imagery in artificial intelligence (AI) systems have been under investigation and discussion for many years now. Kosslyn (1994) provides a comprehensive documentation of this subject from the psychological point of view while Janice I. Glasgow (Glasgow 1993) does this for artificial intelligence. Mental images enable us to picture the world as it could or should be. They enable us to look ahead into possible futures and to relive past events. In our opinion, the most important use appears to be in the prediction of the otherwise unpredictable environment, e.g. *anticipating* future events which requires to be 'faster than reality,' which is according to Atkin (1992) closely related to the development of consciousness. According to Calvin (see Calvin 1989) rehearsal (in the context of his ideas of a 'Darwin-machine' and consciousness) was used primarily as a means of sensorimotor action testing, e.g. throwing effectively a stone at a rabbit. Instead we assume that mental imagery and rehearsal first evolved in a social context, e.g. in order to anticipate the behavior of conspecifics. This can be seen clearly for the synchronization of body movements.

The major hypothesis here is that rehearsal on the basis of mental imagery is developed in coevolution with the complexity of the behaviors of a cognitive system as means to forecast the behavior mainly of conspecifics. This is important to establish and maintain social relationships between individuals. If the cognitive system consists only of very few behaviors which can only be modified in a very restricted way it is easy to forecast the (small set of) possible next behavior(s). This might be done by a kind of look-up table or stimulus-response mechanism. However, if the number of behaviors increases together with the degree of freedom with which these behaviors can be acquired, adapted, and modified, then such mechanisms will not work any longer reliably enough. One solution could be to develop a model of the behaviors of other individuals. But how to acquire it? Again, if the set of possible behaviors is complex enough it will be impossible in practice to construct a model which is accurate enough to forecast the behavior of others. But your own behaviors share this complexity and the problem of forecasting is related to choosing your own next behavior. So, if you are able to come up with a model of your behavior this might serve as a basis for the model of the behavior of other individuals at the same time. The only behaviors which are observable from the outside of a biological cognitive system are body movements. The body images discussed earlier might be related to memorizing movements of other systems. A good guess for an individual might be that similar movements of other conspecifics are related to similar internal sensations.

Predicting movements means always to be faster than real-time which we call on-line prediction, or forecasting. This could be achieved by perceiving few clues from the beginning of a movement which are sufficient to remember the full context of this movement which prepares the system to move itself in an adequate way. This is one possible explanation why body exercises in sports where more than one person is involved, e.g. as competitor or as team member like in baseball, tennis, soccer, or martial arts, lead not only to a higher performance in the movements proper but to a kind of blind confidence of how the other ones or some object like a ball will move.

Rehearsal can be seen as the capability to do off-line forecasting of movements using the same techniques and information as the on-line capability and therefore starts later in the ontogenetic development. The fundamental function of mental imagery, and the human imagination that it supports, could therefore enable us to imagine the possible actions of another human being and our own corresponding movements as response to these actions. Highly elaborated examples of this in human culture are dancing or playing music. These two examples both require to a high extent a synchronization of actions between humans, an aspect of communication which is discussed in more detail in section 7.

6. A Different View on Cognition

Recently several investigations in behavior-oriented robotics point into a direction to describe the behavior of an autonomous robot in terms of attractors in state space (Steels 1994b; Jaeger 1995). One important aspect which inspires researchers from different disciplines to study dynamic systems is 'emergent functionality,' i.e. "... that a function is not achieved directly by a component or a hierarchical system of components, but indirectly by the interaction of more primitive components among themselves and with the world" (Steels 1991). Emergent functionality is an interesting point in development of human infants (Rutkowska 1994) as well as in considerations on consciousness (Atkin 1992).

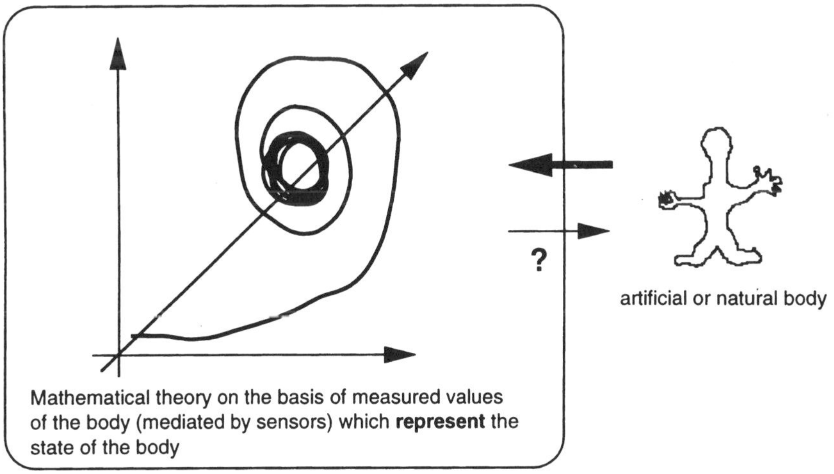

Figure 2a. The traditional separation of body and control architecture which is used in AI, cognitive science and behavior-oriented robotics. State-Space as an disembodied abstraction: Isolated from the physical body.

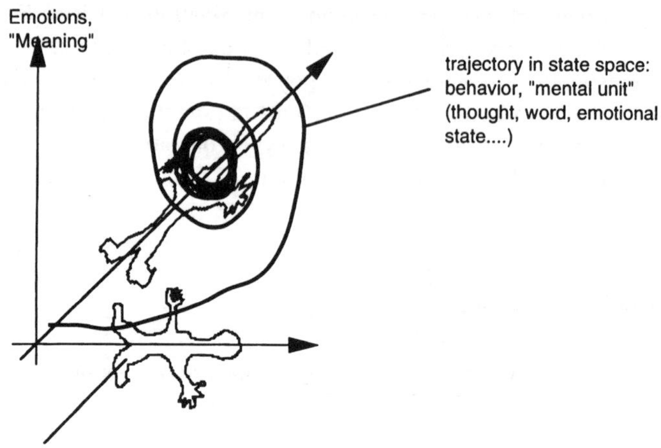

Figure 2b. Agent–environment interaction space as an embodied, physical-psychological unity: Trajectories as perceptions, actions or remembering processes.

So far we have motivated the use of the dynamics systems to investigate artificial as well as natural systems. But we believe that there are severe limitations to the approaches we mentioned so far, namely all investigations are based on the assumption that the state spaces are built out of dimensions which *represent* parameters of the body. In robotic research sensors are used to measure the physical state of the body (external and internal sensors) and these abstractions are used for calculations within a mathematical framework (see fig. 2a). This one-way-direction leads to problems which are frequently discussed especially in AI, including e.g. debates about the best kind of representation (the propositional versus analog debate) and the 'frame problem.' We think that many of these problems arise from the 'representation point of view,' i.e. the assumption that physical states can be measured, modeled and represented and used for further algorithmic or symbolic calculations without providing a direct way back to the physical basis. Fig. 2b) sketches our version of a multi-dimensional state-space as a physical-psychological unity which is an interaction-space of agent-environment and of the processes going on within the agent. The main underlying assumptions and characteristics of the approach shown in fig. 2b) are summarized as follows:

1. There are no abstract, disembodied models in the brain of humans or other natural systems. Physical states cannot be transformed to abstract values in order to build up internal models.[2]

2. The physical-psychological state-space consists of dimensions with links to physical states (location of body parts, proprioception, emotional states etc.). The dimensions of the state-space consist of direct physical links to the body, as they are found in homunculi-like projections in the brain.
3. In the physical-psychological state-space trajectories are perceptions, actions,[3] and 'mental units' of thought and remembering. In this way there is no difference between *storage* and *retrieval* of entities. Instead both refer to a *construction* process. In this state-space all experiences are intuitively related to the body and its history.
4. Remembering means *re-construction* of a trajectory, which is (by definition of the state-space) related to the *actual* physical state of the body. The actual physical state is pushed towards a state which results in a trajectory as close as possible to the remembered one. Therefore there are no two identical results of 'memory retrieval.' Every retrieval process modifies its content![4]
5. If bodily characteristics or the physical links between body and state-space change, the dimensions of the state-space change accordingly.[5] A set of dimensions have links to internal emotional states. Changes in these domains might therefore dominate, suppress or activate other domains and lead to crucial changes of the whole state-space. In the same way as other 'mental experiences' body images both of oneself and of others could be characterized by attractors in the physical-psychological state-space. This might explain why changes in the emotional states might lead to distortion of the conception of one's body. Additionally, certain physical behaviors or habits might influence the emotional attitude towards the body.[6]
6. Learning takes place in the same state-space, i.e. learning is not separated from remembering. In very simple cognitive systems one might think of learning as a reconstruction of 'hard-wired' (innate or learnt via imprinting or conditioning) trajectories. Reacting then means a replaying (remembering and executing) of more or less predefined trajectories (stimulus-response relationships), where adaptation allows the integration of data of the actual situation only to a slight extent. In more complex cognitive systems the amount of 'hard-wired' trajectories is reduced in favor of a greater complexity of the remembered and possibly learnt trajectories. Learning then means that when the cognitive system is confronted with a 'new' situation, a similar past situation is remembered and the new 'knowledge' is fused and integrated into the past experiences. In the case of rehearsal and empathy this can also be done without actually experiencing the situation, but by imagination (self-stimulation) or emotional participation in a

situation. The problem that totally 'new' situations might destroy the consistency of the memory is not given: the cognitive system can only perceive things which it is in principle able to 'think of.'

The metaphor of the *physical-psychological state-space* which changes dynamically in a self-modifying way is meant as a suggestion towards a framework for cognitive architectures. It is a holistic approach towards human and artificial cognition, in viewing phenomena like 'learning' and 'remembering' as different interpretations of the performance of an individual system, i.e. as different aspects which become visible for an internal observer. This is opposed to other structuralist approaches which are well known in AI and cognitive science, namely, assuming different modules located inside a system. In the same way as the questions we ask are based on different underlying assumptations, we get different answers about the nature of the physical phenomenon 'light.' Therefore we argue for a common framework for cognition which can account for the facts that if we do learning experiments with a cognitive system we 'see it learning' and if we let the system memorize we observe remembering processes. These observations do not necessarily verify the assumption of an existence of learning or memory modules!

The next section will outline how symbols, language, and the social environment can be interpreted in this framework.

7. External Symbols and Communication

In the previous sections we often stressed the highly individual character of the bodily and cognitive development. Socially isolated individuals with highly complex and individual cognitive systems, i.e. complex means of remembering and rehearsal processes therefore constantly take the risk to take a cognitive development which leads to a separation from society. If a minimum of consensus on the 'common world view' is not present, social interaction and communication becomes highly difficult and dangerous, since societies usually are very sensitive to 'unnormal' behavior or attitude. Usually they have the tendency to separate 'strangers.'[7] This would lead to *divergence* in the development of cognitive systems, i.e. different individuals would develop in different ways and would more and more lose the common basis which is necessary for social living conditions. In order to compensate this tendency a *convergence* mechanism is necessary. In the case of individuals (e.g. siblings or members of a tribe) which grow up and live under very similar environmental conditions, then the convergence tendency could be provided by the common features of the habitat which would shape their bodily and cognitive development. This mechanism is very susceptible to disturbances, i.e. in cases where the environment changes dramatically on small time scales. This might lead to local and relatively isolated groups. Another convergence mechanism which can be faster, more effective and which can be used across larger distances is communication via language.

The development of communication and language is related to social living conditions (see Dunbar 1993) and is found in one of its most elaborated forms in human societies. Among other factors (like tool-use) one reason for the development of language in human societies seems to be the use of language as a means of effective 'vocal grooming' about social affairs. "In human conversations about 60% of time is spent gossiping about relationships and personal experiences. Language may accordingly have evolved to allow individuals to learn about the behavioral characteristics of their group members more rapidly than was feasible by direct observation alone." (see Dunbar 1993). We think that language does not only function to acquire knowledge about 'behavioral characteristics' of others, but also to get to know the internal 'states' of others, i.e. their feelings, attitudes ('theory of mind') etc. In order to build up a common basis for social interaction and cooperation individuals have to communicate and try to co-adapt their different and unique 'world models,' i.e. their conceptions of the world where the individuals are living in. This includes the social as well as the non-social environment. But how should one communicate internal states? Taken the physical-psychological state-space above this would mean that one has to communicate its dynamical and multi-dimensional characteristics. We assume that this is not possible. But the highly subjective character of this state-space forces agents to agree on *external symbols* whose meaning is socially grounded and has to be continuously updated (via communication and living together). In this way when talking about an object in the world people prefer not to tell each other e.g. how they feel to touch an object, or how they can grasp it best (i.e. subjective features) but they refer to intersubjective features like color or size of the object. In the same line of argumentation grammar of human language can be viewed as a social consensus and not as something which is relevant to the individual's acquisition of language (see Rosenfield 1993). In our view even on a different abstraction level there are no symbols 'in the head,' they are produced by word production actions. The elicitation of word utterance after stimulation of distinct brain areas only shows that there are certain places responsible for the 're-construction' of the word.

We assume that the pressure of social living conditions to communicate, to be cooperative and form alliances has led to the evolution of a 'technical means' (language) to communicate social affairs, individual personal traits and attitudes, i.e. to communicate characteristics of the 'physical-psychological state-space' which enhanced social bonding and the development of individual relationships. Language requires the concentration upon one individual, the adaptation of one's own behavior to the other's behavior, e.g. a huge extent of attention to and 'engagement' in another person. The latter being a prerequisite for empathy.

8. Empathy

> Empathy is the feeling that persons or objects arouse in us as projections of our feelings and thoughts. It is evident when 'I *and* you' becomes 'I *am* you,' or at least 'I might be you.' (Spiro 1992)

According to Spiro empathy, which can be separated from sympathy (more directed towards 'I want to help you') has an esthetic and a personal face. The concept of empathy is discussed in medicine (see Spiro 1992), arguing for the importance of empathy and passion in medical practice), but was first elaborated in the arts. According to Brothers (1989) empathy is nothing 'magical,' but a biological phenomenon, an 'emotional communication' in order to 'read' social signals. "During the evolution of the primate CNS, organization of neural activity has been shaped by the need for rapid and accurate evaluation of the motivations of others." Furthermore, Brothers discusses empathy as "A high-level concept that appears to have great potential utility in bringing together neural and psychological data" and "links data from the disciplines of ethology, child development, psychoanalysis, and neurophysiology." In our view empathy is a social, interpersonal means of 'mind-reading' which can be integrated in our conception of a physical-psychological state space. In order to do this we focus on Godfrey T. Barrett-Lennard's account of empathy, namely the 'cyclic/phasic model of empathy,' from now on referred to as the 'empathy cycle' (Barrett-Lennard 1981, 1993). Specific conditions initiate a sequence of three phases, namely *empathy resonation*, *expressed empathy* and *received empathy*. If the process continues phases 1 to 3 could occur in repeated form. We do not want to describe this model in full detail but only focus on some aspects which are relevant for our framework. As the initial condition Barrett-Lennard describes an 'active openness,' a special condition of being attentive, of "knowing a particular other in their own inside, felt experiencing of self and their world" (Barrett-Lennard 1993). In our framework this could be interpreted as the 'willingness' of allowing a change of the own state space, putting oneself in a similar state as the object of contemplation. This would not only mean to remember a similar former state of ones own, but the attempt to modify the own state towards the one perceived in the other person. In other words: the empathizing person tries to have similar feelings as the other person.

Another interesting point in Barrett-Lennard's empathy cycle is the phase of empathic resonation when the empathizing person *resonates* to the other person, leading to an "immediacy of recognition of the other's felt experiencing and meaning." (Barrett-Lennard 1993). From the viewpoint of a physical-psychological state-space this phase could be understood as a consequence of inner resonance, when the experiences of the other person are dynamically reconstructed in the physical-psychological state-space of the empathizing person.

In a biological sense empathy enhances mind-reading which is necessary to predict the behavior of conspecifics or to handle social matters. The latter is important for establishing close personal relationships to other persons. On the basis of a large number of experiences and in combination with rehearsal empathy provides a powerful mechanism towards coping with a complex social environment. Additionally, since 'resonation' requires remembering and a re-organization of past experiences, empathy allows 'emotional learning,' even if the sources were 'second-hand' experiences.

9. The Study of Physical Artifacts: Social Robots

Although the artifacts which we use for our experiments are far from being as complex as mammals we hypothesize that it might be possible to study incrementally the basic mechanisms along the line of argumentation presented in this paper. So far we have taken very first steps towards artificial social intelligence for robots. We started to implement imitating behaviors in groups of 'individualized' robots (Dautenhahn 1994b, 1995).

Many approaches in artificial life on groups of physical robots, which take into consideration interactions between robots, prefer the simulation of social insect societies, which are anonymous organized societies without individual relationships (Deneubourg *et al.* 1991; Kube and Zhang 1994). The individuals interact only for cooperation, tackling a problem which cannot be solved by a single agent. Other approaches take other agents only into consideration as moving obstacles or as competitors for limited resources. In contrast to these approaches we study 'individualized robot societies,' based on explicit, individual, one-to-one communication and individual interactions between the members of a society. Such social structures are typical for mammalian societies with individual recognition between its group members. As a concrete example for social interactions we study imitation. Arguments why the study of imitation is an important step towards the development of social interactions are given in more detail in Dautenhahn (1995). As a prerequisite for imitative learning we investigate 'keeping contact' and 'following' behavior. The experiments focus on the control of the orientation of the body axis which is, to our mind, important for the control of the body and the development of a 'body conception.' So far we conducted experiments on interactions of autonomous robots in a hilly landscape up to a 'reactive level,' e.g. the robots act upon the actual sensor data without storing data or learning from experiences. Two main advantages of this habitat are (1) the way how it can provide 'meaning' to the actions of the robot and (2) the necessity to control the orientation of the robot's body axis. The former is important since moving up steep hills is highly energy consuming and has therefore a 'displeasurable' meaning.

The robots are controlled using the behavior-oriented approach and a C-based programming language (PDL: Process Description Language, Steels and Vertommen 1993; Steels 1993) for the control of autonomous agents. The PDL programming concept is based on the dynamical system metaphor. The main characteristics of PDL are the concepts of 'quantities' and 'processes.' The quantities belong to sensors, actuators or internal variables. The processes are mappings between the incoming stream of values of the sensor quantities and the outgoing stream of values of the actuator quantities. The processes do not inhibit or activate each other. There is no hierarchy of processes. The influences of the processes on the actuators are accumulated and executed in each PDL cycle. The 'dynamic systems' approach and PDL are not only suitable for implementing the mechanisms described in this paper. Additionally, this approach points to a line of investigation where the behavior of robots is modeled as attractors (Steels 1994b) and can be used for the development of concepts (Jaeger 1995).

Our research issue to study robots which are approaching objects, looking for conspecifics, keeping contact, and imitating are mostly inspired by the idea to have a kind of disposition to 'social grooming' or 'seeking for novelty' (curiosity). Our robots are designed to search for bodily contacts and stimuli in their 'habitat.' This can be seen as opposed to many top-down approaches in robotics which are aiming at *avoiding* objects in the environment. With such a 'cautious' strategy these robots are trying to avoid, or at least reduce, the complexity of the natural environment. For instance, a lot of path planning algorithms try to construct 'free spaces' in order to be able to apply analytic-mathematical algorithms (see overview in Crowley 1987). This can partly be explained by the experimenter's intention to avoid destruction of the robot. But natural agents with a much more flexible body, especially those with an endo-skeleton, are used to behave non-optimally, even adult humans are used to get blue marks while bumping against doors or tables. And this is much more normal for young animals which are trying out their physical relationship to the environment, where bodily contacts play an important role. In contrast to distance sensors, which create a kind of safety-zone around the robot's body surface where disturbances indicate objects approaching or being approached, penetration of the body surface does not only provide information about the external world but affects physically the robot's body, i.e. 'something is done to the body,' e.g. even the slightest contact with other objects can effect direction of movement. Bodily contact (i.e. tactile sensing) is the 'closest' link to the environment.

Using the experimental 'bottom up' approach of studying physical agents (robots) which are moving and interacting in a habitat we hope to approach the answer to the crucial question what exactly the conceptions of 'embodiment' and 'social intelligence' might mean to artifacts.

10. Conclusions and Outlook

In the previous sections we argued for a socially and bodily embedded cognitive psychology for natural as well as artificial cognitive systems. We described *remembering* as a fundamental basis for acting non-reactively and *rehearsal* as an important means for the anticipation of the behavior of conspecifics. We introduced symbols as effective means of communication and described *empathy* as an effective means of reading social signals and motivational states of others. We presented the framework of a physical-psychological state-space and outlined how it could account for the above mentioned phenomena.

In this paragraph we sketch some aspects on the future path to construct artifacts according to this framework. The most important step is to design the physical-psychological space. We envision that in a first step the behavior-oriented approach based upon PDL is sufficient. The crucial point is to come up with a schema for 'pushing the system back' to a remembered state. For doing this we have to identify which parts of the dynamics of a PDL-program can serve as a characterization of an episode and by which operators we can push the system back closely enough. Possibly in parallel we have to develop a learning schema for PDL-programs. All of this should give us the basis for a dynamic memory schema for which we argued above.

The next step is to find a way for a physically changing robot. This is a major prerequisite for an interdependency between adaptation and learning of a PDL-program and adaptation of the mechatronics. A first attempt could be to locate the sensors on some kind of track. Their physical positioning can then be decided upon together with the concrete needs of a robot in a specific environment. A more complicated scheme would be to construct physical enhancements of a robot which enable the robot to perform differently, e.g. a manipulator module. The robot must be on a specific level of development before it can pick up such a module, integrate it into its existing mechatronics, and make use of it on the level of behaviors.

As a consequence of embodiment the scenario plays a prominent role. Usually most of the complexity which in nature forced living beings to become intelligent is stripped away for behavior-oriented robots, e.g. in using mazes. This was a major reason to design the hilly landscape where the same hills and valleys give rise to different behaviors depending on the dynamics of the robot. But this is still too static. The environment must be shapable by the actions of the robots either as side-effects or deliberately (see variations of the hilly landscape in Dautenhahn 1994b). These changes must be detectable by the robots. This is in contrast to most of the current scenarios, e.g. of peg collecting groups of robots.

Last but not least we have to design the psychological part of the above mentioned space. This would require the design of emotional influences. The challenge is that they must be able to color or influence the whole dynamics without introducing the usual hierarchical control schemata. One important aspect in our framework is that the robots should be capable to empathize with one another. This requires to construct the robots in such way that they can recognize each another as an individual. The recognition should not (only) be based upon sending an identification signal or wearing a kind of signature. Again this is interrelated with the emotions, body images, and the required dynamic memory.

In sum, we believe that the 'state-space-approach' metaphor is a promising approach to bring together both engineers, theoretical oriented mathematicians and researchers from natural sciences. Only an interdisciplinary endeavor might be able to test theories of cognition, including the one we proposed in this paper.

Notes

1. Following Daniel Dennett's argumentation (see Dennett 1991: chapter 5) we do not assume that on a neurophysiological level an internal Cartesian Theater exists.
2. Topological projections within the brain (e.g. of vertebrates) are not mediated by 'cables,' but by cells which are living entities with characteristics which are dynamically changing over time.
3. Including word production actions, see section 7.
4. Therefore every re-construction process is a rewriting of our own history. This might explain why memories of the past change during lifetime (i.e. are adapted) and are especially correlated to changes of our body (see Rosenfield 1993: 64f).
5. In the same way after amputations of body parts, e.g. fingers, the sensorimotor projections in the brain change (see also Melzack 1992: 96: "In short, phantom limbs are a mystery only if we assume the body sends sensory messages to a passively receiving brain. Phantoms become comprehensible once we recognize that the brain generates the experience of the body. Sensory input merely modulate that experience; they do not directly cause it."). The 'Feldenkrais method' (see Apel 1992) points into the same line of argumentation, namely that after injuries a totally new state of the whole body system, a different bodily 'awareness' has to be learned, not just a change in a single module.
6. This might be a possible explanation of phantom limbs (see Sacks 1985) or hysterical paralysis (Van der Velde 1985).
7. Only in specific cultural 'niches' (e.g. art, science) human societies more often tolerate 'strange' behavior because it is compensated by other benefits which the persons gave back to society.

References

Apel, U. 1992. The Feldenkrais method: Awareness through movement. *WHO Regional Publications/European Series* 44, 324–327.

Atkin, A. 1992. On consciousness: What is the role of emergence. *Medical Hypotheses* 38, 311–314.

Auchus, M., Kose, G. and Allen, R. 1993. Body-image destortion and mental imagery. *Perceptual and Motor Skills* 77, 719–728.

Barrett-Lennard, Godfrey T. 1981. The empathy cycle: Refinement of a nuclear concept. *Journal of Counseling Psychology* 28(2), 91–100.

Barrett-Lennard, Godfrey T. 1993. The phases and focus of empathy. *British Journal of Medical Psychology* 66, 3–14.

Bartlett, F.C. 1932. *Remembering — A Study in Experimental and Social Psychology.* Cambridge University Press.

Brooks, R.A. 1991. Intelligence without reason. Memo 1293, MIT.

Brothers, Leslie. 1989. A biological perspective on empathy. *American Journal of Psychiatry* 146(1), 10–19.

Bullock, D. 1983. Seeking relations between cognitive and social-interactive transitions. In *Levels and Transitions in Children's Development: New Directions for Child Development*, K.W. Fischer (ed), chapter 7. Jossey-Bass Inc.

Byrne, R.W. and Whiten, A. 1988. *Machiavellian Intelligence.* Clarendon Press.

Byrne, R. 1995. *The Thinking Ape, Evolutionary Origins of Intelligence.* Oxford: Oxford University Press.

Calvin, W. 1989. *The Cerebral Symphony: Seashore Reflections on the Structure of Consciousness.* New York: Bantam.

Chance, M.R.A. and Mead, A.P. 1953. Social behaviour and primate evolution. *Symp. Soc. Exp. Biol. VII (Evolution)*, 395–439.

Cheney, D.L. and Seyfarth, R.M. 1992. Précis of how monkeys see the world. *Behavioral and Brain Sciences* 15, 135–182.

Crowley, J.L. 1987. Path planning and obstacle avoidance. In *Encyclopedia of Artificial Intelligence*, S.C. Shapiro and D. Eckroth (eds), 2, 708–715. John Wiley and Sons.

Dautenhahn, Kerstin. 1994. Trying to imitate — a step towards releasing robots from social isolation. In *Proc. From Perception to Action Conference, Lausanne, Switzerland,* P. Gaussier and J.-D. Nicoud (eds), 290–301. IEEE Computer Society Press.

Dautenhahn, Kerstin. 1995. Getting to know each other — artificial social intelligence for autonomous robots. *Robotics and Autonomous Systems* 16, 333–356.

Deneubourg, J.L., Goss, S., Franks, N., Sendova-Franks, A., Detrain, C. and Chrétien, L. 1991. The dynamics of collective sorting: Robot-like ants and ant-like robots. In *From animals to animats, Proc. of the First International Conference on simulation of adaptive behavior*, J.A. Meyer and S.W. Wilson (eds), 356–363.

Dennett, Daniel C. 1991. *Consciousness Explained.* Penguin Books.

Dunbar, R.I.M. 1993. Coevolution of neocortical size, group size and language in humans. *Behavioral and Brain Sciences* 16, 681–735.

Glasgow, JaniceI. 1993. The imagery debate revisited: A computational perspective. *Computational Intelligence* 9(4), 309–333.

Humphrey, N. 1976. The social function of intellect. In *Growing Points in Ethology,* P.P.G. Bateson and R.A. Hinde (eds), 303–317. Cambridge University Press.

Humphrey, N. 1986. *The Inner Eye.* London: Faber and Faber Ltd.

Jaeger, H. 1995. Identification of behaviors in an agent's phase space. Technical report. *Arbeitspapiere der GMD,* No. 951.

Johnson, M. 1987. *The Body in the Mind.* Chicago: The University of Chicago Press.

Jolly, A. 1966. Lemur social behavior and primate intelligence. *Science* 153, 501–506.

Kaelbling, Leslie Pack, Littman, Michael L. and Moore, Andrew W. 1995. Reinforcement learning: A survey. *Workshop Notes of Practice and Future of Autonomous Agents*, 23 September – 1 October 1995, Monte Verita, Switzerland, Volume 1, 1995.

Kosslyn, S.M. 1994. *The Resolution of the Imagery Debate*. Cambridge, MA: MIT Press.

Kube, C.R. and Zhang, H.Z. 1994. Collective robotics: From social insects to robots. *Adaptive Behavior* 2(2), 189–218.

Kummer, H. and Goodall, J. 1985. Conditions of innovative behaviour in primates. *Phil. Trans. R. Soc. Lond.* B 308, 203–214.

Langton, C.G. 1989. Artificial life. In *Proc. of an Interdisciplinary Workshop on the Synthesis and Simulation of Living Systems, Los Alamos, New Mexico, September 1987*, C.G. Langton (ed), 1–47.

Maturana, H. and Varela, F. 1987. *The Tree of Knowledge: The Biological Roots of Human Understanding*. Boston: New Science Library.

Melzack, R. 1992. Phantom limbs. *Scientific American* 4, 90–96.

Mertens, K. 1987. Zur Interdependenz von Körperbewußtsein und intelligentem Verhalten. *Krankengymnastik* 39(8), 535–542.

Premack, D. and Woodruff, G. 1978. Does the chimpanzee have a theory of mind? *Behavioral and Brain sciences* 4, 515–526.

Quillian, M. Ross. 1968. Semantic memory. In *Semantic Information Processing*. Marvin Minsky (ed), 227–270. Cambridge, MA: MIT Press.

Rosenfield, I. 1993. *The Strange, Familiar, and Forgotten. An Anatomy of Consciousness*. Vintage Books.

Rutkowska, J.C. 1994. Emergent functionality in human infants. In *From Animals to Animats 3, Proc. of the Third International Conference on Simulation of Adaptive Behavior*, D.Cliff, P.Husband, J.-A. Meyer, and S.W. Wilson (eds), 177–188.

Sacks, O. 1985. *The Man who Mistook his Wife for a Hat*. Summit Books.

Slade, Peter David. 1994. What is body image? *Behaviour Research and Therapy* 32(5), 497–502.

Spiro, Howard. 1992. What is empathy and can it be taught. *Annals of Internal Medicine* 116, 843–846.

Stadler, Michael, and Kruse, Peter. 1986. Gestalttheorie und Theorie der Selbstorganisation. *Gestalt Theory* 8, 75–98.

Steels, L. 1991. Towards a theory of emergent functionality. In *From Animals to Animates, Proc. of the First International Conference on Simulation of Adaptive Behavior*, J.A. Meyer and S.W. Wilson (eds), 451–461.

Steels, L. 1993a. Building agents out of autonomous behavior systems. VUB AI LAB Memo 93–05.

Steels, L. and Vertommen, F. 1993b. Emergent behavior. A case study for wall following. VUB AI Lab Memo.

Steels, L. 1994a. The artificial life roots of artificial intelligence. *Artificial Life Journal* 1(1), 75–110.

Steels, L. 1994b. Mathematical analysis of behavior systems. In *Proc. From Perception to Action Conference, Lausanne, Switzerland*, P.Gaussier and J.-D. Nicoud (eds), 88–95. IEEE Computer Society Press.

Stewart, J. 1994. The implications for understanding high-level cognition of a grounding in elementary adaptive systems. In *Proc. From Perception to Action Conference, Lausanne, Switzerland*, P.Gaussier and J.-D. Nicoud (eds), 312–317. IEEE Computer Society Press.

Synnott, Anthony (ed.).1993. *The Body Social*. Routledge.

Thrun, S.B. and Mitchell, T.M. 1995. Lifelong robot learning. *Robotics and Autonomous Systems* 15, 25–46.

Tsotsos, John K. 1995. Behaviorist intelligence and the scaling problem. *Artificial Intelligence* 75, 135–160.

Van der Velde, C.D. 1985. Body images of one's self and of others: Developmental and clinical significance. *American Journal of Psychiatry* 142(5), 527–537.

Part III: Consciousness and Selfhood

Seán Ó Nualláin
Dublin City University and NRL, Canada

1. What is Consciousness?

A burgeoning literature on consciousness is evident in this decade. The writers come from a vast array of different disciplines, from physics to psychology to the philosophy of mind; both mine and Sabah's paper in this section give a brief account of some of the principal theories emerging from each discipline. Making matters worse, Dennett's infamous title "*Consciousness Explained*" has upped the ante on book titles; it has since been "*reconsidered*" (Flanagan) and was to be "regained" until Baars settled on "*In the theater of consciousness.*"

Furthermore, a plethora of different techniques has been suggested for enhancement of consciousness. In these, consciousness is variously identified as mind-body harmonization, information-processing capacity, virtue (a point which Willis Harman makes in the current consciousness academic mainstream), and other processes on which I do not feel qualified to comment.

Nick Herbert (1993: 20–21) gives a list which includes the following:

> … meditation, mantra, chants … psychoactive drugs … martial arts and spirit possession … celibacy, *coitus reservatus* and orgy … self-inflicted pain, fermented grain … gestalt therapy, religious ecstacy, hot baths and cold showers.

Herbert's list is more than slightly tongue-in-cheek, but anyone who has attended the Tucson consciousness conferences has experienced a range of papers of similar variety. It must be asked whether a unified entity corresponding to the word "Consciousness" exists. I shall end this section by giving a qualified "yes"; there is such an entity, but it is of a different nature to anything science has to date encountered, and intuiting it is likewise a qualitatively different act of mind to that to which we are accustomed. Before coming to this rather shocking conclusion, it can rightly be expected that some ground has to be covered. First of all, we need to highlight why science should have (indeed, always has had) a problem with consciousness. In very simple metaphysical terms, science, heretofore successful in dealing with objects, has begun to address subjectivity. John Searle (1992) has recently steered this debate in a direction which has some, perhaps unforeseen (by Searle) consequences.

Secondly, we need explicitly to distinguish physical process and representation from phenomenal feel in order to begin to isolate the focus of our study. Moreover, we need to distinguish between the concept of consciousness implicit in all the disciplines which are putatively currently studying it (from sub-atomic physics through computer science to "experiential" disciplines). Thirdly, we need to look at some proposed solutions to the problems we have raised. One remarkable absentee from Herbert's (and other) lists of possible techniques for realization of consciousness is experiential disciplines which claim to lead one to consciousness itself through a "*via negativa*" claiming we are not currently, but can have a potential to be conscious. These we shall also look at. Finally, a proposal for a "science" of consciousness is given.

1.1. *Consciousness and Science*

"Science" derives from the Latin "*scio*," I know, and its spectacular success since the Renaissance need not be dwelt upon here. My earlier paper in this collection mentioned the prospectus for science outlined by Galileo. It was to deal only with objective ("primary") qualities rather than their subjective ("secondary") correlates; for example, heat was to be measured in the appropriate thermal units, and subjective feelings of warmth were to be ignored. Galileo's methodological prescription has been from time to time interpreted ontologically. Ontology deals with what is; epistemology deals with what we know about it. Ignoring internal experience of any description was extended by philosophical behaviorism to denying the ontologically real status of mental processing. Part of the achievement of Cognitive Science has been to reverse this denial. We shall see later that the role of the observer in modern physics may be the harbinger of a further compromising of Galileo's hard and fast distinction.

As we shall see in the discussion of John Taylor's paper in this volume, Searle has made a further leap. He has decided that it is already time to characterize our phenomenal existence. Consciousness, he argues, has such attributes as familiarity, unity and subjective feel. Taylor speculates on how these inevitably arise from the constraints induced by the computational processes involved, which processes are part of the province of natural science ("*Naturwissenschaft*" in German and practiced by Helmholtz, *inter alia*; the exploration of consciousness by Hegel and others was termed "*Geisteswissenschaft*," and the historical tension between these two resonates in altered form today).

It cannot be too strongly emphasized that the combination of methods envisaged by Searle and Taylor is a break from scientific tradition. In particular Searle attempts to characterize the nature of subjectivity. The evidence used is of a phenomenal kind, based only on inner existence; let us call this approach "inner empiricism."[1] Searle appeals to our consensual validation of his phenomenal reality, and thus the empirical evidence is itself phenomenal, and indeed subjective. The classical scientific objections that inner feelings are fugitive and such thought experiments are unreplicable are open to any would-be critics. Nevertheless, I agree with Searle that we cannot make any substantial progress in studying consciousness without "inner empiricism." However, this method has been developed to levels of sophistication greatly surpassing Searle's in the past, and our science of consciousness must learn from the masters in this area. We shall talk about this later; for the moment, we have established that the consequences of Searle's work include a new type of "inner empirical" data for our science of consciousness, one drawn from consensually-validated phenomenal experience.

"Outer empiricism," on the other hand, is natural science as we know it, insisting on replicable experimentation. Essentially, all the papers in the "architectures for consciousness" section in this volume are, *inter alia*, attempts to study consciousness using the methods of the natural sciences. As mentioned above, Taylor attempts to synthesize the two empiricisms; he proposes a neurophysiological correlate for consciousness and attempts to interrelate the fruits of Searle's inner empiricism with those of the natural science methods. This indeed is how the science of consciousness will advance; it is important to note, however, that this synthesis of *Naturwissenschaft* and *Geisteswissenschaft* is a break from modern western science.

1.2. *Confusions about Consciousness*

Certain further distinctions need to be made to clear the ground for any further discussion of consciousness. One such which has recently gained currency is the distinction between "easy" and "hard" questions. Easy questions arise when one uses the methodology of natural science to investigate (particularly) neurophysiological processes which seem to have phenomenal implications. A classic such example is the enormous opus associated with commisurectomy patients. It is established the left hemisphere is basically linguistic/logical and the right hemisphere spatial; that conflicts can arise between actions lined up by each hemisphere (whether that is qualitatively different from our normal non-commisurectomised state has not been investigated; see below); however, the link from physical process to phenomenal reality has never been made clear. This final linkage is a "hard question" and needs the methodology of inner empiricism (see Taylor's paper for further explication).

I believe the inner/outer distinction better captures the necessary distinctions than hard/easy so it is the former that we will use. In any event, the fact that physical process and phenomenal "feel" are of different ontological kinds has been highlighted once again. A further difficulty arises with the occasional confusing of representation and consciousness. The essence of a mentation is that it should "intend" (roughly speaking point at) something else; existentialist thought in particular identifies consciousness with intentionality; it now becomes impossible to distinguish the representation of x from consciousness of x. This trap we shall also avoid.

A more insidious problem emerges from the sheer variety of disciplines that appeal to one or other notion of "consciousness" in order to meet *explananda* in their own domain of study. It is at this point that one begins to despair of finding any unitary thing as a referent for the word "consciousness"; the remainder of this section is essentially an attempt to salvage the concept. Furthermore, the separate disciplines give accounts of consciousness that conflict with each other, at least superficially. For example, physics attempts to try to explain the "unitary" nature of consciousness, the existence of which unitariness we shortly shall have reason to doubt.

Let us begin with quantum mechanics (QM). The standard view on consciousness emerging here from Penrose's (1994) work insists on the necessity of QM for several reasons. One is the putatively non-computational structure of human mental activity, about which more later; another is the equally putative unitary sense of self, and a final one is free will. Vector state-reduction by an observer has been mainstream physics since the 1920s; Penrose has recently argued, it can be done, observer-independent, by superposition of appreciably different mass distributions. The current story does not bear much retelling here; coherent quantum states appear in the cytoskeleton of individual neurons, somehow cross inter-neuronal boundaries and provide a physical basis for consciousness. The heretofore mysterious workings of anaesthetics can be explained with respect to this process. Other QM versions are different in detail (Herbert 1993) but stress how the randomness in QM can explain free will and how the absence of "things" can somehow supply provender for perceptual ambiguity as phenomenally experienced.

Penrose's conclusions depend, to a large extent, on a hotly-contested interpretation of Gödel. What is certain is that we humans function reasonably well in the world with cognitive structures that we cannot know to be formally sound. Penrose argues that the role of consciousness in these terms is somehow directly to intuit correctness in a non-computational manner. The jury is out on these matters; however, if Penrose makes the mistake of attributing supra-formal powers to consciousness, some of his opponents who attempt to treat consciousness as a species of computation make exactly the same mistake, if for different reasons, as we soon shall see.

Edelman's notion of primary consciousness, commented on in my paper below, identifies consciousness with the ability to assess salience of signals entering from various modalities. It is thus trans-modal and has the capacity to evaluate. Baars' global workspace model (outlined severally in the "architectures for consciousness" section of this paper), is based on the computational blackboard system. Yet blackboards have a notorious shortcoming; it is extremely difficult to intuit which process should be allowed to act next, irrespective of the precise realization (at Marr's Level 1) of the blackboard. Again we see consciousness being attributed a competence not available to the processes in its environment.

Other psychological theories (e.g. Schachter's DICE) focus on consciousness as the seat of voluntary action, as implemented in a single Fodorian module which relays sensory data to a control system in charge of voluntary action. Again, some evaluatory mechanism is implicit. Libet's justly celebrated work (see Taylor's paper) likewise insists on our ability to veto an action once the neural activation corresponding to the onset of that action has reached a certain critical level. This interdisciplinary evidence considered, it is hard not to conclude that perhaps Penrose may be correct in stressing the supra-computational nature of consciousness even if his Gödelian argument is ill-constructed, as may be the case. We have now seen theorists in QM, psychology and neurobiology appeal to concepts of consciousness which have a striking family resemblance.

Dynamical systems theory has of course approached consciousness, as it has approached everything else, but describing it as an emergent phenomenon is unhelpful in the absence of previous explication of conditions of emergence. It begins to seem as though we inevitably will produce far-reaching proposals for a science of consciousness as we produce a coherent concept. However, I need first to fulfil my promise of looking at innerly empirical disciplines, after rejecting one pseudo-synthesis.

1.3. *Inner Empiricism*

Earlier in this volume, the outline of a vision of Cognitive Science, based on an objective (i.e. non-observer-related) concept of information was given. One possible solution presents itself; what if we add a phenomenal aspect to information? The inner and outer empiricisms will then converge; to specify a process informationally is to characterize it phenomenologically. Unfortunately, this leads us to a ludicrous pan-psychism; the entropy of any physical system becomes identified with a particular subjective state.

I believe it is worthwhile to examine the fruits of some of the valid forms of inner empiricism rather than capitulate in this way to pan-psychism. The starting point for these approaches is a phenomenon which has made its way into scientific discourse using the cover term "the self as interpreter." Let's look at Gazzaniga (1988) in his treatment of this phenomenon which is inspired by experiments on commisurectomy patients. (It is not inappropriate to add that some such experiments are morally reprehensible.)

Gazzaniga's notion is that the "interpreter," located as it is in the verbal left hemisphere, maintains a commentary on one's diverse experiences and weaves them into a basically coherent story. To cohere, this story must include as premise the notion that, for the most part, one's actions were initiated by oneself for valid reasons. Often this premise is fallacious; a classic if extreme example is actions initiated due to post-hypnotic suggestion. For example, a patient might be told during a trance state to stamp his foot on a cue like the hypnotist's clicking his finger. However, when the patient does this on cue, he will try and explain his action as due, for example, to a stone in his shoe. Gazzaniga extends this interpreter notion to explain most of our "normal" existence; for example, depression may be initiated by a hormonal imbalance and fixed in place by the interpreter's creating a story to explain the imbalance. The consequences are obvious; much of the time, we are not the authors of our actions, but fallaciously attribute them to ourselves. Let us note Gazzaniga's combination of inner and outer empiricism before finally plunging into the former.

This vital source of data for a science of consciousness has been ignored for a variety of reasons. One is obviously its connection with religion which attenuates its appeal in this secular age. Another is the sheer difficulty of its message, which is shocking for the same reason as Gazzaniga's analysis; we now seem devoid of free will. The religion objection we can dispense with; only one of the four traditions I am about to outline depends on any fideistic notions any more than do Searle or Gazzaniga, on whose empirical basis they also concur. Nor is any commitment made to monism or dualism. What is of great significance is that these traditions regard consciousness and selfhood (which are assumed inextricable) as realizations that emerge from an investigation rather than concepts which can immediately be grasped.

In this, as in other control concepts, the convergences between these assumedly separately-evolving disciplines are impressive. It will be remembered from my earlier paper on this volume that conclusions should be regarded as strengthened if independent lines of study converge on them.

First of all, we need to be convinced that a notion corresponding to Gazzaniga's "interpreter" can be found in the disciplines we are about to look at. I have chosen the Advaita tradition from Hinduism (Mahadevan 1977), the Gurdjieff/Ouspensky (Walker 1957) work (simply referred to as the Work), the Christian mystical tradition as represented by Thomas Merton (1949), and finally Tibetan Buddhism. The Advaita tradition first makes a distinction between "mind" and "self." "Mind" is just a name for thoughts (the analogy computer = mind is not inappropriate) which subsist in and through the medium provided by the thought "I." That this "I" is Gazzaniga's "interpreter" is attested by its ubiquity and spurious nature; it is false selfhood and true selfhood can come into being only after seeing its falsehood. Yet the perception of this falseness may require much (usually meditative) discipline. Once the interpreter has been stripped away, true subjectivity, the true self which is also consciousness, can emerge.

The Work as conceived of by Gurdjieff, whose Weltanschauung is an intriguing mixture of Occident and Orient, starts with a radical assessment of our ordinary "waking" consciousness. Like Gazzaniga, Gurdjieff insists that the attributor of most of our actions to a voluntary, unitary self is incorrect. We are in a state akin to that of the posthypnotic fabricator noted above. Gurdjieff goes even further and now echoes a theme from Tibetan Buddhism; we are fragmented, and indeed consist of a legion of "I's," each of which steps onto the stage, claims to be Prince, Ghost, stage-director and playwright all at once, and then is replaced. For example, one particular "I" may hold the stage at one moment and decide to give up smoking; a second comes along and expresses its disagreement by buying a pack.

If Gurdjieff is right, either consciousness is not the unitary thing Penrose (to take one example) claims or — and this is Gurdjieff's position — we in our current state are not continually conscious. In fact consciousness is a rare achievement (as my paper below suggests in a more conventional context); the

Work is essentially about increasing its incidence and duration. Again, we see the two stages involved; first of all, a recognition of the ubiquity of the interpreter and then a set of techniques designed to point to one's true center of awareness.

Thomas Merton gained his first fame as a Catholic apologist before going on to make an honest living as a social critic in the 60's. The central theme in his inner empiricism is a contrast between the false or empirical self, that with which one navigates socially, and the true self. To achieve the latter is to combine one's true subjectivity and simultaneously a more veridical intentionality. As in Advaita and The Work, consciousness for a person is actually both of these simultaneously. It is difficult to describe Merton's later work as explicitly Christian, or indeed conventionally religious.

Buddhism in the West has unfortunately become to some extent a victim of its own success and it is becoming difficult to encounter true coin. However, the statement "I am composed of simple elements" gives a flavor of the original Tibetan inner empiricism. The notion of an integrated self actually causes suffering, and so should be avoided. Unlike the other sources just cited, Buddhism is explicitly a soteriological religion, so it is less relevant here. However, the parallels in these four streams are impressive.

1.4. *Conclusion: Toward a Science of Consciousness*

The conclusion is inevitable that a science of consciousness be distinct from though overlapping with Cognitive Science (CS). The latter discipline deals with mental processes which can be characterized in bits; the former with an entity whose essence can be arrived at only through a combination of applied experientialism and standard scientific practice. The range of disciplines involved in each is also going to overlap; CS is as stated in my earlier paper, while consciousness research will involve QM and "inner empiricism" also. The notion of consciousness emerging from the discussion immediately above suggests something supra-computational, capable of evaluation of the salience of signals entering from a variety of modalities, and culminating in a subjectivity which yet is pure observation. Given that consciousness is of an ontological nature new to us natural scientists, it should be unsurprising to find its conceptual unitariness is also different in kind to that to which we have become accustomed.

The division of tasks envisaged for the consciousness disciplines, then, is as follows:

1. QM should concern itself with the physical criteria for "observer" status and, together with Philosophy, with the metaphysical implications of vector state-reduction. One of the most embarrassing consequences of philosophy's recalcitrance to live up to its responsibilities here is the "folk physics" espoused by contemporary monists like the Churchlands.
2. Neurophysiology can continue with investigations into localizations of consciousness (as the final papers here do), and its impact on input systems like vision.
3. It is open to Cognitive Psychology and other Cognitive Sciences to consider the information-processing correlates of consciousness. In particular whenever an informational distinction becomes projected into phenomenal space, it is a Cognitive Science phenomenon.
4. Anthropology can inform us about different construals of the world.
5. Computer Science can help us delineate the stage at which consciousness becomes necessary as an *explanandum*; witness the discussion of blackboard architectures here.
6. Philosophers can also help by insisting on linguistic precision.
7. Finally, no progress can be made without "inner empiricism" as described above.

2. Introductory Remarks on the Papers Addressing this Theme

2.1. *Cognitive Science and the Person*

Andrew Brook's paper adds a new dimension to the problems we have already encountered in addressing consciousness and selfhood. He argues that a substantive problem still remains with respect to what Cognitive Science (CS) should be studying. Data given by introspection (a form of inner empiricism) lead to the manifest image of the person i.e. the person as a subject and agent, guided by reason, anticipating her own future, and essentially a unity. This contrasts with current CS analyses like those of Dennett and Fodor which stress fragmentation, be it of the stream of consciousness itself in the case of the former or, less controversially, of the modules which comprise the computational functionality of the brain/mind.

Apart from the obvious ethical issues, Brook argues there is a slew of problems with either approach, particularly once phenomenal data are taken into account. These data are essential to Dennett's case; we can assent to his metaphor of consciousness as a pandemonium of streams of text promoted to further functional roles by a virtual machine only by including phenomenologically-validated consensus. Once these data are included, we are compelled to re-investigate Dennett's anti-homuncular argument thus; how did the homunculus acquire the structure projected on it? What are the *explananda* (in phenomenal terms) for which it provides an explanation? Likewise, Fodor's modularity thesis chips away an encapsulated vision module here, a phonological one there, and leaves us with a large explanatory gap in the characterization of the unencapsulated "horizontal" module. This module contains most of what really interests us about ourselves.

Brook appeals for support to his fellow Canadian philosopher, Charles Taylor, whose work is briefly discussed in my paper. Like Sabah later on, I review the principal current theories of consciousness before converging on a set intersection of the useful suggestions from each of these theories. The link with the innerly empirical disciplines cited above is evident; this summary is a preliminary attempt to put some of their conclusions in cognitive terms. More importantly, with the work of (Charles) Taylor and Cushman as *leitmotif*, the paper continues by pointing out the procrustean bed CS to date has fashioned for the self. At the very best, it confuses the minimalist "punctual" self for the individual, a far richer and politically consequential concept. In so doing, it commits a serious scientific error; at worst, it assents to an increasingly unhumanistic trend in contemporary culture by giving (spurious) scientific imprimatur to a value judgement on the interior life.

On a completely different note, Barnden extends the Johnson/Lakoff arguments on metaphor to the metaphors we use to deal with mind and consciousness themselves. Lakoff has long argued that whole cognitive domains derive their structure from the metaphorical extension of others. For example, we Westerners conceive of a love relationship with the same underlying set of concepts we had impressed on ourselves by the physical experience of going on a journey. Natural language obviously has rich veins to mine for examples of metaphors applied to mind; many of Barnden's examples are from literary work. Several metaphors are investigated in detail; one is that of mind parts as persons (which does not by now surprise us), another that of mind as world-definer. What comes across strongly from Barnden's paper is humans' active attempts to make sense of a world including phenomenal data which indicate that we are both integrated and multiple in the most fundamental possible sense.

2.2. *Architectures for Consciousness*

Katz and Riley begin this section with a warning about all computationally-based theories of consciousness, including those which follow. All such must propose conscious content emerges from a change in the brain's level of activity (algorithmically considered). However, they argue, subjective states can arise from a constant level of activity, and thus can be non-algorithmic. Something about subjectivity is thus ineffable, though the next four papers (coming from a computational standpoint) each address qualia, the "what is it like to be?" issue Hoffman alone makes a stab at isolating the essence of subjectivity qua special relativity. Each of these papers contains a substantial theme and a summary of a long research project.

Newman *et al.* comment that consciousness research is perhaps in better shape than natural language processing (NLP); Sabah courageously takes on both these massive issues. His motive is twofold; NL is absolutely critical for us humans and his past experience in implementing NLP systems has convinced Sabah that consciousness is necessary for them to work properly. This experience suggested that a meta-knowledge, or "reflective" level was necessary, along with unconscious and subliminal levels. He too begins his paper with an analysis of current notions of consciousness like those of Baars, Edelman and Harth. He notes a difficulty with the blackboard architecture on which Baars' "global workspace" is based; there is a difficulty in feeding information from higher levels like the semantic to lower levels like the phonological and so he proposes the "sketchboard" as an extension. However, blackboard systems' need for a scheduler to select the next action in the general case (i.e. irrespective of level), which involves essentially allocating the scheduler homunculus-like omniscience about the system, is not addressed; this problem will remain unresolved even after all the contributions of this volume.

Harth's non-pandemonium, non-homuncular notion of consciousness requires that the brain's observation of its actions is as fundamental a neural process as, for example, Edelman's neuronal group selection or re-entry. In a computational project of great intrinsic interest, Sabah is bringing together features of Edelman's, Harth's, and Baars' work. Finally, the mixture of inner and outer empiricism we are later to see in John Taylor's work leads Sabah to discuss how the computational constraints lead to the phenomena of selectivity and exclusivity, *inter alia*.

Newman *et al.*'s work can be read as a state-of-the art *envoi* from the "Global" Workspace theorists and so merits our closest attention. In the wake of works purporting to "explain" consciousness, it is refreshing to note the minimal commitments ethos in the statement that consciousness **reflects** the operation of a global integration and dissemination system. The current stories about this system converge on an elegant "wagon-wheel" analogy; the thalumus is the hub; the nuclei reticularis thalami (nRT) are the sleeve around the hub and they get to "gate" (in the "logic gate" sense) many crucial neural highways. The sensory modalities at the bottom rim of the wheel project to modality-specific nuclei in the hub. As they do, they also give off projections to cell assemblies like those in the superior colliculus, which perhaps might explain the efficiency of the "egocentric" cognition which occurs there. The wagon wheel metaphor includes highly modular spokes carrying afferent and efferent signals with the nRT, the "seat of awareness," crucial for these processes.

We discuss this metaphor further below; it is unquestionable that the nRT are crucial, and the survival of consciousness after excision of even half the cortex points to such a localization. Moreover, as Hoffman comments later, the data from anaesthesia confirm that the dissemination system (labelled ARAF by him, ERTAS by Baars) is involved in deluging the brain with input. Where Newman *et al.* may overstate their case is in the litany of AI systems they appeal to as corroboration, particularly when gating is used. In particular, their "multi-expert" blackboard requires that each expert include within itself the functionality of a scheduler; the invocation of PDP and other implemented systems does not address this crucial objection.

As neuroscience, the work reviewed is superb. The nRT complex gates interaction between specific thalamic nuclei and the cerebral cortex under the control of the reticular core and the prefrontal cortex. Processing must be global as well as local; a tangential intracortical network spreads the waves in oscillations at 40Hz across the cortex to this end. The content of consciousness can be characterized with respect to the dissemination system; however, the insistence (present also in Taylor) that the nRT is the site of subjectivity, because it is the source of "knowing that we know" violates a crucial distinction between epistemological fact (knowledge) and ontological reality (subjectivity). In short, two problems remain; the issue of subjectivity itself, which Hoffman handles with the Lorentzian contraction, and the computational scheduling problem.

Newman *et al.* make much reference to Taylor's PDP work on basal ganglia (bg) and prefrontal influences on the nRT. Taylor himself is concerned with the elucidation of phenomenal data with current scientific methodology. In particular he espouses the "relational mind" view of consciousness; conscious experience is shaped by the entering into relation by past stored experiences with current input. His paper stresses control and processing strategies; for example, the (putative) existence of a single strand of consciousness requires a global control structure, explicable by the role of nRT. The dissemination procedure and 40 Hz oscillation are also *à la* Newman *et al.*

However, Taylor is after bigger game; none other than "what it is like to be" itself. The twin themes of the relational mind and the gating nRT are brought to bear on this, and we've seen what Searle's phenomenal data are addressed *en route* in what is extremely valuable work. Of primary interest to us is that the two issues raised about Newman *et al.* are still unresolved; here, however, we also have an exciting extension of the blackboard model, and PDP simulations.

The final say is left to William Hoffman, whose research described here goes back three decades. It is nothing less than an attempt to consider all the heretofore troublesome phenomena of mind with the geometry of systems; consciousness (treated similarly to Baars), affect (a limbic system output) and social factors are all to be addressed. The geometry of systems is a co-ordinate free synthesis of the structures of differential geometry and topology and control theory. The paper is enormously difficult but repays the effort; because of this and its assault on the questions which provoked the workshop in the first place, this introduction will be lengthy compared to others.

Hoffman includes the notion of simplicial objects. Hoffman argues that the simplicial structure for concepts in information-processing psychology leads to an isomorphism between it and dialectical psychology. The latter provides us with a passport to social factors; for Hoffman, moreover, its treatment of many issues in cognition (including, for example, the surprising adult failure at Piagetian formal operations) is unparalleled. It is assumed that cognitive acts have a 2-stage structure; first, discrimination and classification occur, followed by synthesis with context. The range of reference includes memory, learning and creativity; the Hegelian dialectic can include subject and environment (including social environment) in an encompassing schema.

Memory is assumed to divide into the separate constructs of the perceived field, working and long term memories and knowledge itself. It is by dint of this analysis that the mathematical symmetric difference operation is introduced and the isomorphism with information-processing psychology highlighted.

So far, we have focussed on two aspects of Hoffman's work; the search for a mathematical formalism to unify all cognitive phenomena of mind in a way consistent with neuroscientific knowledge, and the use of category theory to capture an isomorphism between dialectical and information-processing approaches to mind. The seemingly paradoxical simultaneously local and global operation of the brain is addressed through integration of the differential operations. The analog nature of the mind's operations, its realization as flows in neural tissue, needs the type of co-ordinate-free math that has been introduced.

Hoffman builds his model of consciousness from the ground up; it is first of all self-awareness, and is not to be identified with qualia but rather our observation of experience. Though these are self-acknowledged parallels with GW theory, the basis (i.e. special relativity phenomena) for subjectivity is firm. Using the notion of Lie groups, Hoffman demonstrates how perceptual invariances occur; each level of analysis up to cognition is given its own formal treatment.

Hoffman's work needs more mathematical expansion than can be handled here. However, the outline principles can be explained simply. The first states the role of simplicial structures in information-processing treatment of cognition. The second points out the importance of the symmetric difference operation in treatment of cognitive processes by dialectical psychology. The third comments that the application of this operation to memory yields a new entity. The fourth stresses its role in generating new knowledge. Finally, the fifth posits the isomorphism we have noted necessary to put information-processing and dialectical psychology on the same footing.

This work is consonant with some being done at McGill, also using category theory, which attempts to supply a firm basis for cognitive science to advance. The converging of senior members of the academic community on this formalism must give us pause. Might it be that this work by Hoffman, encompassing GW theory, dialectical psychology and current neuroscience while simultaneously providing a firm mathematical basis for all the phenomena of mind, will give us the impetus we need to proceed to the next stage of the development of our science(s), as we desperately need to do?

Notes

1. This term was invented by Jacob Needleman: see his (1982) "The Heart of Philosophy," published by Harper Collins, NY.

References

Gazzaniga, M.S. 1988. *Mind Matters; How Mind and Brain Interact to Create our Conscious Lives.* Boston: Houghton Mifflin.
Herbert, N. 1993. *Elemental Mind.* New York: Penguin.
Mahadevan, T.M.P. 1977. *Ramana Maharshi.* London: Allen and Unwin.

Merton, T. 1949. *Seeds of Contemplation*. Norfolk, CAN: New Directions.
Penrose, R. 1994. *Shadows of the Mind*. New York: University Press.
Searle, John. 1992. *The Rediscovery of the Mind*. Cambridge, MA: MIT Press.
Walker, K. 1957. *Gurdjieff*. London: Unwin.

Reconciling the Two Images

Andrew Brook
Philosophy and Interdisciplinary Studies
Carleton University

1. The Problems Facing Cognitive Science

The problems currently facing cognitive science have been a subject of much discussion lately. As a number of speakers observed at the workshop on Cognitive Science Education at the meetings of the Cognitive Science Society in 1994, even after forty years of work, cognitive science is still far from having a unified research program. In this regard, the neurosciences make an interesting comparison. Though much more recent as a self-identified activity, their major international organization has over ten times the members of the Cognitive Science Society and its members aren't constantly worrying about its progress. The problems run very deep: sometimes it is difficult to know how to take or assess the very claims that cognitive science makes.

Doubtless there are many reasons why cognitive science is facing problems but one of them, I think, is this. Using the terms introduced by the philosopher Wilfred Sellars thirty years ago (Sellars 1963), we have two images of the person. One is the image we have of persons as highly unified centers of representation and intention. This is the image that dominates in everyday life and in philosophy, the *social sciences*, law, etc. The other is the image of persons as consisting of vast assemblies of postulated tiny units of some kind: neurons, syntactic structures, stimulus-response arcs, or whatever. This is the image that dominates in the sciences of the person, including cognitive psychology and cognitive science. Sellars called the former the *manifest image* and the latter the *scientific image.* One of the reasons for the current difficulties facing cognitive science, it seems to me, is that the scientific image of cognitive science has not as yet made much contact with the manifest image of everyday life and the social sciences.

Why is this a problem, it will be asked. Why can't the two activities be perfectly happy going their separate ways. After all, each has its own proper domain of discourse and a lack of good bridges from one domain of discourse to another is not always a problem, not by any means. It is a problem in the special case of cognitive science, however. We live in the manifest image. It is where we start when we set out to think about ourselves and others. Thus, if we do not know in some measure how to connect claims made within the scientific image to our manifest-image experience of ourselves and others, to that extent we do not know how to take or assess those claims at all.

To understand the kind of connection between the manifest and the scientific image of the person, it might be helpful to have an example. In the last decades, we have developed many marvelous inference and decision systems. Now ask, what do these systems tell us about the activity of me and other people making a decision: the effort to think the situation through and reach a decision, then the effort (the same or a different kind of effort?) to adhere to the decision in the face of some unforeseen complication or — a different kind of case — against a strong desire to do otherwise? Very little; the operating principles of inference systems, etc., make almost no contact with decision-making as we find it in persons.

This gap appears in many different places. Cognitive science has had less impact on social policy, interpersonal relationships, rules and structures for regulating interpersonal practice, the law, and many other aspects of social life than might have been expected. Where it has had an impact, it has often not been in the way one might have expected, by illuminating important aspects of individual or social life, but by providing new and powerful prosthetics, in the way word processors have had an impact on writing. (Education is a partial exception, and one could think of others.) Again, one reason, in my view, is that no way has been found to apply the conceptions of human cognition developed within cognitive science to people as we encounter them in everyday life, society, philosophy, the law, etc. More simply, we have no idea how to use these conceptions to understand ourselves. Note that the problem is not about introspection. How to apply cognitive science's notions of a cognitive system to others is as much a mystery as how to apply it to ourselves.

To forestall a possible misunderstanding, let me hasten to add that I am not setting the stage for any form of mysterianism. Flanagan (1991) calls philosophers like Nagel and McGinn who urge that consciousness, subjectivity or whatever is somehow beyond the ken of rational modeling New Mysterians. To the contrary, if there is a problem here, it is a problem to be solved, not a mystery to be venerated. Philosophers, Charles Taylor for example, have taken up some aspects of the issue.

2. The Manifest and Scientific Images of the Person

We seem, then, to have two broad conceptions of the person and no good way to connect them. As I said, Sellars took a large first step towards getting these two conceptions clearer thirty years ago when he introduced the notions of the manifest image and the scientific image in his very well-known paper, 'Philosophy and the scientific image of man' (1963). To repeat, the manifest image is the image of the human person of ordinary moral, social, and interpersonal life. What makes the scientific image special is that it starts from some postulated tiny unit of analysis and views the person or the human cognitive system as a vast assemblage of these tiny units. The leading theory in the scientific image nowadays is that persons are a vast assemblage of neurons tied together in complex biochemical and informational relationships.

Sellars treats these two images of the person in concert with a similar pair of images of the rest of the natural world but we will consider them in isolation here. He urges that the two images are radically different from one another.

'The' scientific image is in fact more than one image. A number of scientific pictures have had an effect on our conception of persons, including: the neurological picture, computational, information-processing pictures such as the condition-action rule of classical production systems, and more recently the connectionist, distributed-representation picture. If we extend our considerations to include historical ones, the number becomes even larger. In addition to the neural, informational and weighted node units of more recent theorizing, there is the neural hydraulics of Descartes, the neuron and its quantities of energy of Freud's Project for a Scientific Psychology, and behaviorism's stimulus-response arc.

These pictures are very different from one another in many ways, of course, but they all have a similar general structure, one that makes them all quite different in much the same way from the manifest image. Thus no harm is done by talk of 'the' scientific image.

To draw out how the manifest and the scientific images are different a bit more clearly, let us start with the point that in the manifest image, the basic unit of analysis is not some tiny postulated entity. The basic unit of analysis in the manifest image is in fact nothing less than the whole person, a being conceived as able to observe, make decisions, identify itself with things, enter into relationships with others, govern itself by standards, and so on. Much of social science and virtually all of practical, interpersonal or social activity take the whole person as the basic unit of analysis in this way. They start from the person as a unit and focus on what moves these units, how they relate to other such units, how they relate to things in the nonpersonal world, and so on. More generally, the manifest image starts with the person as manifest in everyday life.

By contrast, the scientific image takes as its basic building block unobservable entities and unobservable processes of one kind or another, postulated to exist in order to explain features of the manifest (Sellars 1963: 19). The idea behind this image is that the large basic unit of the manifest image consists of vast numbers of the smaller basic unit of the scientific image.

It is important to understand that these two images are not images of parts of the person. As Sellars puts it, they are not two halves of a single picture. Rather, they are,

> two pictures of essentially the same order of complexity, each of which purports to be a complete picture of man-in-the-world. (1963: 4)

3. The Nature of the Manifest Image

In broad outline at least, the scientific image is fairly straightforward and what we have said already is enough to introduce it. The manifest image is more complicated.

At root, the manifest image of something is simply the image we have of that thing in everyday life. Sellars' picture of the manifest image is more complex than this, however, because he also includes in it everything we can discover about the thing without postulating simpler, unobservable objects to explain it, everything we can discover by observing correlations, etc. Thus with persons, we have been able to discover, for example, that anger is correlated with insults, contentment with good relationships, adult disturbance with childhood traumata, and many other things by observing and organizing correlations. Moreover, many of these discoveries go well beyond what was contained in our original commonsense conception.

Sellars sometimes equates the manifest with the observable (p. 19). Even if we include the introspectible in the observable as Sellars does (p. 19), this characterization is too restrictive. There is a great deal in our manifest image of persons that is not observable or introspectible, very straight-forward things like character and levels of ability, for example. Neither is observable or introspectible, yet both are clearly aspects of our manifest image of the human person.

So what does characterize our manifest image of the person? One natural suggestion is that the manifest image is the arena of psychological explanation in the language of intentionality. The scientific image would then be characterized as the arena of mechanistic, i.e., non-teleological explanation.

The idea at the heart of psychological explanation is the familiar idea that objects of perception, desire, fantasy, belief, memory, etc. can have meaning for people. As a result, people can think and feel and do things for reasons, not just as the result of mechanistic causes. Reasons in turn have intentionality; they are about something, have an object.

Psychological explanations in terms of reasons are very different from standard mechanistic explanations and it would certainly be true to say that part of the difference between the two images is captured by this difference. But there is more to the manifest image than this. At least one kind of mechanistic explanation also plays a role in the manifest image, namely patterns of correlation, as we just saw. They are not postulational and we make constant use of at least of the correlations we have observed in everyday life, which is what makes them manifest-image, yet they are clearly not intentional. The difference between psychological and mechanistic explanation does not exhaust the difference between the manifest and scientific images.

Sellars has a suggestion that may take us further: he says that the manifest image is simply the framework in which we encounter ourselves (p. 6). That is to say, it is the framework within which we experience, reflect on, relate to, and interact with ourselves and one another. This suggestion leaves room for two things that the previous suggestion about psychological versus mechanistic explanation left out. One was the point that beliefs about correlations are also a part of the manifest image. The second is this. In addition to a picture of psychological content and attitudes, which is what the distinction between the two kinds of explanation focuses on, our manifest image of persons also gives us a picture of the thing that has this content, takes up these attitudes. More succinctly, the manifest image is not just an image of content and attitudes, it also contains an image of the subject of content and attitudes. And it contains an image of an agent, a being that forms intentions, makes decisions, and originates actions.

The subject and agent of the manifest image has a quite specific character. We can capture at least part of what was missed by the suggestion that the manifest image is the realm of psychological or intentional explanation if we can describe it.

Probably the most important feature of this part of our manifest image of the person is the one just sketched:

1. Persons are subjects and agents — centers of representation, imagination, reflection, and desire, originators of intentions, decisions, and actions.

By 'action' we mean not simply behavior but behavior that is the result of the formation of an intention and the taking of a decision.

As pictured in the manifest image, to be a subject of perception and representation and an agent of intention and action requires, in turn, that:

2. Persons guide themselves by reasons: states and events that work by providing motivation — motivation to accept or reject beliefs, motivation to feel in one way or another, motivation to act or not to act in certain ways. In the manifest image, we picture persons as beings whose motions are generally caused by motives, not by nonintentional causal forces.

Finally, as pictured in the manifest image,

3. Among the most important motivators or reasons for action are emotions — fears, feelings of affection, feelings of gratitude, resentments, hostilities — and biologically-based desires — hungers, lusts, feelings of discomfort, and so on.

These three features capture a lot of what is distinctive about persons as we picture them in the manifest image. As thus conceived, when a being has these features, it also has a distinctive constellation of powers, dispositions, and abilities. So the next question is, what are the other things that go with these three features like? It would be impossible in a short space to give anything like an exhaustive list of them but here are a few representative examples.

3.1. *Making an Effort*

Sometimes it is easy to reach a decision, but sometimes it takes effort — effort to understand a situation, effort to figure out what to do, effort, sometimes, to resist temptation and keep to a decision ('I really should reread three more pages of this paper but I would so much like to quit for the night'), and even effort to overcome obstacles, both human and non-human. Moreover, try as hard as we might, sometimes we do not manage to do what we decided to do. What is making an effort like? Are all the exertions of effort just sketched of one kind or a number of different kinds? What is the agent who makes these various efforts like? It is hard to say.

3.2. *Unity of Focus*

Making choices requires something else that is central in our manifest image of the person. I will call it *unity of focus*. To see what it is like, start with the better known unity of consciousness (UC). We can define UC more formally as follows:

> The unity of consciousness (UC) = *df.* (i) a representing in which (ii) a number of representations and/or objects of representation are combined in such a way that to be aware of any of these representations is also to be aware of other representations as connected to it. (For more on UC, see Brook 1994: Ch. 3.)

This is one form of UC. It is a matter of being aware of a whole group of representations at the same time. Another consists of being aware of oneself as the common subject of those representations.

Clearly UC in both forms is an important part of the manifest image of the person. I think a kind of unity found on the volitional side is even more central to what we conceive a person to be, however, what I call unity of focus. We conceive of persons as beings able to focus their intentional resources on courses of action. They can focus on a number of considerations at the same time and weigh up their implications. They can focus on a number of alternative courses of action at the same time and assess them against one another in the light of desires, moral beliefs, wishes for other people, etc. They can bring these considerations together to form an intention and choose a course of action. And they can focus their intentional resources on carrying out that course of action, against obstacles, conflicting desires, and so forth. The unity of focus involved in these activities is something more than just unified consciousness of representations.

3.3. *Anticipating a Future as One's Own*

I have a striking ability to imagine a future person as myself, to imagine me having the experiences and doing the actions that I suppose he or she will have or do, even when I suppose at the same time that I am connected to that person by all manner of unusual connections and/or lack of connections. Williams explored this phenomenon in a series of interesting thought-experiments in (1973: esp. Ch. 3).

Unified awareness of self and anticipating a future as one's own requires that one refer to oneself. Reference to self is worth a quick sketch even in a short account because it has some very interesting features.

3.4. *Reference to Self*

Persons can refer to themselves. Moreover, when I refer to myself, I know that it is to myself that I am referring. The indicators that we use to perform these acts of referring have some unusual semantic properties, properties different from those found in virtually any other indicator, so unusual that Castañeda (1966), for example, does not think that they are true indicators at all. He calls them quasi-indicators. Shoemaker calls the kind of reference in question *self-reference without identification*:

> My use of the word 'I' as the subject of [statements such as 'I feel pain' or 'I see a canary'] is not due to my having identified as myself something [otherwise recognized] of which I know, or believe, or wish to say, that the predicate of my statement applies to it. (1968: 558)

That is to say, I am aware of myself, and of myself as myself, without inferring this from any other feature of myself. If so, that the referent is myself is something I know independently of knowing anything else. If that is so, in turn, I must be able to refer to myself as myself independently of 'noting any quality' in myself, as Kant put it (1781: A355). If all that is true, finally, the first-person pronouns are semantically quite unusual.

All of (1) to (4) seem to be either parts of our manifest image of the person or natural extensions of it.

Two things are striking about the resulting conception of the person. One is its central importance in practical and especially interpersonal and social life. The other is how little anybody, philosopher, cognitive scientist or anybody else, has managed to say about it. We have made especially little progress with capturing any of (1) to (4) in a postulational, mechanistic model. That is as true of contemporary neuronal, computational, and connectionist models as it was true of the models of earlier times. This gap, to bring us back to where we started, is one reason, in my view, why cognitive science has had as little impact on activities that revolve around the manifest image of the person as it has had.

Much the same gap may underlie the sorry state of functionalism as a high-level model of the mind at the moment. In much the way that we do not know how to map any mechanistic model onto manifest image phenomena, we cannot map the claims of the various flavors of functionalism onto manifest image phenomena.

4. Bridging the Gap: Dennett and Fodor

To attempt to address this gap or these gaps between manifest phenomena and our various models of the mind, various strategies are unfolding. The 'mind-mapping' work so central to current neuroscience (MRI imaging, etc.) is one. The rapidly-increasing empirical research on consciousness of the past few years is another. Two others are found in the work of two contemporary philosophers, Daniel Dennett and Jerry Fodor. I will restrict my comments to the last two approaches.

Dennett's strategy for addressing the gap between manifest image phenomena and what we have been able to capture in our scientific image models is to try to eliminate the manifest image, at least as anything worth taking seriously. The manifest image is merely that: an image, an interpretation we have constructed because it is very economical way of making some sense of certain complex patterns of behavior. Thus all the philosophers and others who thought that they were doing metaphysics of the mind have merely been engaged in arcane studies of the self-image we have collectively constructed of ourselves. Oh dear! Unfortunately, I do not think this strategy can get us very far.

There is a feature of the manifest image underlying all the aspects of the manifest image sketched above that I have not yet mentioned. As we picture ourselves, we are each of us a kind of homunculus, a conscious entity who is at the center of and the unifier of our representational and volitional world. Dennett (1991) has mounted the most thorough attempt to date to undermine this idea, to persuade us that there is nothing in people to correspond to this picture. Thus his approach to the gap, over-simplified a bit, is to urge that the problem is our sense that there is a problem. There is no problem. We think there is only because we misinterpret 'phenomenology' in certain natural ways, phenomenology being the way we appear to ourselves and to others.

The problem with Dennett's strategy, in my view, is that it merely shifts the problem. The puzzle now becomes the phenomenology itself: what could produce such a sense as our sense that each of us is an homunculus? And how could we have been led so astray — how could a conception of such profound, even ineliminable social importance as our manifest conception of the person be at its core nothing more than a huge mistake? (This in outline is all I think most eliminative strategies achieve; they merely shift the problem.)

By contrast, Fodor *is* willing to mechanize the manifest image, or at least to try. His approach is to chip away at the edges: sort out a syntax module here, a vision module there, and postpone the rest to a better day. The problem with this approach is that, while the chips are eminently worth knocking off and we have learned a great deal about the abstract and in the case of vision even the neural structure of some encapsulated, nonconscious subsystems by doing so, the central person, the big homunculus, has not been touched. It remains as much of a mystery as ever. All Fodor's and others' discoveries about the various nonconscious cognitive subsystems in a person have done is to shrink the range of the unified subject and agent, shrink the homunculi. They have done little or nothing either to discharge it or tell us what it is like.

Why do we need to 'reduce' the homunculi? Why not adopt Dennett's strategy and just ignore them or explain them away? For this reason. On the one hand, we cannot dispense with it; we could as soon dispense with the language of love, law, psychology, motive, feeling, and representation. On the other hand, as the Churchlands and others have pointed out, the manifest image does not take us very far. In fact, it works at all only so long as the person being explained is functioning well. Introduce any amount of cognitive or emotional damage or breakdown and the manifest image instantly 'claws the air,' as Paul Churchland has put it (1984: 46). And so on. Abandoning the manifest image is not one of our options (on this point I do not agree with the Churchlands) but abandoning the drive to capture it in a mechanical model that illuminates and helps us understand it is not an option either.

In the light of these needs, it is striking that most researchers in cognitive science right now are either manifest image extenders or manifest image ignorers. There is very little work on reducing the manifest image going on. Philosophers are prominent in the first group: they spend most of their time trying to find ways to capture manifest image phenomena in extensions and developments of the same kind of phenomena. These folk worry hardly at all about how their ideas might be realized in a computational system or any other kind of mechanical, well-characterized postulational system.

The second group includes most of the rest of cognitive science, the researchers who spend their time developing ever deeper and broader postulational models. Few of these people give much thought to how their models can capture manifest image phenomena. Sometimes one finds both sides in a single researcher, quite unconnected. I am thinking of cognitive psychologists I have known, for example, who work with pure postulational models in their labs, yet delight in tossing around notions such as the unity of consciousness outside it.

What is missing on both sides is an effort *to bring the two images together*, to find postulational, computational accounts that illuminate phenomena of our everyday experience of ourselves. And that, I contend, is one reason why cognitive science is in the state it is in. In the absence of robust roots connecting our scientific image models to our manifest image experience of ourselves, we not only do not know how to assess these models, we do not even know how to take them. As Dennett once said on a related issue, a meeting of minds would seem to be in order.

References

Brook, A. 1994. *Kant and the Mind.* Cambridge: Cambridge University Press.

Castañeda, H.-N. 1966. 'He': A study in the logic of self-consciousness. *Ratio* 8, 130–157.

Churchland, P. 1984. *Matter and Consciousness.* Cambridge, MA: Bradford Books/MIT Press.

Dennett, D. 1991. *Consciousness Explained.* New York: Little, Brown.

Flanagan, O. 1991. *The Science of the Mind*, 2nd ed. Cambridge, MA: Bradford Books/MIT Press.

Fodor, J. 1983. *The Modularity of Mind.* Cambridge, MA: Bradford Books/MIT Press.

Freud, S. 1895. Project for a scientific psychology. In *Standard Edition of the Complete Psychological Works of Sigmund Freud*, Vol 1, J. Strachey (ed. & transl.), 283–398. London: Hogarth Press and Institute of Psycho-Analysis.

Kant, I. 1781. *Critique of Pure Reason.* Transl. by Norman Kemp Smith as: *Immanuel Kant's Critique of Pure Reason.* London: Macmillan, 1963.

Sellars, W. 1963. Philosophy and the scientific image of man. In *Science, Perception and Reality*, 1–40. London and New York: Routledge and Kegan Paul.

Shoemaker, S. 1968. Self-reference and self-awareness. *J. of Phil.* 65(20), 555–567.

Taylor, C. 1985. *Philosophical Papers*, 2 vols. Cambridge: Cambridge University Press.

Williams, B. 1973. *Problems of the Self*, esp. Ch. 3, The self and the future. Cambridge: Cambridge University Press.

Consciousness and Common-Sense Metaphors of Mind[1]

John A. Barnden
Computing Research Laboratory and Computer Science Department
New Mexico State University

1. Introduction

At the workshop on which this volume is based, consciousness was a focal issue, because of its importance to any fully-fledged science of the mind. In the present paper, I take the line that such a science needs to carefully consider *people's common-sense views of the mind,* not just what the mind really is.[2] This is for more than one reason. First, such views are of course themselves an aspect of the nature of (conscious) mind, and therefore part of the object of study for a science of mind. Secondly, the common-sense views allow people to interact with each other in broadly successful ways — to predict each other's mental states and behavior with enough success for most mundane communications and other interactions to proceed fairly smoothly. Therefore, it is reasonable to look to the common-sense views for some rough guidance as to the real nature of the mind. The three words "some," "rough" and "guidance" are, naturally, to be stressed here. But, thirdly, to the extent that common-sense views are inaccurate, and perhaps even in gross conflict with the true nature of the mind, one interesting scientific question is: why do we hold such views, given our access to our own minds? Why should introspection be limited in a way that allows inaccurate views to hold sway? Finally, I shall claim that metaphors of mind can lead us to useful considerations concerning the phenomenal quality of consciousness.

I adopt two major working hypotheses:

1. Mundane natural language discourse that refers to mental states often reflects rich common-sense views of the mind. For instance, when someone

says "Perhaps some part of John realized that Sally was right," the idea of *part* of a person having a mental state reflects a prevalent, conceptually rich, common-sense view of the mind (which will be discussed below). Talk of parts of people having thoughts is not just an idiomatic form of words with no underpinning in the thinking of the speaker.

2. Common-sense views of the mind, at least those reflected in natural language discourse, but plausibly also those that are not, are highly metaphorical. The study of metaphor is, for this reason by itself, central to the study of mind (quite apart from other possible reasons for its being central). The part-of talk in (1) is metaphorical. (The metaphor is one I call MIND PARTS AS PERSONS. It will be discussed below.)

These claims involve a further, broader claim to the effect that metaphor is not primarily a linguistic matter, but rather a conceptual one. A metaphor is, metaphorically, a *view* of one "target" subject-matter, T, that uses descriptive resources from another "source" subject matter, S. Roughly, at least, the view is a matter of "seeing T as S."[3] Such a view can, importantly, be *manifested* in ordinary natural language discourse. But it can also be manifested in other external ways (see, e.g., Woll 1985; Kennedy, Green and Vervaeke 1993). And it can affect purely private thoughts about the target T. The argument that metaphor is not just a linguistic phenomenon has been championed by Lakoff (1993b), and rests on the open-ended and systematic way in which mundane, non-literary metaphorical expressions can be varied and still be easily understandable. See Barnden *et al.* (1994b) for some evidence that sentences like some of those listed below do indeed contain non-frozen manifestations of metaphors as opposed to frozen (canned) forms of language that may once have been non-frozen. The arguments are similar to those used by Lakoff (1993b).

Thus, I shall assume that when people use metaphorical sentences such as "Perhaps some part of John realized that Sally was right," they are, typically, thinking in terms of the metaphor, or in other words describing the thing to themselves in terms of the metaphor, at the time of utterance. The metaphor is not just a convenient, superficial tool for verbally describing something that the describer is actually thinking of in other terms.

I should say straight away that I am not claiming that someone who holds a metaphorical view of T as S at some moment necessarily does so consciously, or really believes an identification of the two domains. For instance, I am not claiming that if John says "The idea was deep in the recesses of my mind," then he really believes that his mind is a three-dimensional space in which there are recesses. Rather, the metaphor is, for John at the particular time in question, a convenient way of (consciously or unconsciously) thinking about his mind. But notice carefully that John's lack of belief in the metaphor need not be salient for him, and the metaphorical view may even seem non-metaphorical to him. Also, even if he is conscious of taking the metaphorical view and consciously believes the view to be a false one he may not have any non-metaphorical way of thinking the same thought (though he may have a number of other metaphorical ways of doing so).

There have been previous discussions of (a) metaphors for mind and, more particularly, consciousness (see papers cited in section 2.10) and (b) the role of metaphors for mind in introspection. However, work of type (a) that has looked at a wide range of metaphors has not had the particular goals of the present paper (as laid out in the first paragraph of this Introduction), and work of type (b) has largely focused on a small range of metaphors — notably COGNIZING AS SEEING (mind's eye), for instance in Banks (1993), Dennett (1991), Gopnik (1993), Jaynes (1982) and Rorty (1980), and MIND AS PHYSICAL SPACE, especially in Jaynes (1982). For other related work, see Shoemaker (1994) for an argument that introspection should not be seen as a type of perception. See Dennett (1991) and Jaynes (1982) for accounts of consciousness that are centrally informed by a consideration of inner speech (a phenomenon which I encompass under the metaphor of IDEAS AS INTERNAL UTTERANCES).

Thus, the present work points towards a more broad-ranging study of metaphor than heretofore as a way of illuminating the nature of consciousness.

And it does only point towards such a study. The paper is preliminary, and concentrates on analyzing the nature of a range of important metaphors of mind, briefly discussing the extent to which they can be used to describe or qualify states of consciousness, and pointing to important questions about the nature of consciousness that study of the metaphors raises. The paper represents a very recent offshoot from some research that was not originally aimed at elucidating the nature of consciousness. This research is reported in Barnden (1989a, b; 1992, 1995) and Barnden *et al.* (1994a, b, 1995). It has recently led to an implemented, prototype computer program that reasons in a specialized way on the basis of metaphors of mind in discourse. The article is not concerned with this program, but rather with the concepts underlying the research.

The research just mentioned involves a rather unusual meld of the field of metaphor and the field of representing and reasoning about mental states. Although cognitive linguists have considered metaphors of mind in some detail, they have not developed detailed schemes for mental-state representation/reasoning. On the other hand, investigators in the latter area, mainly philosophers and artificial intelligence researchers, have largely ignored the involvement of metaphor in mundane language about mental states. I believe, however, that metaphor is a crucial consideration for mental-state representation/reasoning, at least when it is concerned with mental states as reported in natural language (and this is indeed the concern in much of the work in philosophy and artificial intelligence on mental-state representation/reasoning). The reasons why it is crucial are brought out at greater length in Barnden (1989a, b, 1992) and Barnden *et al.* (1994a, b, 1995).

Metaphor is often used to convey details or qualifications that would be difficult to convey by other means. In the mental realm, this is evident in talk of "fringes" of consciousness, ideas "surfacing" in the mind, things being in the "recesses" of one's mind, "seeing" mental things "clearly" or "obscurely," and so forth. In using the word "consciousness," I will mean the type of consciousness that one has when one is "consciously thinking" about something, or consciously feeling an emotion, or feeling pain or pleasure. Indeed, it is the feeling of being conscious with which I am most concerned, and which, together with qualia in general, I take to be the central, unsolved problem of consciousness. Not all of this article directly addresses this problem, however.

2. Some Metaphors of Mind

When we speak or write, we often describe mental states (and processes) using metaphors of mind. Some examples are as follows. They are all closely modeled on examples found in real text and speech that I have encountered.[4]

1. "*One part of Mike knows that Sally has left for good.*"
2. "*Part of Mike was insisting that Sally had left for good.*"
3. "*John was leaping from idea to idea.*"
4. "*Veronica caught hold of the idea and ran with it.*"
5. "*In one part of his mind, Bob was thinking about the party.*"
6. "*George put the idea that he was a racing driver into Mary's mind.*"
7. "*Peter hadn't brought the two ideas together in his mind.*"
8. "*That belief was firmly fixed in his mind.*"
9. "*In the recesses of her mind, Cynthia knew she was wrong.*"
10. "*His desire to leave was battling with the knowledge that he ought to stay.*"
11. "*Martin had a blurred view of the problem.*"
12. "*It was crystal clear to Susan that Tom was unfaithful.*"
13. "*Things didn't smell right to Kevin.*"

14. "*In John's mind, terrorism is getting worse every day.*"
15. "*Sally told herself, 'Mike is untrustworthy.'*"
16. "*Sally said to herself that Mike was untrustworthy.*"

We will now look at the metaphors manifested in these sentences. We will comment in this section on the extent to which they address conscious states of mind, but will leave further discussion of the light they may eventually throw on consciousness to section 3.

The above examples are all in the third person, but second-version versions occur occasionally, while first-person versions are common and especially important for the present paper.

2.1. *Mind Parts as Persons*

Sentences (1) and (2) manifest a metaphor of MIND PARTS AS PERSONS. Under this metaphor, a person's mind is viewed as having "parts" that are themselves people — or, at least, complete minds — having their own thoughts, hopes, emotions, and so forth. I will call these parts "inner persons." Different inner persons can have conflicting mental states, or a mental state held by one can fail to be held by another. In addition, the inner persons can communicate in ordinary language, as the word "insisting" in (2) indicates. Another example of this phenomenon is:

17. "*Half of me whispered that I'd drive all the way there.*"[5]

MIND PARTS AS PERSONS in general is not specifically directed at consciousness, but particular manifestations can explicitly bring in consciousness, as in:

18. "*It was as if his consciousness didn't want him to be without anxieties.*"[6]

Here the agent's consciousness is being reified as a thinking, desiring entity.

But even without an explicit mention of consciousness, it is often natural to interpret the sentence as implying consciousness. For example:

19. "*Part of you wants to talk about your personal problem but part of you hates the idea.*"[7]

Also, if this sentence is indeed interpreted as describing a conscious mental state, it also shows illustrates the point that, under MIND PARTS AS PERSONS, several different parts of a mind can be conscious, and that none of them needs to have clear dominance over the others. Notice further that in (18) the mind-part is identified as a special component of the mind that has long-term existence, whereas in (17) and (19) the parts do not have any special nature and might well be very short-lived.

MIND PARTS AS PERSONS is related to the multiple-selves metaphors discussed by Lakoff (1993a).

2.2. *Ideas as External Entities*

Sentences (3) and (4) manifest a metaphor of IDEAS AS EXTERNAL ENTITIES, under which an agent is conceived as being *within* a space populated with some of his/her own ideas, hopes, etc. (I use the word "IDEAS" broadly in the name of the metaphor.) That is, some of the agent's ideas are viewed as external to the agent. The ideas are typically conceived of as concrete physical objects. The ideas can move around, or be active in other ways, and the person can move relative to the ideas, or physically manipulate an idea as in (4). Other manifestations of the metaphor, illustrating various types of activity on the part of the person or ideas, are:

20. "*The idea came to him to replace the statue in the garden.*"[8]
21. "*The facts, the truth slammed back at him once more, the reality that she had murdered someone.*"[9]
22. "*He would have shunned the idea if he could. He would have escaped from the knowledge that ...*"[10]
23. "*An idea had come to him out of the air, out of nothing, an idea of stupendous magnitude, a total solution. It felled him so that he spoke to her in a tone of vagueness, hesitantly, unable to find the ordinary simple words.*"[11]
24. "*But I was cheap. Cheap enough that the idea of paying the detective more after what he'd done so far tugged at me.*"[12]
25. "*I set the notion aside, but I had a feeling it was going to stick to me with a certain burrlike tenacity.*"[13]
26. "*She laughs and shrugs, as if to shake off gloomy thoughts.*"[14]

All these examples plausibly refer to conscious ideas (no. 20 less clearly so than the others), even though consciousness is not explicitly mentioned in any of them. This is partly because the physical interactions between person and idea that are described in the sentences are largely of types that would normally involve a conscious state of mind on the part of the person. For example, one is normally conscious of setting something aside or of someone or something tugging at one. Again, if an idea "strikes" you "forcibly" then you are likely to be vividly conscious of the idea, because physical things that literally strike you forcibly typically lead to your having vivid conscious awareness of the object (unless you are knocked unconscious, of course!).

2.3. *Mind as Physical Space*

As a contrast to IDEAS AS EXTERNAL ENTITIES there is the extremely prevalent metaphor of MIND AS PHYSICAL SPACE. Sentences (5) to (9) are manifestations of it. Under this metaphor, a person's mind is a physical region within which ideas, thinkings, hopings, etc. can lie at various positions. and such entities can move in and out of the region, as in (6). The person is not within the space; rather, the space is at least partially within the person. An important aspect of the metaphor is that ideas that have not been "brought together" (cf. sentence 7) are likely not to interact. For example, conclusions are unlikely to be drawn from them as a group. Another significant aspect is that if one's mind is "full of" something, there is little "room" to think of anything else.

Different positioning of ideas in the mind-space can be used to convey differing degrees of consciousness. For instance, things that are to the "back" or one "side" of one's mind, or are in the "far reaches" or "recesses" of one's mind, are less within consciousness than are things at the "front" or that are "uppermost." Indeed, they may even be out of consciousness. Examples in which thoughts are "deeply buried" strongly indicate unconsciousness. The notion of the "fringes" of consciousness is also relevant here, and was the target of a special issue of the journal *Consciousness and Cognition* (namely 2(2), 1993). Notions of conscious thoughts as occupying a special subregion within the mind-viewed-as-physical-space is evident in phrases such as "pop up into consciousness" and "seep into [one's] consciousness."[15]

A complication is that the word "mind" is ambiguous as between meaning the whole mind, including unconscious aspects, and meaning only the conscious mind. Both cases are common. The latter case is explicit in the following sentence from D.H. Lawrence's *Women in Love*, quoted in Cohn (1978: 49): "All this Gudrun knew in her subconscious, not in her mind." That the mind can, on other occasions, include things that are not in consciousness is graphically illustrated by examples like "His unconscious mind had known what his conscious mind had refused to know."[16] See also Bruner and Feldman (1990).

The metaphor has been studied by other researchers (e.g., Gentner and Grudin 1985; Hoffman, Cochran and Nead 1990; Jakel 1993). Jaynes (1982) adopts the extreme stance that people's use of MIND AS PHYSICAL SPACE (combined with COGNIZING AS SEEING) concerning their own mental states is an essential aspect of self-consciousness. MIND AS PHYSICAL SPACE has often been called MIND AS CONTAINER, but in this name the word CONTAINER often means no more than a bounded region, as opposed to a physical container like a box (Lakoff, personal communication). Some sentences do cast the mind as a physical container, however. I take MIND AS PHYSICAL CONTAINER to be a special case of MIND AS PHYSICAL SPACE.

In sentences (6, 7, 8), it is possible to replace "mind" by "head" without changing the meaning much. (The replacement seems less acceptable in (5) and (9).) I still take the modified sentences to manifest MIND AS PHYSICAL SPACE, where the mind-space is conceived of as being within the head-space. It is not clear, however, that this spatial inclusion carries over to cases where the mind is mentioned but not the head. I take this issue up again below.

2.4. *Ideas as Physical Objects*

It is common for the MIND AS PHYSICAL SPACE metaphor to be accompanied by the metaphor of IDEAS AS PHYSICAL OBJECTS, which is manifested most clearly in (8). However, in (9) there is no particular reason to think that Cynthia's realization is being viewed as a concrete physical *object;* rather, her realizing is an *event* that is localized in the "recesses."

Of course, IDEAS AS EXTERNAL ENTITIES is a special case of IDEAS AS PHYSICAL OBJECTS, when the external entities are physical objects. Sentence (10) manifests a metaphor of IDEAS (etc.) AS BATTLING ENTITIES. This metaphor is a special case of IDEAS AS PHYSICAL OBJECTS. Notice that the sentence does not strongly suggest MIND AS PHYSICAL SPACE. It would also be possible for the sentence to lie within a context that suggests that the ideas are external to the agent.

The IDEAS AS PHYSICAL OBJECTS metaphor has, in general, nothing special to say about consciousness. Nevertheless, when an idea is cast as an especially "visible" physical object, it is especially present to consciousness. I say more on this in section 2.9.

2.5. *Cognizing as Seeing*

Sentences (11) and (12) manifest a metaphor of COGNIZING AS SEEING. This is an extremely common metaphor, and is manifested most simply and palely in phrases like "see that [something is the case]."[17] Other common phrases or words manifesting it include: see as, see how, see through, in [one's] view, view as, looks like, look to, look forward to, shortsighted, lose sight of, blind to, focus on, outlook, viewpoint, flash of insight, flicker of recognition, bright, brilliant, having an eye toward. Under "cognizing" I include understanding, believing, thinking, predicting, and so on. Further examples of the metaphor's manifestations are:

27. "*Primakov looks at Saddam, and he sees a longstanding Soviet relationship that he wants to preserve if he possibly can.*"[18]
28. "*Iacocca concedes cost cutting slowed development of the LH, but sees a silver lining.*"[19]
29. "*There may be a flicker of recognition, but it's snuffed out immediately. Darkness is preferable to threatening self-awareness.*"[20]

Notice that at least the first two of these examples, as well as (11) and (12), portray the agent as looking outside him/herself; the object of sight is external to the agent. Thus, there is an implicit use of IDEAS AS EXTERNAL ENTITIES. On the other hand, many other manifestations of COGNIZING AS SEEING accompany MIND AS PHYSICAL SPACE instead. I will take up this point in section 2.9.

Forms of COGNIZING AS SEEING, especially when allied with MIND AS PHYSICAL SPACE, play a prominent role in the literature on consciousness. This is perhaps most topically so in Dennett's (1991) discussion of the "Cartesian theater" view of consciousness (which he attacks). Under this view, the mind contains a homunculus that sees the contents of consciousness as if on a stage. But the more common notion of the "mind's eye" is of course a reflection of COGNIZING AS SEEING. Rorty (1980) places great blame on the mind's-eye metaphor, and the related metaphor of the mind as mirror of nature, as perverters of the philosophy of mind. Sweetser (1990) contains a discussion of UNDERSTANDING AS SEEING from a linguistic standpoint. See also Richards (1989). Gopnik (1993) addresses COGNIZING AS SEEING in conjecturing that folk psycho*psychology* (the study of the relationship between mental states and our experiences of them) may arise from the folk psycho*physics* of visual perception.

Visual images, which are normally considered to be an aspect of conscious as opposed to unconscious mentation, are usually described by analogy to external pictures, diagrams, etc. That is, visual images are often metaphorically viewed as pictures, etc. It is common in ordinary language to talk of "picturing" things or of having "pictures" of objects or situations in one's mind.

Aside from one major type of exception (discussed in the next paragraph), manifestations of COGNIZING AS SEEING almost always imply consciousness. This is because literal seeing is normally a conscious matter, at least in the ordinary, common-sense view. It would seem possible to say "He unconsciously saw that he was wrong" — presumably because we realize that we sometimes literally see things without consciousness — but such examples seem to be rare in real discourse.

A phrase like "Martin's view of the issue," however, does not strongly connote consciousness on Martin's part, possibly because one can have a (literal) view of something without looking at the thing at all, consciously or otherwise. For example, "Martin's view of the mountain is the best in town" does not of itself imply that Martin has yet looked at the mountain.

2.6. *Cognizing as Physically Sensing*

COGNIZING AS SEEING is an important special case of a more general metaphor, namely COGNIZING AS PHYSICALLY SENSING. I will return to this matter below. A further, less common, special case is COGNIZING AS SMELLING, which is manifested in (13). A COGNIZING AS BEING TOUCHED metaphor touch is manifested in examples like (21–26), along with IDEAS AS EXTERNAL PHYSICAL OBJECTS. A metaphor of COGNIZING AS HEARING is evident in sentences like "*It sounds a good idea to me*," but its manifestations appear to be considerably more frozen than other physically-sensing metaphors such as COGNIZING AS SMELLING and COGNIZING AS SEEING.[21]

Manifestations of COGNIZING AS PHYSICALLY SENSING generally suggest consciousness, because physical sensation is, common-sensically at least, a quintessentially conscious matter.

2.7. *Mind as World-Definer*[22]

Sentence (14) manifests the metaphor of MIND AS WORLD-DEFINER. Other examples are:

30. "*He was, in her mind, an unquestionably beautiful child.*"[23]
31. "*I'd just invented a date at the Ritz, hadn't I? It wasn't true. It hadn't been arranged. Only in my mind.*"[24]

32. *"His six officers have not fired a shot in his memory."*[25]

In a manifestation of this metaphor, a belief (or other mental state) of an agent X is reported by describing the believed situation (e.g. terrorism getting worse every day) and adding a qualifier like "in X's mind." This is analogous to the use of qualifiers like "in the novel," "in the movie" and "in the painting," where the "in" takes us into a fictional world. Let us call a novel, movie, play, painting, etc. a *world-definer*. (This is an intentionally bland term, chosen to abstract away from a multitude of different ways in which a world can be defined; and I intend "defined" in a loose way — I do not mean to imply a complete, clearcut or unique definition.) I claim that (14) and the examples just above cast the agent's mind as a world-definer. Notice, for one thing, that in all the given examples the in-so-and-so's-mind qualifiers can be meaningfully replaced by phrases like "in the novel," "in the movie" or "in the newspaper report."

The metaphor is a tricky one to analyze because of various complications. All the examples so far given of the metaphor convey a state of belief on the part of the agent. However, qualifiers like "in his mind" can indicate states such as planning ("He was writing the letter in his mind") or merely entertaining or imagining ("In his mind, he was scoring goal after goal"). Which state is conveyed is delicately dependent on the nature of the target situation (e.g., terrorism getting worse) and syntax. On the question of syntax, compare the goal-scoring example just given with "He was scoring goal after goal in his mind." The latter is more susceptible to an interpretation in which the goal-scoring is a metaphorical description of successful problem solving events. I have not sorted out these matters fully in my own mind [sic] and so will not go into them further here.

Another complication is that it is often, but not always, reasonable to assume that the agent has a visual image of the target object or situation. For instance, it is reasonable to impute visual images to the agent in (30). Also, in "In John's mind, Sally was winning the race," John's mind is analogous to a play or film in which Sally is winning, and we can hypothesize that the agent has a visual image of a showing of this film or play. (Under this view there are two distinct things: the film/play; and the showing of it.) An example where a hypothesis of visual imagery is less called for would be "In Sally's mind, Clinton is morally weak."

Cases where the agent can be expected to have a vivid visual image (intermittently at least) — such as "In Sally's mind, Mel Gibson is sexy" — naturally convey consciousness of the imaged state of affairs. In other cases, the extent of consciousness is much more unclear. In the case of "In Sally's mind,

Clinton is morally weak," it is not clear to what extent or intensity Sally has conscious occurrent thoughts about Clinton being weak. It could well be that Sally has a only a background belief that he is weak, much like a background belief that, say, apples are fruit.

However, in those cases where consciousness is clearly conveyed, the responsibility is not entirely with the content. It also lies in part with the "In X's mind" qualifier itself. To see this, compare "In Sally's mind, Mel Gibson is sexy" with "Sally believes that Mel Gibson is sexy." With the latter sentence, it could easily be that Sally has never seen that star — perhaps she has merely heard from a reliable friend that he is sexy. This possibility seems less likely with the former sentence.

This contrast can be neatly explained through a much more general claim: *viz.* the claim that "In X's mind, P" causes us to attribute to X a belief in or other cognition about a whole plausible, rich scenario involving P, whereas this is much less the case with "X believes that P." Moreover, the richness comes from standard information, images, etc. we ourselves have concerning P. These claims follow from the analogy with "In the novel, P" or "In the picture, P." Unless a novel is of a special genre such as a fairy tale, we assume that normal relevant facts about the real world hold in the novel unless contradicted. For instance, in interpreting "In the novel, the detective meets Queen Victoria" we assume that standard facts about Queen Victoria, meetings with royalty, and detectives hold in the novel unless contradicted. We may also form a visual image of the meeting and assume (tentatively) that extra features of the image, such as the nature of the room containing the meeting, hold in the novel. Similar observations hold for "In the picture, the detective meets Queen Victoria" (unless the picture is of a strange sort). By analogy, I claim we do a similar thing in interpreting "In X's mind, the detective met Queen Victoria." We attribute standard facts to X as part of his/her belief state, and we attribute to X any visual image we ourselves might have of the putative meeting, because the image depicts a plausible form that the whole believed scene could take. By contrast, "X believes that the detective met Queen Victoria" does not strongly suggest that X has a visual image of the situation, or any beliefs about extra features such as the room.[26]

A related observation is that MIND AS WORLD-DEFINER sentences describe holistic, unified mental states as opposed to highlighting specific, discrete ideas as is often the case in manifestations of MIND AS PHYSICAL SPACE and IDEAS AS EXTERNAL PHYSICAL OBJECTS. Even in sentences such as "*These ideas were seeping into his consciousness*" (MIND AS PHYSICAL SPACE), where there is a liquid quality to the ideas (individually or as a mass), there is no strong suggestion of unity among the ideas.

A final complication: In a sentence like "The embarrassment he had caused her was in John's mind all day long" it is tempting to see a manifestation of MIND AS WORLD-DEFINER. However, note that the "in his mind" qualifier is not being used adverbially to qualify a situation description but is rather used predicatively of something (the embarrassment). I therefore suggest that the sentence actually manifests MIND AS PHYSICAL SPACE, and the phrase "the embarrassment he had caused her" is being used metonymically to refer to *some idea* that John had of the embarrassment. This hypothesis is backed up by the fact that we get natural variants by referring to particular locations within the mind, for instance by using "in the back of John's mind" in place of "in John's mind." Such localization is not natural for examples like (14) — the sentence "In the back of John's mind, terrorism is getting worse every day" sounds peculiar.

2.8. *Ideas as Internal Utterances*

Sentences (15) and (16) manifest the metaphor of IDEAS AS INTERNAL UTTERANCES, as long as these sentences are not describing an out-loud speaking event. This metaphor casts a thinking event as an event of "internal speech." Internal speech is not *literally* speech.

Sentence (2) involves the same metaphor, combined with MIND PARTS AS PERSONS, since what an inner person says is an idea of the overall person's. Conversely, sentences (15) and (16) can (in many contexts) be taken to involve MIND PARTS AS PERSONS: one part of Sally is saying it to another part.

Sentences (2), (15) and (16) use a speech verb, and no speech verb seems to be barred from being used to describe an internal-speech event. For instance, recall example (17), where "whisper" is used. And a modifier like "to herself" is not even needed, if it is clear from context that a thought is being reported. But, on the other hand, the use of a speech verb is not necessary for the metaphor to be manifested. Consider:

33. "*Everyone here immediately thought, 'Wow, I'm going to have access to the world's largest market.'* "[27]
34. "*What bothered him was that he didn't want to live the rest of his life with the thought, 'There's Norm Schwarzkopf, the Butcher of Baghdad.'* "[28]

The use of quotation and speech-based mode of expression within the quotation strongly suggest that the agent's thoughts are being likened to speech.[29]

The examples above that do use a speech verb also happen to manifest MIND PARTS AS PERSONS. However, an involvement of that metaphor is not necessary, as shown by:

35. "*Sharon pulled herself out of her jeans, the words 'How could he? How could he?' jumping about her wearied brain.*"[30]
36. "*'I should never have got engaged,' I groaned inwardly.*"[31]

We should be aware of a complication and ambiguity arising with IDEAS AS INTERNAL UTTERANCES examples. It appears to be common for people to experience some of their own conscious thoughts as internal speech. Therefore, it is often reasonable to take a sentence like (15) as reporting a situation in which the agent herself would say she was using internal speech. Nevertheless, there is no compulsion to take the sentence in this way. It could be that the author of the sentence is describing a mental state merely *as if* it were one of those in which the agent herself would say she was using internal speech. A similar point is made by Cohn (1978: 76). Chafe (1994) remarks that in the use of IDEAS AS INTERNAL UTTERANCES in a past-tense self report, such as "*I said to myself, 'Time to go!'*" the original thought may not have appeared to the agent as internal speech, but the agent reconstructs it as such at a later time. There can thus be a sort of double metaphor in IDEAS AS INTERNAL UTTERANCES manifestations: one metaphorical step is in describing a thought as if it were internal speech (even though it isn't really); the other step is in the metaphorical relationship of internal speech to real speech.

Cohn (1978) also points out (e.g. 62–63, 95–98) that, in novels, manifestations of IDEAS AS INTERNAL UTTERANCES (which she calls "quoted (internal) monologue") are often more fragmentary than real dialogue, and the words used can have more flexible or idiosyncratic meanings. It may be that novelists are obeying an intuition that "internal speech" differs from real speech in important ways, much as visual images can be less distinct and more fragmentary than visual perceptions or external pictures.

In addition, even when the quoted expression is a properly formed discourse chunk, and we can assume the agent would agree that (s)he was experiencing inner speech, we should not assume that the agent's actual internal speech is identical to the quoted expression. That expression could still only be an idealization. This observation is backed up by the experimental study of Wade and Clark (1993), supporting the contention that direct quotation is in general used merely to depict *some features* of an utterance (actual or hypothetical, out-loud or mental); the use of a direct quotation to depict an utterance *verbatim* is just a special case.

In any case, virtually all manifestations of IDEAS AS INTERNAL UTTERANCES appear to indicate conscious, occurrent thoughts. This is presumably because almost all real speech is consciously produced and, when interpreted at all, consciously interpreted. However, the degree of consciousness can be modified, as in the (17) and through the use of "murmur" in:

37. "*Perhaps somewhere in the back of his brain there ran a murmur telling him that* … "[32]

Also, the consciousness can be attenuated and/or made non-central, so to speak, by the speech being ascribed to just one mind subregion as in (37), or to one inner person, as in (17) again and in:

38. "*Did part of you think, 'Yes, I'm flattered'?*"[33]

2.9. *Relationships between the Metaphors*

The above metaphors are not independent of each other, both in that there are some taxonomic relationships between them and in that some are often smoothly combined in the same sentence.

We have already noted that COGNIZING AS SEEING and COGNIZING AS SMELLING are special cases of COGNIZING AS PHYSICALLY SENSING, and that MIND AS PHYSICAL CONTAINER is a special case of MIND AS PHYSICAL SPACE. Also, IDEAS AS INTERNAL UTTERANCES and MIND PARTS AS PERSONS are special cases of MIND AS PHYSICAL SPACE. Internal utterances are "internal" to the agent's mind, and inner persons inhabit the agent's mind. Nevertheless, the particular locations of the utterances or inner persons are typically not salient or significant. Sentence (35) is an example where the location of an internal utterance is significant. Although MIND PARTS AS PERSONS examples rarely mention a specific location of an inner person, there is an implicit positional factor of great significance: the inner persons are often taken to be in verbal communication with each other, so that presumably they are assumed to be in relatively close physical proximity.

As for combinations of metaphors, we have already seen that IDEAS AS PHYSICAL OBJECTS and MIND AS PHYSICAL SPACE are often combined, as are MIND PARTS AS PERSONS and IDEAS AS INTERNAL UTTERANCES. This is extremely important, as it allows communication and other interaction between inner persons to take on the full complexity of real linguistic and social interaction. As one small illustration of the implications of this, consider (2) again. A person normally only insists that something, X, is the case when someone in the same conversation has objected to X. (This is a common-sense observation about social interaction.) Thus, in the example we can take it that some other inner person has probably objected to X. Thus, the agent is in a strongly conflicting state of mind. One must be careful to note that most of the examples of combinations of MIND PARTS AS PERSONS and IDEAS AS INTERNAL UTTERANCES given above only cast the internal utterances as occurring *between* inner persons. Distinctly different is the combination in (38), where internal speech occurs *within* an inner person. The first type of combination is a "parallel" mixing of metaphors: two different metaphors are used to get at different aspects of an overall situation that are, so to speak, side by side: the division into inner persons, and their interaction. The second type of combination is "serial" mixing (or "chaining") in that a source-domain item of one metaphor (namely an inner person) is itself described metaphorically in terms of the other metaphor (IDEAS AS INTERNAL UTTERANCES).

Manifestations of MIND AS PHYSICAL SPACE are often enriched by COGNIZING AS SEEING. Some examples of this are:

39. "*It seems to me in some dark recess of my mind that ...* "[34]
40. "*All animation departed from her face, and it was as if those eyes turned inwards to contemplate the workings of her mind.*"[35]
41. "*Cheryl's name and a kind of vague picture of her had been in his mind.*"[36]
42. "*All these things flowed through U Po Kyin's mind swiftly and for the most part in pictures.*"[37]
43. "*His youthful prank of being a policeman had faded from his mind.*"[38]
44. "*Blotting out the whole business from his mind, he had avoided everything that might be associated with it.*"[39]
45. "*Now his mind had curiously blanked, emptied but for the presence in it of a small black Scottie dog.*"[40]

From such examples it is clear that the mental "space" need not just be a repository for ideas, but can also be a space within which visual perception is an important consideration. If something that is/was within the mind cannot (now) be seen, then it is not available to the agent's consciousness. Something which is seen is in the focus of consciousness. Something in a "dark" part of the mind might be visible, but only dimly, and is therefore not prominent to consciousness. Note how "dark" intensifies the effect of "recess" in (39). An idea that is cast as a large object is thereby that much more visible. A "flash" of insight is strongly present to consciousness because sudden and bright.

It is also plausible that inner "vision" is part of the reason why phrases like "back of the mind" and "front of the mind" mean what they do. Why should things in the back be less prominent than things in the front? One reason is no doubt that in many physical contexts, items in/at the front of something (an person's body, a car, a stage, a line of people) are more prominent or interactive in some sense.[41] And in some of these cases, at least — a stage, for instance — the extra prominence comes from higher visibility. So "front" and "back" may have general associations with higher and lower importance/prominence and often also with higher and lower visibility.

But there may be an additional visual effect that is more special to the PHYSICAL SPACE metaphor for mind. If we assume that a person often conceives of his or her "inner self" as a homunculus sitting within the space of the mind (Banks 1993), looking frontwards out into the world as well as looking at the contents of the mind, then mental things the self-homunculus can easily "see" are those towards the front, and things he/she/it cannot so easily see (or does not see without special effort) are towards the back. The suggestion of a *forwards*-looking homunculus also accounts for why the front-back axis of the mind is aligned with that of the head, as evidenced by "at back of X's mind" having a similar effect to "at the back of X's head." (But the front of the head is not used in mental metaphor examples, to my knowledge.)

We also get an interesting illumination of the import of the phrase "back of the mind." Sometimes this seems to indicate a lack of consciousness, whereas at other times it indicates only some fringe type of consciousness. Consideration of the seeing-homunculus suggests that what "back of the mind" conveys is that the thing is for most of the time not seen by the homunculus, but that the homunculus may nevertheless remember its existence and be able to turn round to see it. That is, "back of the mind" arguably conveys *potential and/or intermittent consciousness*: a low level of consciousness of the thing most of the time but the possibility of fuller consciousness of it when the agent is not concentrating on more salient things.

Finally, IDEAS AS EXTERNAL ENTITIES and MIND AS PHYSICAL SPACE are not altogether easy to distinguish. It is not clear whether the mind is always viewed as a part of the space within the head, or sometimes the other way round. It might seem natural for the mind-space to be viewed as being within the head. However, some people say that, especially when they close their eyes, the mind-space they are aware of is *larger* than the head. If this is allowed for, one could say that in the case both of MIND AS PHYSICAL SPACE, in some manifestations, and of IDEAS AS EXTERNAL ENTITIES, some part of the mind forms a space that is outside the person. The real distinction is that in IDEAS AS EXTERNAL ENTITIES the whole agent, body and all, is conceived of as being within the idea-populated external space, and no space within the person's head is taken into account, whereas in MIND AS PHYSICAL SPACE the space does not contain the person's body and at least overlaps the interior of the head. Another complication is that sometimes MIND AS PHYSICAL SPACE seems to involve a self-homunculus inside the mind-space. To the extent that this entity is person-like, it is then in much the position of a real person who is being thought of by means of IDEAS AS EXTERNAL ENTITIES.

2.10. *Other Metaphors*

The above example sentences and metaphors merely scratch the surface of the possibilities for metaphorical description of mental states, events, and processes. Not only can each metaphor be manifested in an indefinitely large variety of ways, using different extents and types of coloration from their source domains (PHYSICAL SPACE, UTTERANCES, etc.), but also there are other metaphors. For more examples and analysis of metaphors of mind, see for example Asch (1958), Belleza (1992), Casadei (1993), Cooke and Bartha (1992), Fesmire (1994), Gallup and Cameron (1992), Gentner and Grudin (1985), Gibbs and O'Brien (1990), Hoffman, Cochran and Nead (1990), Jakel (1993), Johnson (1987), Katz *et al.* (1988), Lakoff (1993b), Lakoff, Espenson and Schwartz (1991), Larsen (1987), Leary (1990), Lehrer (1990), Pollio (1990), Richards (1989), Roediger (1980), Smith (1985), Sweetser (1987), (1990), Tomlinson (1986), and Weitzenfeld *et al.* (1992).

3. Relevance to Consciousness: Part A

The discussion of various metaphors in section 2 well illustrates the fact that manifestations of metaphors of mind often help to indicate or qualify the extent, type, or level of the agents' consciousness. Now, the metaphors we have been discussing are used by people who typically are not cognitive psychologists. Therefore, there is no a priori reason to think that a metaphorical description captures some scientific truth about the mind (or about consciousness in particular). For instance, there is no reason to think that mind "parts" correspond to modules in some scientific, information-processing account of the mind.

However, different ways of metaphorically describing mental states point to different real types of mental state (including different types of consciousness), even though the metaphorical descriptions taken individually may not help much with the construction of scientific accounts of those types. To the extent that people have found the metaphorical descriptions useful in ordinary life it is worth considering what scientific reality about consciousness etc. could underlie them.

To take the mind-parts case again, it does make a difference to describe *one part of* Veronica as believing that the recipe was wrong, rather than just saying that Veronica believed it was wrong. In particular, different parts can have differing beliefs. Questions that then arise include:

- What scientific sense can be made of the notion that one part of someone believes something but another part lacks that belief, or even believes something inconsistent with it?
- In particular, what does this say about the nature of consciousness, when both parts are conscious?
- Is there, in ordinary, healthy people, a common state where either the whole mind or the conscious mind is well modeled as including components that are about as independent from each other as different people are? In other words, just how accurate or inaccurate is the MIND PARTS AS PERSONS account?
- If MIND PARTS AS PERSONS does reflect different mental components postulated by some scientific theory, do those components arise dynamically (so that the mind could have different components at different times) or do they exist over the long term? This question is obviously affected by whether the mind "parts" are special and long-lived, as in (18), or non-special and possibly short-lived, as in (17) and (19).
- Again, if the metaphor reflects different scientifically-postulated components of the mind, how well is communication between them modeled by natural language communication (as revealed for instance by the word "insisting" in sentence 2 or "whispering" in sentence 17)?

This last question is clearly similar to questions we could ask in connection with IDEAS AS INTERNAL UTTERANCES. For example:

- How well does IDEAS AS INTERNAL UTTERANCES reflect one mode of conscious thought? Are conscious thoughts in that mode really as well structured even as spoken language (let alone written language)?

For any given broad metaphor, such as MIND AS PHYSICAL SPACE, there is a wealth of possible manifestations that draw on different aspects of the "source" domain (PHYSICAL SPACE in this case). For instance, one can talk about ideas being "at the back" of a mind, "in the recesses" of the mind, "to one side," "hidden away," "buried" and so forth. Although these particular terms have a rough similarity of effect — they all suggest that the ideas in question are not playing a central, active role in the person's thoughts or behavior — we should at least entertain the possibility that they allude to importantly different mental states. For instance, perhaps things that are at the back of a mind are more liable to become central than ideas that are in the recesses. Thus, metaphorical language about the mind could reveal fine differences between mental states (including conscious or partly conscious ones) that would be crucial to account for in any good theory of consciousness.

Similarly, what reality underlies the difference between, say, "blotting [some idea] out" (a manifestation of COGNIZING AS SEEING) and "burying [some idea] in a corner of [some] mind"? Both stop the idea being consciously attended to. Or is it a mistake to think that either phrase means any more than "stopping [some idea] being consciously attended to permanently or for a long period"? Such questions can be multiplied indefinitely for each individual metaphor.

4. Relevance to Consciousness, B: Metaphorical Self-Description

Sentences (1) to (16) are all in the third person, but first person versions are also common. Thus, it is common to hear people say things like "Part of me wanted to stay at home," "I thought, 'Shall I stay at home?' " or "I'd pushed the idea to the back of my mind."[42] The claim I wish to pursue is that the first-person case may provide valuable insights or avenues of research into the nature of consciousness, over and above those I suggested in the third-person case.

First, the sheer fact that people use metaphors such as the ones above to think about their own mental states, including conscious ones, may tell us something about the nature of the mind's own capabilities for conscious self-inspection. If it is true that the typical person X generally uses a metaphor of A as B because X has more effective or richer understanding of B than of A, then, when A is X's own mind and B is, say, PHYSICAL SPACE, we are confronted with the possibility that X is more at home with thinking about PHYSICAL SPACE than about thinking directly about his own mind without metaphorical intervention. This is an interesting proposition in view of the high degree of intimacy people might a priori be thought to have with their own minds.

Secondly, the particular metaphors of mind used for self-description may give us insight into the detailed nature of the self-inspection capability. What is it about this self-inspection capability that leads someone to cast her mental state in terms of mind "parts," or vision, or "recesses"? Possible inaccuracies in the metaphors that rest on such notions may be especially revealing here. Suppose the mind in reality is *not* very accurately describable as having (temporarily) a number of "parts" that are about as independent of each other as real people are. (Recall the list of questions at the start of section 3.) Then the use of MIND PARTS AS PERSONS in self-description could betray a limitation in the self-inspection process: perhaps the metaphor is the closest that the self-inspection capability is able to get to the realities.

Notice the difference between the point being made here and one made above. Above, I claimed in effect that metaphors of mind may be *accurate* enough as tools for third-person mental description for them to provide useful pointers towards the nature of the mind. The point I am now making is that scientific investigations aimed at exposing *inaccuracies* of first-person uses of the metaphors may provide insight. Of course, the third-person point does carry over to the first-person case as well, so that accuracies in first-person metaphorical descriptions could still reveal things, or guide research, about the mind.

A complicating factor is whether the self-inspection we have been discussing is conscious. When someone says something about himself verbally, using a metaphor, it is a good guess that he is thinking about himself *consciously* in terms of that metaphor. But there could be unconscious usage of the same metaphor during unconscious self-inspection episodes (that are not allied to the making of verbal statements).[43] Now, the descriptive tools available to unconscious thought may be different to those available to conscious thought. Therefore, any evidence about self-inspection capabilities to be drawn from verbal self-description by people is of less clear relevance to unconscious self-inspection than to conscious self-inspection.

In the previous section we commented on such things as the difference between something being in the back of a mind and being in the recesses, or between blotting an idea out and burying it in a corner. We queried whether there is any real difference within these contrasts or whether instead we just have alternative modes of expression for the same state of affairs. Now consider what happens in the first-person case. If an agent thinks in terms of, say, COGNIZING AS SEEING and MIND AS PHYSICAL SPACE to *think* consciously about his own mind, rather than just to describe it to other agents, then there is a case for saying that those metaphors are partially constitutive of his consciousness at that time. Therefore, if at one moment the agent perceives himself as being unable to see something in his mind clearly because it is obscured by something else, and at another as being unable to see something in his mind clearly because it is in a dark corner, then that sheer fact creates a difference between his states of mind at those two moments. To put the point more strongly, to the extent that visual images are analogous to ordinary visual perceptions, subtle differences between different conditions of ordinary vision can have analogues in imagery. Equally, to the extent that internal speech is similar to external speech, subtle differences between external utterances can have analogues in internal speech.[44]

5. The Phenomenal Quality of Consciousness

I now turn to an issue that gets closer to the central problem of consciousness, namely the *feel* or *phenomenal quality* of being conscious. This feel may include ordinary perceptual qualia, but, more importantly for this section, can include the phenomenal quality of the "perception" of one's own mental state.[45] I contrast the matter of the feeling of consciousness to other aspects of consciousness — such as the sheer self-inspection quality of consciousness — as does Goldman (1993). After all, many artificial intelligence computer programs that are presumably not conscious have self-inspection aspects. See, e.g., Lenat (1983). There is no mystery to self-inspection in itself.

Now, it is reasonable to suppose that:

a. The IDEAS AS INTERNAL UTTERANCES metaphor is commonly used in mental-state description because conscious thinking often *feels* like making and/or hearing natural language utterances.

I contrast this proposition with the proposition that:

b. people use that metaphor because they have worked out or learned about a structural analogy between conscious thinking and utterance-making/hearing.

By a structural analogy I mean a bundle of correspondences between the parts/aspects of one subject-matter and the parts/aspects of another. (This is the main type of analogy studied in artificial intelligence and cognitive psychology: see Hall 1989 and Vosniadou and Ortony 1989.) Propositions (a) and (b) are not inconsistent with each other: conscious thinking could both feel largely like utterance-making/hearing *and* bear a structural analogy to it. But, if there is any truth to (a) then it is an important aspect of the *feel* that consciousness has.

Claim (a) can be broadened to a conjecture about metaphors other than IDEAS AS INTERNAL UTTERANCES. For instance, perhaps we are prone to using MIND PARTS AS PERSONS in self-description because the mental state thereby described *feels* somewhat like having several inner persons (or at least sub-minds akin to whole minds) inside oneself. Equally, perhaps when one talks of the "side," "back," "recesses," etc. of one's mind one really *feels* (or internally *sees)* that one's mind is a physical region and that the ideas in question are at a particular position within it. Perhaps one *feels* that ideas are physically interacting when one is drawing conclusions from them. Analogous statements could be made about other metaphors of mind, and possibly about all of them, though we do not need to claim universality here. Vision-based metaphors are a clear case, because thinking that is based on visual imagery feels much like seeing. These points are related to observations in Asch (1958), amounting to a claim that metaphorical talk about psychological states or attributes is partly grounded in feelings.[46]

Relatedly, there is a case for saying that the use of sentences like:

47. "*Experts feel that the economy is slowing down*"

hints at our recognition of thought as having a *feeling* for us, not just being a (somewhat-)logical structure or process. The type of feeling here would presumably be more akin to emotional feeling than ordinary physical sensation such as sight, smell, indigestion, or physical pain. However, the line is difficult to draw, and we also need to account for sentences such as:

48. "*Experts feel in their guts that the economy is slowing down.*"

The exact way metaphor is involved in sentences such as (47) and (48) is unclear. If thought literally includes feelings of one sort or another, then (47) is arguably literal and (48) manifests an EMOTIONAL FEELING AS PHYSICAL SENSATION metaphor (and, consequently, the COGNIZING AS PHYSICAL SENSATION metaphor as well). In addition, Goldman (1993) discusses the possibility of different mental states (believing, hoping, etc.) having a different feeling from each other. It is surely reasonable to say that this is part of the common-sense view of mental states. This could be true even if Nelkin (1994) is right in repudiating difference of feel as an *essential* difference between mental states.

The conjectures in this section are friendly to Johnson's (1987, 1991) claim that we understand abstract matters, including mental states and processes, through metaphorical projection from bodily experience. However, his account appears to rest on structural analogy as in (b) above rather than on direct feeling as in (a).

6. Conclusion

The purpose of this article has been to advocate the detailed study of metaphors of mind in the study of consciousness (and indeed mind in general). Such metaphors are used by us within our minds when we think about minds (including our own), and so are part of the object of study. Apart from this, any loose scientific accuracy in the metaphors is worth uncovering, and the undoubted inaccuracies they involve present an interesting research issue in themselves. Additionally, metaphors of mind can give us clues to the nature of the phenomenal quality of consciousness. This does not solve the problem of consciousness — the problem of why and how there are any feelings in the universe at all — but at least it leads us into the right ballpark of discussion.

Notes

1. The work was supported in part by grants IRI-9101354 and CDA-8914670 from the National Science Foundation.
2. The topic of common-sense views of the mind is closely allied to the topic of "folk psychology." However, I will not be addressing the particular concerns that have gyrated around the latter term (see, e.g., Churchland 1989; Goldman 1993; Ramsey, Stich and Garon 1991; Stich 1983), so I will continue to use the former.
3. Cooper (1986) strongly objects to this characterization, largely because the notion of "seeing as" is itself unclear and metaphorical.
4. I have amassed about a thousand discourse chunks exemplifying metaphors of mind. They are in a publically accessible database. Please contact the author if interested.
5. Mona Simpson, *The Lost Father*. New York: Vintage Books, 1992, p. 265.
6. Ruth Rendell, *The Bridesmaid*. London: Hutchinson, 1989, p.172, with minor adaptations.
7. Adapted from a lecture by Ann Dale, Wycombe Abbey School, England, 10 May 1995.
8. Ruth Rendell, *ibid.*, p.264, with minor adaptations.
9. Ruth Rendell, *ibid.*, p.255, with minor adaptations.
10. Ruth Rendell, *ibid.*, p.162.
11. Ruth Rendell, *ibid.*, p.142.
12. Mona Simpson, *ibid.*, p.432.

13. Sue Grafton, *I is for Innocent*. London: Penguin Books, 1993, p.182.
14. Jane Rogers, *Mr Wroe's Virgins*. London: Faber and Faber, 1991, p.91.
15. Latter example is from *My Story* magazine (Editions Press Ltd, Gibraltar), May 1995, p.35.
16. Janna Trollope, *The Rector's Wife*. UK: Corgi Books (Black Swan ed.), 1993, p.235.
17. It may be tempting to regard "see that" as an example of a frozen manifestation of the metaphor. However, the phrase can be productively and systematically varied in ways that suggest the metaphor is live even here. Consider, for example: "he could only see in a blurred way that…" In any case, the major points made in this paper are tolerant of the possibility that some phrases are only frozen manifestations of the metaphors of interest.
18. *Newsweek,* 19 November 1990, p.25.
19. *Newsweek*, 6 January 1992, p.30–32.
20. *Cosmopolitan*, March 1994, p.192.
21. See also Richards (1989) for discussion of a range of physically-sensing metaphors, and other metaphors. However, Richards is reluctant to accord metaphorical status to some sentences I would say are metaphorical.
22. This metaphor supersedes the IDEAS AS MODELS metaphor that I have discussed elsewhere.
23. George and Weedon Grossmith, *Diary of a Nobody*. London: Penguin, 1965, p.212, with minor adaptations.
24. *My Story* magazine (Editions Press Ltd, Gibraltar), May 1995, p.46.
25. *Newsweek*, 19 November 1990, p.59. Whether or not the memory is normally regarded as part of the mind, I allow this type of example under the heading of MIND AS WORLD-DEFINER, taking the word "MIND" broadly.
26. According to Jackendoff (1983), belief reports are cognitively similar to statements about states of affairs depicted in stories and pictures. Here the belief reports can use "In X's mind" or "X believes that," among other devices. My claim is that mental state reports using "in X's mind" are cognitively much *more* analogous to story or picture reports than are "X believes that" statements.
27. *Newsweek*, 20 May 1991, p.42–45, with minor adaptations.
28. *Newsweek*, 11 March 1991, p.32–34.
29. It is common, especially in fiction, for speech-like thoughts to be portrayed in much more implicit ways, without even the use of a mental verb let alone a speech verb. See, for instance, Cohn (1978) and Wiebe (1994). However, the more explicit forms are enough for the present paper.
30. *My Story* magazine, May 1995, p.17.
31. *My Story* magazine, May 1995, p.8.
32. R. Barnard, *Little Victims*. UK: Corgi Books, 1993, p.129, with minor adaptations.

33. Program presenter Sue Lawley to interviewee in Desert Island Discs program, Radio 4, England, 12 May 1995.
34. In lecture by John Locke (Harvard University) at Psychology Department, Sheffield University, England, 27 April 1995.
35. Ruth Rendell, *ibid.*, p.162, with minor adaptations.
36. Ruth Rendell, *ibid.*, p.55, with minor adaptations.
37. From Orwell, *Burmese Days*, p.7, quoted by M. Jahn, 1992, "Contextualizing represented speech and thought." *J. Pragmatics* 17: 347–367.
38. G.K. Chesterton, *The Man Who Was Thursday: A Nightmare*. London: Penguin Books, 1986, p.66. In this example and the following two, the item apparently stated to be in someone's mind (the prank) is actually a non-mental object. However, we can take the reference to the prank to be a metonymic reference to some idea or image of the prank. Similarly for the "whole business" and the dog in the following two sentences.
39. Ruth Rendell, *ibid.*, p.181.
40. Ruth Rendell, *ibid.*, p.250.
41. This accords with the discussion of the notions of front and back in Allen (1995).
42. It is plausible that the third-person use of metaphors of mind is parasitic on their first-person use. A number of psychologists have favored the notion that we ascribe mental states to others by analogy to our own mental states: see, e.g., Beckwith (1991) and Frye (1991). However, the truth of this claim is not essential for the present paper, though certainly friendly to it.
43. By unconscious self-inspection I mean merely a situation in which an agent X has unconscious cognitive states/episodes (thoughts, reasonings, etc.) concerning the mind of X or the mental states/processes of X. This does not necessarily involve cognitions about the conscious "self" of X. However, it might do so; there would be no contradiction, any more than there is in supposing that a person (or computer program) X could unconsciously think about the conscious self of some *different* person Y.
44. Of course, anything spoken out-loud can also be inwardly imagined or rehearsed. This is not the type of internal speech I wish to allude to. I am focusing on internal speech that arises as a natural part of thinking about some issue.
45. Nelkin (1989) may be right in saying that neither believing nor introspection of believing *essentially* involves any feeling. But that does not mean they cannot do so.
46. Of course, it is conceivable that a person's use of some particular metaphor of mind arises solely from immersion in a language community that uses that metaphor; moreover, the person's learned usage might conceivably then *lead* to him or her having corresponding feelings.

References

Allan, K. 1995. The anthropocentricity of the English word(s) *back*. *Cognitive Linguistics* 6(1), 11–31.

Asch, S.E. 1958. The metaphor: A psychological inquiry. In *Person Perception and Interpersonal Behavior*, R. Tagiuri and L. Petrullo (eds), 86–94. Stanford, CA: Stanford University Press.

Banks, W.P. 1993. Problems in the scientific pursuit of consciousness. *Consciousness and Cognition* 2(4), 255–263.

Barnden, J.A. 1989a. Towards a paradigm shift in belief representation methodology. *J. Experimental and Theoretical Artificial Intelligence* 2, 133–161.

Barnden, J.A. 1989b. Belief, metaphorically speaking. In *Procs. 1st Intl. Conf. on Principles of Knowledge Representation and Reasoning*, 21–32. San Mateo, CA: Morgan Kaufmann.

Barnden, J.A. 1992. Belief in metaphor: Taking commonsense psychology seriously. *Computational Intelligence* 8, 520–552.

Barnden, J.A. 1995. Simulative reasoning, common-sense psychology and artificial intelligence. In *Mental Simulation: Evaluations and Applications*, M. Davies and T. Stone (eds), Oxford, 247–273. UK: Blackwell.

Barnden, J.A., Helmreich, S., Iverson, E. and Stein, G.C. 1994a. An integrated implementation of simulative, uncertain and metaphorical reasoning about mental states. In *Principles of Knowledge Representation and Reasoning: Proceedings of the Fourth International Conference*, J. Doyle, E. Sandewall and P. Torasso (eds), 27–38. San Mateo, CA: Morgan Kaufmann.

Barnden, J.A., Helmreich, S., Iverson, E. and Stein, G.C. 1994b. Combining simulative and metaphor-based reasoning about beliefs. In *Procs. 16th Annual Conference of the Cognitive Science Society*. Hillsdale, NJ: Lawrence Erlbaum.

Barnden, J.A., Helmreich, S., Iverson, E. and Stein, G.C. 1995. Artificial intelligence and metaphors of mind: Within-vehicle reasoning and its benefits. To appear in *Metaphor and Symbolic Activity*. Also appeared as *Memoranda in Computer and Cognitive Science*, No. MCCS-95–282, Computing Research Laboratory, New Mexico State University, Las Cruces, NM.

Beckwith, R.T. 1991. The language of emotion, the emotions, and nominalist bootstrapping. In *Children's Theories of Mind: Mental States and Social Understanding*, D. Frye and C. Moore (eds), 77–95. Hillsdale, NJ: Lawrence Erlbaum.

Belleza, F.S. 1992. The mind's eye in expert memorizers' descriptions of remembering. *Metaphor and Symbolic Activity* 7, 119–133.

Bruner, J. and Feldman, C.F. 1990. Metaphors of consciousness and cognition in the history of psychology. In *Metaphors in the History of Psychology*, D.E. Leary (ed), 230–238. New York: Cambridge University Press.

Casadei, F. 1993. The canonical place: And implicit (space) theory in Italian idioms. In *Philosophy and the Cognitive Sciences*, R. Casati and G. White (eds). Kirchberg am Wechsel, Austria: Austrian Ludwig Wittgenstein Society.

Chafe, W. 1994. *Discourse, Consciousness, and Time: The Flow and Displacement of Conscious Experience in Speaking and Writing.* Chicago, IL: The University of Chicago Press.

Churchland, P.M. 1989. Folk psychology and the explanation of human behavior. In *The Neurocomputational Perspective*, P.M. Churchland (ed). Cambridge, MA: MIT Press.

Cohn, D. 1978. *Transparent Minds: Narrative Modes for Presenting Consciousness in Fiction.* Princeton, NJ: Princeton University Press.

Cooke, N.J. and Bartha, M.C. 1992. An empirical investigation of psychological metaphor. *Metaphor and Symbolic Activity* 7, 215–235.

Cooper, D.E. 1986. *Metaphor*. Oxford: Basil Blackwell.

Fesmire, S.A. 1994. Aerating the mind: The metaphor of mental functioning as bodily functioning. *Metaphor and Symbolic Activity* 9, 31–44.

Frye, D. 1991. The origins of intention in infancy. In *Children's Theories of Mind: Mental States and Social Understanding*, D. Frye and C. Moore (eds). Hillsdale, NJ: Lawrence Erlbaum.

Gallup, G.G., Jr. and Cameron, P.A. 1992. Modality specific metaphors: Is our mental machinery 'colored' by a visual bias? *Metaphors and Symbolic Activity* 7, 93–101.

Gentner, D. and Grudin, R. 1985. The evolution of mental metaphors in psychology: A 90–year perspective. *American Psychologist* 40, 181–192.

Gibbs, R.W., Jr. and O'Brien, J.E. 1990. Idioms and mental imagery: The metaphorical motivation for idiomatic meaning. *Cognition* 36, 35–68.

Goldman, A.I. 1993. Consciousness, folk psychology, and cognitive science. *Consciousness and Cognition* 2, 364–382.

Gopnik, A. 1993. Psychopsychology. *Consciousness and Cognition* 2(4), 264–280.

Hall, R.P. 1989. Computational approaches to analogical reasoning: A comparative analysis. *Artificial Intelligence* 39, 39–120.

Hoffman, R.R., Cochran, E.L. and Nead, J.M. 1990. Cognitive metaphors in experimental psychology. In *Metaphors in the History of Psychology*, D.E. Leary (ed), 173–229. New York: Cambridge University Press.

Jackendoff, R. 1983. *Semantics and Cognition.* 4th printing, 1988. Cambridge, MA: MIT Press.

Jakel, O. 1993. The metaphorical concept of mind: Mental activity is manipulation. Paper No. 333, General and Theoretical Papers, Series A, Linguistic Agency, University of Duisburg, D-4100 Duisburg, Germany.

Jaynes, J. 1982. *The Origin of Consciousness in the Breakdown of the Bicameral Mind.* Paperback ed. Boston, MA: Houghton Mifflin.

Johnson, M. 1987. *The Body in the Mind.* Chicago, IL: The Chicago University Press.

Johnson, M. 1991. Knowing through the body. *Philosophical Psychology* 4(1), 3–18.

Katz, A.N., Paivio, A., Marschark, M. and J.M. Clark. 1988. Norms for 204 literary and 260 nonliterary metaphors on 10 psychological dimensions. *Metaphor and Symbolic Activity* 3, 191–214.

Kennedy, J.M., Green, C.D. and Vervaeke, J. 1993. Metaphoric thought and devices in pictures. *Metaphor and Symbolic Activity* 8(3), 243–255.

Lakoff, G. 1993a. How cognitive science changes philosophy II: The neurocognitive self. Paper presented at 16th International Wittgenstein Symposium, Kirchberg am Wechsel, Austria.

Lakoff, G. 1993b. The contemporary theory of metaphor. In *Metaphor and Thought*, 2nd edition, A. Ortony (ed). New York: Cambridge University Press.

Lakoff, G., Espenson, J. and Schwartz, A. 1991. Master metaphor list. Draft 2nd Edition. Berkeley, CA: Cognitive Linguistics Group, University of California at Berkeley.

Larsen, S.F. 1987. Remembering and the archaeology metaphor. *Metaphor and Symbolic Activity* 2, 187–199.

Leary, D.E. (ed). 1990. *Metaphors in the History of Psychology*. New York: Cambridge University Press.

Lehrer, A. 1990. Polysemy, conventionality, and the structure of the lexicon. *Cognitive Linguistics* 1, 207–246.

Lenat, D.B. 1983. EURISKO: A program that learns new heuristics and domain concepts. The nature of heuristics III: Program design and results. *Artificial Intelligence* 21, 61–98.

Nelkin, N. 1989. Propositional attitudes and consciousness. *Philosophy and Phenomenological Research* 49(3), 413–430.

Nelkin, N. 1994. Patterns. *Mind and Language* 9(1), 56–87.

Pollio, H.R. 1990. The stream of consciousness since James. In *Reflections on The Principles of Psychology: William James after a Century*, M.G. Johnson and T.B. Henley (eds), 271–294. Hillsdale, NJ: Lawrence Erlbaum.

Ramsey, W., Stich, S.P., and Garon, J. 1991. Connectionism, eliminativism, and the future of folk psychology. In *Philosophy and Connectionist Theory*, W. Ramsey, S.P. Stich and D.E. Rumelhart (eds), 199–228. Hillsdale, NJ: Lawrence Erlbaum.

Richards, G. 1989. *On Psychological Language and the Physiomorphic Basis of Human Nature*. London: Routledge.

Roediger, H.L., III. 1980. Memory metaphors in cognitive psychology. *Memory and Cognition* 8(3), 231–246.

Rorty, R. 1980. *Philosophy and the Mirror of Nature*. Oxford: Blackwell and Princeton, NJ: Princeton University Press.

Shoemaker, S. 1994. Self knowledge and 'inner sense.' *Philosophy and Phenomenological Research* 54(2), 249–314.

Smith, M.B. 1985. The metaphorical basis of selfhood. In *Culture and Self: Asian and Western Perspectives*, A.J. Marsella, G. DeVos and F.L.K. Hsu (eds), 56–88. London: Tavistock.

Stich, S. 1983. *From Folk Psychology to Cognitive Science: The Case against Belief.* Cambridge, MA: MIT Press.

Sweetser, E.E. 1987. Metaphorical models of thought and speech: A comparison of historical directions and metaphorical mappings in the two domains. In *Procs. 13th Annual Meeting of the Berkeley Linguistics Society*, J. Aske, N. Beery, L. Michaelis and H. Filip (eds), 446–459. Berkeley, CA: Berkeley Linguistics Society.

Sweetser, E.E. 1990. *From Etymology to Pragmatics: Metaphorical and Cultural Aspects of Semantic Structure.* Cambridge, UK: Cambridge University Press.

Tomlinson, B. 1986. Cooking, mining, gardening, hunting: Metaphorical stories writers tell about their composing processes. *Metaphor and Symbolic Activity* 1, 57–79.

Vosniadou, S. and Ortony, A. (eds). 1989. *Similarity and Analogical Reasoning.* Cambridge, UK: Cambridge University Press.

Wade, E. and Clark, H.H. 1993. Reproduction and demonstration in quotations. *J. Memory and Language* 32, 805–819.

Weitzenfeld, J., Reidl, T., Chubb, C. and Freeman, J. 1992. The use of cross-domain language by expert software developers. *Metaphor and Symbolic Activity* 7, 185–195.

Wiebe, J.M. 1994. Tracking point of view in narrative. *Computational Linguistics* 22, 233–287.

Some Consequences of Current Scientific Treatments of Consciousness and Selfhood

Seán Ó Nualláin
Dublin City University and NRC, Canada

1. Theories of Consciousness Revisited

We have seen that among the consciousness theories currently ricocheting around the sciences of mind are computational theories (Dennett 1992; Johnson-Laird 1988), cognitive theories (Baars 1988; Jackendoff 1987), analytic philosophers' approaches (Searle 1992; Flanagan 1992) and finally the neurobiological speculations of two Notel laureates (Edelman 1992; Crick 1994). They are worth another brief glance for a number of reasons, apart from the intrinsic interest of the subject. First of all, as we have seen, they seem bewilderingly diverse at first glance in their approaches and conclusions. Secondly, they converge on a notion of self which has massive and mainly negative political consequences. We may note, also, that they are almost all monistic — sometimes aggressively so — despite much evidence that monism/materialism lacks any knockout scientific evidence (Margenau 1984; Eccles 1987; Libet 1985) and/or is quite as intellectually incoherent as dualism (Searle 1992).

Dennett's "Consciousness Explained" may be the most ambitious title since Sartre's "Being and Nothingness." Unlike Sartre, Dennett owns up: his is not an explanation, but a new set of metaphors. To be precise, consciousncss is a pandemonium of streams of text, some of which streams get promoted to further functional roles by the activity of a virtual machine. Self is similarly fragmented, but there exist centers of "narrative gravity" which are as close to coherent selves as one can get. Despite the shock induced by its initial statement, that this view is by no means as indefensible as it might initially seem. However, perhaps due to the claim implicit in the title of the book, Dennett has been attacked, with increasing violence as one goes left to right, by Korb (1993), Searle, Crick (*opera cit.*) and Edelman (1993), the latter pair of whom doubt that he has even addressed the issue. We have seen one glaring omission in Dennett which has, perhaps charitably, been ignored by his other critics: the virtual machine is playing a homuncular role, and no plausible computational structure is proposed for it.

Such is also the case for Johnson-Laird (1983, 1988), whose consciousness is that particular class of virtual machines called operating systems. It will come as no surprise to those who know his work that Johnson-Laird roots selfhood, Consciousness and free will in the notion of "mental models." Yet his theory is inadequate, and falls in the face of a Gödelian argument. (The details of the latter are outside our current scope. Essentially, it requires the impossible; that a system should simultaneously model the world and itself, including its own fully updated model of the world). Of more importance to us here is its equating the self with Unix, VM and — who-knows? — MS-DOS 2.0.

Cognitive theory tends to ground its description of consciousness in terms of the distinctions between objects in the internal and external world which comprise it. The Cognitive approach has been termed a "paradigm shift" by Roger Sperry. Bernard Baars' (1988) which we've looked at in terms of neuroscience already, is the best worked-out such theory. He adduces a set of requirements for conscious experiences, i.e., that they be internally consistent, accessible by a self-system, involve perceptual/imaginal events of some duration, be globally broadcast through the central nervous system… but how much of everyday "conscious" mental life can pass these tests? In fact, Baars may, perhaps accidentally, have stumbled on a crucial notion: Consciousness is best thought of not as a continuous process, but as an occasional achievement in mental life. With that, it's apposite to look at his view on self, which frankly is unlikely to inspire a new generation of Romantic poets; self is identical with some of the more profound levels of the "context hierarchy," a hierarchy which structures our quotidian intercourse with the world. Self is nothing *an sich*. Again, he may be nearer the mark than immediately seems the case; were he explicitly to distinguish the "cognitive self" from other manifestations of selfhood, he would have a palpable hit.

Ray Jackendoff (1987) completes Baars' mission of describing consciousness in terms of the distinctions which he adjudges comprise it, and his analysis of linguistic, musical and spatial cognition in these terms is wonderfully precise. However, he holds an inessentialist position on consciousness: it has no causal role, being merely an early warning system. For Jackendoff, as for his mentor Lashley (*op. cit.*, 277), no causal mental process ever is conscious (though he is now willing to grant an informational role to attention.) This inessentialist position is counter-intuitive enough to require a great deal more evidence than

Jackendoff supplies. His notion of self is minimalism itself; in these terms, it is an occasional fleeting visitor in the field of consciousness.

The philosophical study of consciousness was once the preserve of phenomenologists, and perhaps the best index of its newly-achieved respectability is the extended argument offered by the American philosophers Searle and Flanagan that consciousness soon will yield its secrets to science. Searle (correctly, in my view) argues that a real Science of Mind must include consciousness and takes Cognitive Science to task for that reason; we have now seen how a division of tasks between consciousness and CS might be done. Flanagan (1992: 219) agrees with Searle that a theory of consciousness is in fact a theory of mind. He excoriates views like mine and those of David Chalmers (1996). He argues consciousness *will* soon be explained in Darwinian terms; its content is the cognitive process with most degrees of freedom. It is above all a stream (175), and is causal in allowing us focus on the present and veto actions in the manner Libet (1985) so brilliantly elucidates. It is coextensive with the classes of mental states which involve awareness (Flanagan 1992: 31). Fine, but that does not explain consciousness, nor does Flanagan's argument, for all the violence of its epithets, attack the consequences arising from the crucial issue: Consciousness is ontologically of a different nature to anything science has so far approached. Charles Taylor (1989: 33–34) makes the same point about selfhood. It behooves us to be very careful.

Especially so when one is attacking the topic with the armory of the hard sciences. Penrose's (1989) "The Emperor's New Mind" is famously about everything under and over the sun, but its most substantial line of argument is on consciousness. AI will not succeed precisely because it is based on algorithms: Consciousness is non-algorithmic. In fact, it is based on physical processes corresponding to wave-function breakdown, and the very laying down of neural pathways as a result of conscious experience is best handled by a non-recursive formalism like tiling theory. Penrose, as we saw already, finds solace in the work of Hameroff (*op. cit.*) which finds a biological basis for consciousness in quantum-mechanical events, in particular the persistence of coherent states in the cytoskeleton of neurons. All well and good, and this may be the beginnings of a mathematics of consciousness; however, switching disciplines rapidly, Penrose goes on to claim that consciousness is contact with a Platonic form. It may well be, but it is as many other things besides as the objects of which we can become aware.

Gerald Edelman's is a neuroscientific approach to consciousness which strikes this writer as a step in the right direction. We saw that he distinguishes between the "Consciousness" of animals like the snake, which is the ability to detect salience among varied signals entering through the senses (we'll call this "C1" and the "C2" of humans which involves socially constructed selfhood.

Some convergence may be seen among these theorists in selfhood and the implicit distinction between consciousness and sentience. Let us attempt to salvage what we can from them in outlining a theory of Consciousness and Selfhood.

2. Outline for an Adequate Theory of Consciousness and Selfhood

It is necessary to begin with a few disclaimers. This theory, like any other such, is as much prescription as description: apart from its being related to the (scanty) available scientific evidence, it must also be tried on like clothing to see if it resonates with one's phenomenal experience. Secondly, only an outline is relevant here. Thirdly, we do not as yet have the scientific tools fully to study either consciousness or Selfhood: the best we can do is make intuitive leaps based on tentative hypotheses.

The burden of this section is the provision of notions of selfhood and consciousness which are coherent in experiential terms. As prescription, in the sense just indicated, they should work. Before providing these notions, I wish again briefly to indicate a higher level of the debate, a concept of consciousness *per se* which can unify the perspectives of the philosophers, neurobiologists and physicists we have just looked at who seem to me to be in an "elephant in the dark" situation. (Again, the disclaimer; Consciousness cannot be mentally grasped in any way other than experientially. The *veridicality* of the grasp is attested to by the range of data which seem explicable). First of all, Consciousness is not a single entity, the same for humans as for animals, in experiential terms. It manifests itself quite differently for us, where we construct a self with linguistic mediation, than for animals who lack such symbolic mediation. The consciousness of bats is the "C1" noted above, which we might more correctly have called "primary awareness." We now allow the possibility that there are indeed higher levels of consciousness, about which most of us know little. Let us return to the main business of this section.

First of all, some troublesome aspects of Selfhood should be looked at. As we act, we have a coherent, if tacitly experienced, notion of self. However, in introspection, this pillar of experience dissolves. If we attempt to identify an essence of self separate from its contents, we find ourselves grasping at air. Should we attempt to divide ourselves into subject and object in order progressively to isolate the former, we find the regress is infinite. This aspect of subjectivity, the view afforded as Consciousness gazes inward, Taylor (1989) labels the "punctual" self and he traces it back to Locke. It is the self emerging from the Cognitive Science work. One of the tasks of the latter part of this paper is to demonstrate it as only one of many possible configurations of self, and to show that there is considerable danger in regarding it as anything else. Moreover, if we are truly honest, we find within ourselves a multitude of "selves" bobbing up and down in the mental ether.

In fact, the nature of selfhood is as varied as the objects with which one can identify: self is, in this sense, a principle of identification, as we note again below. (Yet Oriental mystical traditions, in a view with which I have sympathy, treat it as also identical with the Absolute (Needleman 1982: 44). This is the kind of paradox we have seen the later Thomas Merton seized on so brilliantly, arguing that because of its very emptiness, the self could "contain" the Universe as it relaxed into its origin). In cognitive terms, there is a growing consensus that the role of self is to preserve a distinction between subject and object; its misidentifications are corrected (in fact, much recent work on the topic uses an immunological metaphor for the self). The primal such experience is the child's making a distinction between herself and her physical surroundings, and thus releasing herself from the "Uroboros" of indifferentiation between self and world. This point merits some amplification.

The analysis of egocentrism is the royal road to the study of Consciousness. Egocentrism should not be understood as a wilful placing of oneself at the center of the Universe: rather, it is an indifferentiation between subject and object. The child who in the state often described as "Narcissism without Narcissus" fails to distinguish physical self and world is egocentric. The (adult) crashing bore who goes through life without discerning the spaces clearing around him is similarly egocentric. In each case, there is an unauthentic projection of self: "I am the world" in the first case, and "People find me fascinating" in the second. We saw a similar process apparent in Gazzaniga's notion of the interpreter; most of us seem to settle on a granularity of selfhood which is neither that of the bore nor the psychopath (in that we control our "moral" behavior). The differentiation needed in the child's case is a universal human experience and the identification of a certain set of phenomenal events with oneself is near-perfect for most humans; however, we all know to our cost that some adults never achieve the latter differentiation. Once achieved, it becomes a pillar of conscious experience. Selfhood and Consciousness are crucially interrelated, then, first of all, in that self-authentication enriches the distinctions available to Consciousness. Their interrelation can be seen also in the great differences which exist between people in the degree of integration of one, and quality of distinctions available to the other. That said, I wish now to list a set of characteristics of each of these entities.

2.1. *Selfhood*

1. Develops from initial fragmentation and misidentification to a degree which varies from individual to individual
2. Includes at least the following distinct phenomena distinguished by previous authors:
   a. The "cognitive" self, which preserves subject-object distinctions.
   b. The "punctual" self, that which reveals itself to introspection.
   c. A consensual socially and linguistically constructed self, which will be the main topic discussed in the later part of this article, in particular by Cushman and Taylor. Cushman refers to it as the "individual," who is the recipient of political rights. We shall find that it changes configuration from culture to culture, subject to some parameters analogous to those imposed on individual languages by universal grammar. Just as no language consists solely of prepositions, neither will a socially-constructed self have the attribute that everyone is assumed continually unable to control his behavior. A fair summary of the thrust of this is that Cognitive Science and AI may cause great harm by identifying the cognitive self as the individual: unfortunately, we currently lack the terminological resources to expose this identification as the category error it is, so we need to attend to what Taylor has to say.
   d. Merton's "empty" self, quite distinct from the one Cushman is about to outline. Merton is concerned with establishing that even the cognitive self, when viewed correctly, is identical with the Absolute. He is arguing against Buddhist notions of "no self" by showing that they are a special case of his more encompassing view.
   e. The "*persona*" which is a role, initially adopted perhaps on an "as if" basis. R.D. Laing's (1970) brilliant early work demonstrated how authenticity and ontological survival demand that the *persona* must reflect the deeper interests of the self.
   f. This in turn leads us to Jung's (1933) notion of the "Self" which is the full authentication of one's essential nature. Jung's "Ego" rather resembles the cognitive self.

3. The integration of self involves also moral self-mastery, which emerges from the confrontation of self and environment in both its social and physical aspects. Again, the degree to which this is achieved varies between individuals and cultures.

2.2. *Consciousness can usefully be regarded in the following way:*

1. It is an occasional achievement.
2. It corresponds in metaphysical terms to a differentiation of subject and object.
3. Since this differentiation results in an appropriate identification by the cognitive self, we can claim that conscious events are remembered.
4. It involves a tacit experience of self; thus its simultaneously subjective nature.
5. It develops: we can augment its incidence and duration, as well as the number of distinction on which it builds. Perhaps this is the primary role of will.
6. It can have several different types of content:
   a. Intentional objects.
   b. Oneself in relation to an encompassing context.
   c. The distinctions thrown up as the products of the computational mind.

3. What is at Stake?

It may be as well to remind the reader of precisely what is at stake in this discussion. At first glance, the following issues seem to have emerged.

1. The existence of a "soul" or dualist "self-conscious mind" (Eccles 1987).
2. Moral responsibility.
3. Consciousness and its relation to the world.

Yet the situation is even more fraught. Several of the theories we have reviewed, including those of Baars, Johnson-Laird and the present author, have found it essential to include will and selfhood in their discussion of Consciousness. There is even more at stake than we might immediately discern.

Cushman (1990) is concerned with the notion of selfhood, as it has historically unfolded. For Cushman, the concept of self is historically-conditioned: "By the self I mean the concept of the individual as articulated by the indigenous psychology of a particular cultural group" (*ibid.*). Moreover: "The self, as an artifact, has different configurations and different functions depending on the culture, the historical era, and the socioeconomic class in which it exists" (*ibid.*).

We may note in passing that for Karl Jaspers, this chameleon quality is the most perplexing feature of self. For Jaspers, "*Existenz*" is "that capacity of our self for free decisions in virtue of which the self is inexhaustible by scientific knowledge, not because it is too complex to be fully described, but because there are no limits to what it can make of itself" (Passmore 1966: 473). Let us return to Cushman.

The major argument in Cushman's article is that the present configuration of self is the "empty self" as distinct from, for example, the Victorian "sexually restricted self" (Cushman 1990). The "empty self" is prone to abuse by "exploitative therapists, cult leaders and politicians." Psychology is to be castigated for playing a role in "constructing the empty self and thus reproducing the current hierarchy of power and privilege" (all quotations from Cushman 1990).

It is regrettable that Cognitive Science and AI may also be to blame in this regard. The viewpoint criticized by Cushman can be seen also in Baars (1988) and reaches its nadir in Minsky (1987). We have seen that, for Baars, self is merely a level in the "context hierarchy." For Minsky, the concept is more fugitive: "We construct the myth of ourselves" (Minsky 1987: Section 4.2). Yet the constructor of the myth must also be in some sense identical with the myth! The only framework in which a notion like Minsky's works is the one like Jasper's and makes the self, as we have seen, inexhaustible by scientific knowledge. Of even more significance are the cultural consequences of such an empty self theory, which are negative in the extreme. The final point that must be made is that Minsky has chosen one from a possible multitude of configurations of self and proposed it as the only valid one. His theory of self, as stated, is not just ethically questionable, but scientifically incorrect.

The neurophysiologist Blakemore (1988) in a popular book based on a TV series, insists that we consider our notions of consciousness, selfhood and free will simply as convenient fictions, genetically engendered (p.272). This is not a scientific hypothesis of any description: it is simply an assent to the prevailing "empty self" culture dressed in evolutionary clothes. Dennett (1990) similarly attempts to deprive us of "intrinsic intentionality," while offering as little cogent argument as Blakemore and Minsky.

When it comes to discussing notions like self, Cognitive Science (or whatever replaces it as the forum for the interdisciplinary study of information-processing in biological and computational systems) and AI must begin to examine their own culturally-conditioned origins. The alternative is that they merely assent to the *Zeitgeist*, while erroneously claiming objectivity and scientific corroboration. The consequence is not just incorrect scientifically but destructive ethically.

4. Consciousness and Selfhood as Treated in Non-analytic Philosophy

It is apposite now to delve into consciousness and Selfhood as treated in "continental" and non-analytic philosophy in general. However, instead of a historical treatment of the area, which would require several volumes we will look at just one issue: how can we say that consciousness constitutes the world, and yet avoid solipsism?

The answer is that we cannot say so at all if we reduce consciousness to mere *qualia* (Ayer 1982: 219). Consciousness assumes, *a priori*, the existence of one's own and other bodies.

> It is through my body that I understand other people and things.
> (Ayer, *op. cit.*, p.220)

In his account of Merleau-Ponty, Ayer (*op. cit.*, pp.216 221) agrees with the French philosopher that a child must start from a concept of Being itself which includes:

1. the embodiment of self and others;
2. the Consciousness of others;
3. the existence, a priori, of objects set apart in space.

in order for any cognitive development to occur. Some initial differentiations are therefore imperative.

Nor is it by any means a solecism in philosophy to insist on this primordial intent toward Being: it can be found *inter alia* in Brentano (Passmore 1966). Baars' formulation, for all its rich range of evidence, limits itself to Consciousness as a process and lacks this connection with Being itself.

An enormous literature has grown up about the historicity of the notion of selfhood (Cushman, *op. cit.*; Taylor 1989). The major current difficulty with the concept is the poverty of the current philosophical vocabulary used to discuss it. This confusion of tongues we've noted also in Cognitive Science and AI: section 3 above resolves it.

While castigating the dearth of philosophical vocabulary as noted above, Taylor (1989) points to three characteristics of selfhood which yet endure:

1. The notion of depths within the self;
2. The affirmation of ordinary life which emerges;
3. The notion of selfhood as somehow still embedded in nature.

The argument of this article is that Taylor's critique and positive recommendations are correct, and it is appropriate now to give his work the attention it deserves. He is concerned most of all with modern identity. This obviously requires that he finds Cushman's chameleon "self" or "individual" (Taylor, like Cushman, unfortunately does not linguistically distinguish the two) a sympathetic construct: in fact, his notion resembles that suggested by the "universal grammar" analogy above. Interestingly, Taylor's view of subjectivity is a "situated cognition" one: to be a self is above all to be a perspective on moral issues.

The viewpoint of this article is that the consensually-validated "self" of Cushman and Taylor is constructed from achievements of the basic building block of subjectivity, the cognitive self: the manner in which it is done is outside our present scope. As mentioned above, we are after much bigger game, for Taylor argues that our ethical notions are bound up in our notion of the good, and our notion of the good in the self:

> ... to be a self requires a notion of the good ... our visions of the good are tied up in our understandings of the self. (Taylor 1989: 93–95)

He is concerned with salvaging the moral yearnings hidden from many who have considered modern identity, which for Taylor (503) crystallizes in:

> ... a sense of self defined by disengaged reason as well as by the creative imagination, modern understandings of freedom and dignity and rights, ideals of self-fulfillment and expression, demands of universal benevolence and justice.

We have indeed journeyed far from the punctual self, or the minimalist vision offered by Cognitive Science and AI. The affirmation of the human which Taylor seeks is as removed from our current scope as the details of the construction of the modern identity: what we have hopefully learned from him and Cushman is the dangers of the images of Selfhood and Consciousness currently emerging from our disciplines. Cushman argues that the current configuration is the "empty self," Taylor that once the impoverished vocabulary is enriched, the modern identity is a great deal more substantial than we had thought. Moreover, our notion of the Good and consequently our grasp on the moral depends on it.

References

Ayer, A.J. 1982. *Philosophy in the Twentieth Century*. New York: Bantam.

Baars, B. 1988. *A Cognitive Theory of Consciousness*. Cambridge: Cambridge University Press.

Blakemore, Colin. 1988. *The Mind Machine*. London: BBC.

Chalmers, D. (1996). *The Conscious Mind.* New York: Oxford University Press.

Crick, F. 1994. *The Amazing Hypothesis*. New York: Charles Scribner's sons.

Cushman, P. 1990. Why the self is empty: Toward a historically situated psychology. *American Psychologist* 45(5), 599–611.

Dennett, D. 1990. Evolution, error and intentionality. In *The Foundations of Artificial Intelligence*, D. Partridge and Y. Wilks (eds). Cambridge, UK: Cambridge University Press.

Dennett, D. 1992. *Consciousness Explained.* London: Allen Lane.

Edelman. 1992. *Bright Air, Brilliant Fire*. New York: Basic Books.

Edelman. 1993. Quoted in the consciousness wars. *Omni* 16(1), 51.

Eccles, J. 1987. Brain and mind, two or one? In *Mindwaves*, C. Blakemore and S. Greenfield (eds). Oxford: Basil Blackwell.

Flanagan, O. 1991. *The Science of the Mind*, 2nd Edition. Cambridge, MA: MIT Press.

Hameroff, S.R., Kasniak, A.W., and Scott, A.C. (eds). 1995. *Toward a Scientific Basis for Consciousness*. Cambridge, MA: MIT Press.

Jackendoff, R. 1987. *Consciousness and the Computational Mind.* Cambridge, MA: MIT Press.

Johnson-Laird, P. 1983. *Mental Models*. Cambridge, UK: Cambridge University Press.

Johnson-Laird, P. 1988. *The Computer and the Mind: An Introduction to Cognitive Science*. London: Fontana.

Jung, C.G. 1933. *Modern Man in Search of a Soul*. London: Kegan Paul, Trench, Truber and Co.

Korb, K. 1993. Review of "Consciousness Explained." *Psyche* 1(1), Dec '93.

Libet, B. 1985. Unconscious cererral initiative and the role of conscious will in voluntary action. *Behavior and Brain Sciences* 8, 529–566.

Margenau, H. 1984. *The Miracle of Existence*. Woodbridge, CN: Ox Bow Press.

Merton, T. 1993. *Zen and the Birds of Appetite*. Boston: Shambala.

Minsky. 1987. *The Society of Mind.* Guildford, UK: Heinemann.

Needleman, J. 1982. *Consciousness and Tradition*. New York: Crossroad.

Ó Nualláin, S. 1995. *The Search for Mind: A New Foundation for Cognitive Science*. Norwood: Ablex.

Passmore, J. 1966. *One Hundred Years of Philosophy*. Middlesex: Penguin.

Penrose, R. 1987. Minds, machines and mathematics. In *Mindwaves*, C. Blakemore and S. Greenfield (eds). Oxford: Basil Blackwell.

Penrose, R. 1989. *The Emperor's New Mind*. London: Vintage.

Taylor, C. 1989. *Sources of the Self*. Cambridge, UK: Cambridge University Press.

Idle Thoughts

B.F. Katz & N.C. Riley
School of Cognitive and Computing Sciences
University of Sussex

1. Introduction

The computational metaphor has been without peer in explaining human psychology. It has enabled cognitive scientists to model the full range of human behavior from inference to induction to imagination. Its successes have been peppered with doubts and problems, but have been sufficiently forthcoming that many naturally desire to extend the metaphor to cover subjective states in addition to objectively observables such as overt behavior. The purpose of this paper is to suggest that not only is this extension unwarranted, but that current, computational models of mind point to the opposite conclusion, viz., that the subjective character of experience is not the result of computation.

Our argument is as follows:

Premise 1.	Subjective states can be the products of stationary brain states.
Premise 2.	No algorithm can run during a stationary state.
Conclusion.	Therefore, subjective states are not necessarily the products of a computation.

Premise 1 states that subjective experience can arise when the brain is in a steady state, i.e., the activity level of its neurons is constant. In a system as large as the human brain, with 10^{11} such computational units, it is extremely unlikely that no change is occurring. What is meant here is that the units subserving the subjective experience, whatever that may be, are firing at a more or less constant rate. This premise is justified in section 2, by first noting that neural systems that are not trivial feedforward networks compute by relaxing into a steady state, and that the subjective state can be derived by noting which units are highly active when the network has relaxed. Further support for this premise comes from consideration of the relation between positive affective states and neural states.

Premise 2 states that algorithms depend on the notion of change. Computation cannot occur in static systems. In conjunction with Premise 1, this fact enables us to conclude that there exist subjective states that are not products of an effective computation that the brain is running. Therefore, it is a mistake to extend the computational metaphor to qualitative experience. Some other quality of the brain must be causally responsible for these states.

2. Subjective States can be the Products of Stationary Brain States

In this section, the first, and key premise will be defended. It will be argued that current theory suggests that subjective, cognitive states can be a function of stationary brain states. It will then be discussed how affective theory has come to a similar conclusion. The section concludes with possible objections to this view.

2.1. *Cognitive States*

The fundamental mode of computation of a connectionist system is to relax into a steady state such that the constraints embodied by the network are maximally satisfied. In the case of feedforward network, such that used in a backpropagation learning system (Rumelhart, Hinton and Williams 1986), relaxation occurs in a single forward pass through the system. Once recurrent connections are added to the network, however, relaxation takes place over many cycles.

This is true of the brain, which has numerous recurrent connections. It is also true of constraint networks, such as that modeling the perception of the Necker cube (Rumelhart, Smolensky and McClelland 1986).

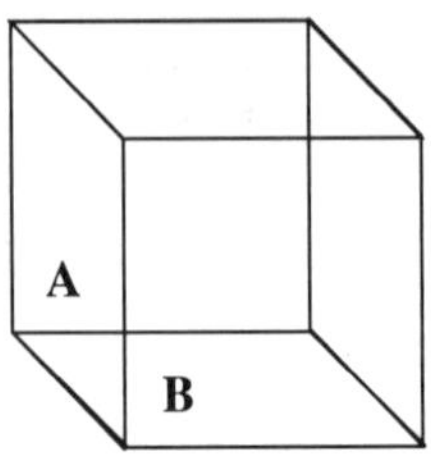

Figure 1. The Necker Cube

The Necker cube is shown in Figure 1. The subjective view of this two-dimensional projection oscillates between the state where face A is in front, and face B in the back, to the opposite view. It is extremely difficult to see the cube as a two-dimensional object, and more importantly for the purpose of this discussion, it is difficult, if not impossible, to see a third state intermediate between the two views. The Rumelhart *et al.* model (1986) reveals why this is the case. The model consists of a set of excitatory connections within a set of units subserving one of the two views, and inhibitory connections between the units from the different sets. Given a small initial bias toward one view or another, the network effectively acts as a winner-take-all network, forcing all the units subserving one view to be on, and the units subserving the competing view to be off.

At no time are units for both views simultaneously in a state of high activity; thus only one view is admitted to awareness at time. Moreover, this view is persistent, and corresponds to a steady state of the system. That is, according to the model, there is no change in activity in the relevant units once the network has relaxed. It is only when the cube flips in the mind's eye that the network changes state. Thus, the subjective state of seeing one view or the other corresponds to a stationary brain state.

It must be admitted that the Necker cube model is a simplified view of a complex process which imposes volume on a flat object. No one yet knows exactly what is happening in the brain when the Necker cube is processed. But the conclusion does not depend directly on the details of this model, and follows from a weaker set of assumptions. The minimal requirements are that (a) there are steady brain states for a short but non-zero amount of time, and (b) awareness (of something) takes place during this time. Requirement (a) follows from the fact that effective computations can only take place if relaxation occurs. Requirement (b) follows from the fact that the alternative is difficult to entertain. If (b) were not the case, then we would temporarily "black out" during these stationary states. But, in fact, it is precisely during this time that we need to be aware, because this is the time that the network has made its decision. In other words, if (b) were not the case, then there would be no subjective awareness of the computational decisions our brains have made, just of the processes leading up to these decisions. In fact, our knowledge of such processes is poor, as protocol analysis reveals. It is the final result that we "see" in our mind's eye, e.g., one of the two alternating views of the Necker cube.

2.2. *Affective States*

This section will treat one dimension of affect only, albeit an important one, viz. hedonic tone, or the degree to which one is responding positively or negatively to a stimulus. Two of the more prominent theories of hedonic tone attempt to relate this variable to cortical activity. The first, attributed to the pioneer in aesthetic research, D.E. Berlyne, states that hedonic tone is an inverted U-shaped function of the arousal potential of a stimulus (Berlyne 1971). Berlyne based his theory on physiological assumptions which have not been supported since his theory was first proposed. Nevertheless, the intuition behind Berlyne's proposal is clear. A stimulus that lead to low arousal will not be enjoyed, but overarousal is irritating and unpleasurable. More formally, Berlyne demonstrated that hedonic tone is an inverted U-shaped function of a number of properties of the stimulus, such as complexity.

Martindale (1984) and Katz (1993, 1994) have proposed an alternative formulation that proposes a monotonic relation between hedonic tone and cortical arousal. They have shown that this relation best captures the aesthetic ideal of unity in diversity in neural terms. High activity network states will be those in which the diverse elements of a complex stimulus are simultaneously active, and therefore unified by the network. Low network activity implies either a stimulus of low diversity, or one in which the diverse elements of the stimulus cannot be entertained at once. The monotonic theory explains non-monotonic relations between a stimulus variable and hedonic tone by showing that network dynamics are responsible for such a result, rather than the hedonic measure.

The key point for the purposes of this paper is that if either the monotonic or nonmonotonic measures (Berlyne's) turns out to be correct, then problems arise for an algorithmic view of subjective states. The reason for this is that both measures are functions of the stationary state of the system. That is, they do not depend on the path by which the brain has achieved its current state of activity; they are only concerned with the level of such activity at a given time. If that activity level is maintained, then the hedonic tone will also be maintained at that level. I.e., stationary brain states can give rise to subjective affective states.

2.3. *Possible Objections*

We consider first the objection that this first premise is just wrong, i.e, that there must be processes occurring in the brain for a subject to enjoy most conscious experiences. Consider a subject watching an object moving across their visual field: any cognitive mechanism that tracks the object must use a continuous process. It is not possible for a static brain state alone to be responsible for a tracking function. On this view, the Necker cube evidence is contrived to deflect attention away from other brain processes necessary to produce subjective states. This objection can be met by stating that there are a relatively large class of processes, such as that responsible for the perception of the Necker cube, in which relaxation is essential. Furthermore, there are invariant subjective experiences associated with intrinsically dynamic processes such as tracking (e.g., the experience of the moving object is more or less constant, although the visual field as a whole is evolving).

Another possible objection to the claim that the subjective states can be the product of stationary brain states will now be considered. The intuition behind this objection is that a computational process has taken place in order to achieve the steady state. Could not this process be the causal basis for the qualitative experience? The problem with this view is that the trajectory by which the steady state is reached will determine the nature of this experience, rather than the end result. That is, two different means of approaching the same steady state will result in two different experiences.

But this runs counter to the fact that computational decision in current neural models depends not on trajectory, but on the end state. Or, to put it in alternative way, suppose that there was some device that could activate the neurons in one's brain that are ordinarily active during the steady state associated with one view of the Necker cube. This device is not that far-fetched, although with current technology would depend on invasive stimulation. The process view then says that nothing would be experienced, because no algorithm was run in the brain in order to achieve this state.

It seems counter-intuitive that such brain stimulation would have no effect on qualitative experience. Let us concede this temporarily, however, and deny the causal efficacy of relaxed states in producing qualitative experiences. By this view, jumping a cognitive system to any stage of the dynamic process could never constitute introduction of a subjective state where such a state is defined as a section of continuous neural activity. This objection successfully eliminates the premise but it also commits itself to some radical theory. If all of our conscious brain states are processes, then an explanation is required of how the information (or content) of this state can ever be made available to the rest of the system. Within connectionist theory, a relaxation into the pattern of activation producing the conscious experience of a car also activates links to other relevant concepts: traffic wardens, ignition, furry dice, et cetera. However, content addressability of this sort should also be present on the process model of consciousness posited in objection to our premise, at least as an emergent property. If the dynamic process was independent of the rest of the system then how can any *part* of the process bear the content of the *whole*?

The way round this is to treat the process not as independent but as a sub-process of a larger one, i.e., cognition is a dynamic process *all the way down.* There have been recent suggestions that cognition would be most fruitfully studied as state space evolution in a dynamical system (Van Gelder 1991). Typically, such a theory is strongly anti-representational. Van Gelder's agenda proposes cognitive modeling using low dimensional dynamic systems as a replacement paradigm for the Turing Machine or the Neural Network in cognitive psychology. Such a model, based on differential equations, may also introduce steady states. In order to divorce these states from the contents of consciousness, one must assume that these steady states are non-representational. Otherwise, as before, one's awareness would not consist of the end-product of the computation.

Alternatively, one can conceive of a dynamic system, perhaps more radical than van Gelder's, whereby the only convergent processes are non-static (e.g., strange or chaotic attractors). While not inconceivable, systems depending on these sorts of processes alone are unlikely to have evolved given that simpler solutions are available.

3. No Algorithm can run during Stationary States

The second premise of the argument is that no program is running during stable states. This is as true of a Turing Machine as it is of a neural network. Of course, transitions from one state to another are what constitute a program in discrete rule-based mechanisms such as a Turing Machines, so we are accustomed to say that when a machine is in state S then a program is running. However, our issue is temporal. Imagine adding a feature to the operation of a Turing Machine ensuring that whenever it reached state S it remained in that state for three seconds before moving to the next state. During these three seconds no input is being transformed into output. In virtue of the S's being an intermediate state of the program we can, in a weak sense, claim that the program is still running during these three seconds, but there is stronger sense in which it is not. The three seconds of stability have no effect on the output of the program. Equivalently, a neural network that has settled into a stable state is no longer functionally equivalent to any machine running a program, Turing or otherwise.

4. Conclusion

In summary, if current connectionist models of mind are accurate, then the steady-state of the brain yields up both the decision that has been computed, as well as the subjective experience corresponding to this decision. This means that subjective experiences are not necessarily the products of computations, because the program is no longer running during the steady state. In effect, Cognitive Science, in pursuing ever more detailed models of computation, has hoisted itself on its own petard, if it also possess ambitions to describe subjective experience in addition to observable behavior. Something has to give. Either it must move to a more radical, dynamic view of the mind, or it must be augmented by a physicalist/energistic notion of consciousness. We find ourselves attracted, not strangely, to the latter option.

References

Berlyne, D.E. 1971. *Aesthetics and Psychobiology.* New York: Appleton.

Katz, B.F. (1993). A neural resolution of the incongruity-resolution and incongruity theories of humour. *Connection Science* 5, 59–75.

Katz, B.F. 1994. An ear for melody. *Connection Science* 6, 299–324.

Martindale, C. 1984. The pleasure of thought: A theory of cognitive hedonics. *Journal of Mind and Behavior* 5, 49–80.

Rumelhart, D.E., Hinton, G.E., and Williams, R.J. 1986. Learning internal representations by error propagation. In *Parallel Distributed Processing,* vol. 1, D.E. Rumelhart and James L. McClelland (eds). Cambridge, MA: MIT Press.

Rumelhart, D.E., Smolensky, P., McClelland, J.L., and Hinton, G.E. 1986. Schemata and sequential thought processes in PDP models. In *Parallel Distributed Processing,* vol. 2, D.E. Rumelhart and James L. McClelland (eds). Cambridge, MA: MIT Press.

Van Gelder, T. 1991. What might cognition be if not computation. *Indiana University Cognitive Science Research Report 75.*

Consciousness

A Requirement for Understanding Natural Language

Gérard Sabah
Language and Cognition Group (LIMSI)
CNRS, Paris

A psychologist can no more evade consciousness than a physicist can side-step gravity. (Baars)

1. Relations between Natural Language and Consciousness

1.1. *Generalities*

In my previous publications concerning the relations between Natural Language Understanding and architecture (Sabah 1990b, 1990c, 1993, Sabah *et al.* 1988; Sauvage, *et al.* 1989), CARAMEL stands for "*Compréhension Automatique de Récits, Apprentissage et Modélisation des Échanges Langagiers*," which translates into English as "*Automatic Story Understanding, Learning and Dialogue Management.*" This indicates that the model only intended to tackle Natural Language Understanding. Even if the model had a very ambitious goal (as the system applied to many different kinds of applications — e.g., dialogue, story understanding, abstracting), it was not included into a general model of reasoning and intelligence.

However, I have considered consciousness as a crucial aspect for natural language understanding for a long time, and the studies cited above have shown how reflectivity and distributed artificial intelligence allow for computer programs to represent their behavior and reason about these representations in a dynamic way.

Non-controlled processes appeared also to be necessary in this kind of program for computer efficiency reasons as well as for cognitive ones. Therefore, I proposed a blackboard extension (the *Sketchboard* — or *carnet d'esquisses* in French) that implies a different kind of relation between processes and allows for reactive feedback loops at different levels between processes that do not know each other.[1]

In order for these two categories of processes (controlled and uncontrolled) to collaborate, consciousness clearly has a central role to play. I propose here a new general cognitive model that is based on the assumption that language is the necessary ability that allows for intelligence.

To take this point into account, the name CARAMEL now means (in French): *Conscience, Automatismes, Réflexivité et Apprentissage pour un Modèle de l'Esprit et du Langage* (in English: *Consciousness, Automatic processes, Reflectivity and Learning for a Model of Mind and Language*). This model takes many inspirations from various sources whose essential points are summarized below.

Baars (1988), with a psychologist's point of view, offers an "economical" conception of consciousness. A general workspace, specialized unconscious processes and contexts (conceived as hierarchies of goals) are the only three main components of his theory. From him I retain the following ideas:

- a blackboard as a workspace where conscious data are written;
- the hierarchy of interpretative contexts and the handling of interruptions;
- the competition between several unconscious processes; and
- the model of voluntary control and attention.

Harth (1993) is opposed to Cartesian dualism as well as to the more recent radical pluralism ("a million witless agents instead of one clever homunculus"). For him, mental images are not replicas of world objects, they are combined with previous knowledge. Top-down ways allow higher knowledge to modify messages that come from senses and to inject into them additional information. This process is active as soon as sensorial input begins (and not at the end, as one would assume if considering it as an advanced function of the brain). Therefore, there is no homunculus scrutinizing the state of the brain. The brain itself acts as an observer of these first levels and influences them in order to maximize the recognition: the brain analyses, recreates and analyses again its own productions, in a truly "creative loop."
From this theory, I took:

- the idea of feedback between unconscious processes;
- the a priori evaluations of unconscious processes; and
- consciousness acting at the first levels of treatment rather than at the end.

Eccles (1992) is practically alone (with Popper) in basing his theory on dualism! He distinguishes three worlds among which there exist recursive relations (1–material, 2–consciousness-states, 3–objective knowledge). With his micro-sites hypothesis, based on quantum physics, he tries to explain how mental events can act upon neural events (mind being compared with a quantum probabilistic field allowing for activating pre-synaptic net vesicles). Though his presentation lacks precise information on the real "implementation," he develops the interesting idea (close to Harth's) that conscious mind is not only engaged in a passive reading of the brain activity, but has a) a proper searching activity (e.g., attention, pulsion, necessity) and b) the role of unifying the whole set of information. The main parts of his theory that are useful for my new model are:

- the division of data and knowledge among three separated but communicating worlds; and
- the asymmetry of the brain and the articulations between language and consciousness.

Edelman's vision of consciousness (1989, 1992) is based on a theory of the brain functions, in its turn based on a thesis of evolution and development.

The core of his approach is the TNGS (Theory of Neuronal Group Selection) based on three principles: ontogenetical selection; secondary synaptic reinforcements or decay; interactions among cerebral maps by a bi-directional re-entry.

He exhibits the neurobiological functions that have allowed the emergence and the evolution of more and more elaborated characteristics of the human mind. He also demonstrates how these characteristics are an explanation of consciousness. They are: 1) neural specializations allowing the distinction of internal signals from world signals, 2) perceptual categorization, 3) memory as a process of continuous re-categorization with the possibility of representing the activation order, 4) learning (links between the categories and the essential values of the individual), 5) concept acquisition (categorization of the brain activity through global maps) 6) primary consciousness (allowing to connect internal states that result from previous perceptual categorizations, to present perceptions — what he calls : the remembered present), 7) an ordering capability which results in a presyntax (the basis for symbols) 8) language and 9) higher order consciousness. Here, I have been mainly inspired by:

- the definition of unconscious processes as basically producing correlations,
- the definition of semantics as correlations between concepts, sensory input and symbols,
- the memory model as a categorization of processes rather that as a zone for storing representations, and
- the role of language for symbol manipulations.

I will first give some theoretical and practical justifications for taking into consideration unconscious as well as conscious processes, and then propose a model of memory that is in agreement with this aspect. Secondly, I will recall the basic elements on which is based our global model of understanding (the Sketchboard for unconscious processes and a model of reflective agents for conscious processes — the old Caramel). Finally, I will propose a computer model that shows how these two levels can be linked through a process having a strong analogy with consciousness.

1.2. *Necessity for Two Kinds of Processes*[2]

Understanding is not only based on logical criteria, it is also the emerging result of non rational cognitive processes that cannot be described in an algorithmic way. While we think, either we are able to say something about the way some mental operations are performed (these will be called "conscious processes"), or we are only able to realize what their results are ("unconscious"). As clearly enlarged on (Baars 1988), these two kinds of processes differ on the following points (see Table 1):

Consciousness-processes capabilities	Unconsciousness-processes capabilities
1. Computationally inefficient (errors, low speed, mutual interferences between conscious computations. 2. Great range of various contents, great ability to relate conscious contents to each other as well as to their unconscious contexts. 3. Internal consistency, seriality and limited capacity.	1. Highly efficient in their own tasks (low number of errors, high speed, little mutual interference). 2. limited range, each processor is relatively isolated and autonomous. 3. Diverse, can operate in parallel, great capacity together.

Table 1. Relations between Conscious and Automatic Processes

Even if non controlled, unconscious cognitive processes are supposed to be exhaustive and not to depend on the cognitive load (Newell 1990). From my point of view, this may not be true at every level. As soon as they draw near to consciousness, unconscious processes have to limit themselves, as they have to compete in order to take control over a very limited working memory. We propose here a smart and simple explanation for this transition between subliminal perception and conscious perception.

In order to control the analysis rules, and to avoid combinatorial explosion, classical rational approaches use metaknowledge (Pitrat 1990). However, if it seems difficult to foresee some conflicts only by using logical reasoning, it seems even more difficult to solve them by these means alone.

For example, during syntactic parsing, the best interpretations usually follow the *minimal attachment* principle (don't postulate, in the syntactic tree, a potentially useless node) and the *differential closing* principle (if it is grammatically possible, attach any new element to the current phrase). A big problem with such very general principles is that we are unable to easily detect the exceptions (e.g., the study of corpora allows to set up general rules but does not enable specific cases). It is the reason for which these regularities cannot be used as formal parsing rules. However, they may be explained as an emerging consequence of the competition between interpretative processes: usually, the interpretations that follow *minimal attachment* and *differential closing* are the most simple to build, and, as a consequence the first ones to be consciously perceived.

Most certainly, rational thought is also an important part of understanding, but it should intervene only *after* such a spontaneous perception of meaning (this distinction allows us to distinguish between "true" ambiguities, due to the communicative situation — and that should be solved by a dynamic, rational planning — and artificial ambiguities that usually remain implicit without a deep linguistic analysis).

Our hypothesis is that data in working memory is transferred to short term memory (i.e., become conscious — cf. § 4.1 and 4.5) when a given threshold of persistence is overshot, depending upon their accessibility and their lasting time. This transfer is viewed here as the "feeling of understanding." This process generally concerns only one interpretation at a time Such an interpretation is perceived globally as an already constituted entity (*pop-out*) and may further become the basic element of a logical simulation of rational thought.

As psychological experiments about semantic priming have shown (e.g., Meyer and Schvaneveldt 1971), knowledge structures are characterized by variable and dynamic accessibilities. This property has to be taken into account in an ergonomic model of understanding. Indeed, the cost associated to each elementary interpretative operation is closely linked to the time necessary to attain pieces of knowledge in the memory. Therefore, interpretations which are coherent with the most accessible knowledge are the most likely to become the best ones: since they are more rapidly constructed, the cognitive system that constructs them can devote more attention to them. Thus, the system naturally prefers the interpretations that are the closest to the activated knowledge, i.e., the most relevant considering the current state of the context.

Due to the fact that the attention of a process may be distributed among several interpretations, the analysis is performed in a parallel way. However, this capability is not very flexible, since the computed entities are stored in a limited-capacity memory. The cognitive system should use this memory in the most efficient way, which is performed by an automatic non-conscious optimization: the attention is focalized on the most relevant interpretations.

This allows us to take advantage of the dynamic properties of the memory: frequently accessed knowledge becomes more and more accessible, which makes the relevant interpretations more and more likely, which in turn makes the corresponding knowledge more and more accessible, and so on (*positive recursive feedback*). The space occupied by less accessed knowledge is quickly used for a more useful usage, and the associated interpretations slow down or disappear (let us remark that an explicit evaluation of relevance is no longer necessary, since it is implicitly performed with regard to the state of knowledge evolution: dynamic accessibility of knowledge and parallel analysis are sufficient to explain how the most relevant interpretations emerge). The interpretations are not compared on a structural basis, but through their competition to occupy memory.

Individually, each possible interpretation is computed in a bottom-up (data-driven), sequential way. Nevertheless, context implies that the whole system converges towards a resulting interpretation, very often a unique one. Indeed, the interpretations are developed at different speeds, depending upon the plausibility of the chosen solution, i.e., lastly depending upon the accessibility of the knowledge they are based on. The state of the cognitive context acts as a set of hypotheses that favor the most relevant interpretations. This predictive mechanism differs heavily from classical top-down analysis.

Thus, while trying to make explicit the subliminal processes underlying language ability, we want to define a more realistic model of understanding (which may substantially differs from a linguistic analysis!). Here, we do not want to account for an explicit, formal reasoning process, but for spontaneous, non controlled inferences that allow information to go from the subliminal, perceptual level to the conscious level.

1.3. *A Model of Memory*

As mentioned above, memory plays a crucial role towards understanding, since it contains mental structures created or used by the interpretative process. In computer science, it is usually thought that memory is a homogeneous space where knowledge is stored, a position that has been widely criticized (Edelman 1989; Rosenfield 1992; Vignaux 1992): several aspects of language, not very often studied in natural language processing, are more easily explained when associativity of mental structures and their dynamic aspects are considered. In other words, computational functionalism may be appropriate to model rational thought, but, as I already said, rational thought should not be considered as the only component of verbal communication.

I present below several *functional* structures into which memory can be divided. I insist here on the term *functional*, since these structures should not be taken for the biological underlying structures, even if some of these functions may be localized somewhere in the brain (our objective here is to model some cognitive capabilities of the mind, without any commitment to neural nets). We propose a model based on three distinct kinds of memory (Figure 1).

Some basic properties of the model find their origin in psycholinguistic studies. First, two kinds of memory can be distinguished in human beings: a very active but short-spaced memory, whose content is constantly renewed and where the interpretations take place (*short term memory*), and a more stable memory, which keeps worthy of interest results of these operations (*long term *memory*). This division is particularly clear with people whose hippocampus has been injured (Kandel and Hawkins 1992): they are still able to remember their past (before the injury) and they seem to behave normally, but, as soon as they stop their current activity (which is occupying their attention) they completely forget it (e.g., they are unable to remember who they were speaking with and what they were speaking about). Therefore, the hippocampus seems to be necessary for the cognitive system to update its long term memory, an absolutely necessary function for learning to take place.

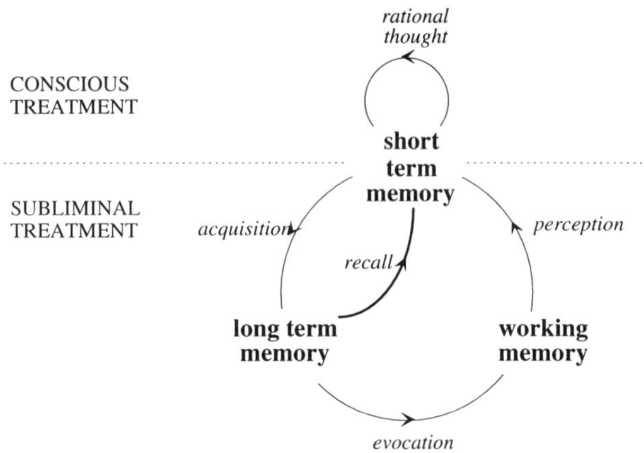

Figure 1. General organization of memory. Knowledge in long term memory is evoked by linguistic entities (associative recall). Then they are transferred in the working memory where interpretations set up coherence towards the cognitive context (which corresponds to the recognition of the utterance cohesion). A coherent interpretation becomes conscious and appears in the short term memory. This conscious perception triggers an automatic acquisition process and a rational, controlled treatment (cf. § 2.2.2 and 4.5 for a more detailed presentation).

Volatile memory in its turn may be divided into two parts, one very limited in space and conscious (*short term memory*, Miller 1956), and a subliminal part, a little bit larger (which we named *working memory*). In order for mental structures to be reused, they have either to be still in the short term memory or to have been stored in the long term memory.

The subliminal level has been less studied than the conscious level, but several psychological experiments have shown its existence, particularly concerning lexical access (Swinney and Hakes 1976): all the meanings of a polysemic word are activated in a parallel way, even if, in a real context, only one of them is usually selected and consciously perceived (for every meaning — be it relevant for the following discourse or not — the same priming effect on close semantic data can be demonstrated).

I propose below a computer model (implemented in Smalltalk 80) capable of simulating this kind of behavior.

2. The "Sketchboard" (a Memory for Subliminal Processes)

2.1. *Neurobiological Evidence and Computer Science Point of View*

First, it should be noted that — in contradiction with what seemed to be implied by our previous model (Sabah and Briffault 1993) — there is no homunculus scrutinising the state of the brain: the brain analyses itself, creates and examines again its own productions in a "truly *creative loop*" (Harth 1993). This is particularly obvious for the vision process, where the existence of top-down paths has been demonstrated at the neurobiological level very early (Ramón y Cajal 1933), and is still a topical question (Kosslyn 1980; Kosslyn and Kœnig 1992). It has also been shown that these paths actively participate in the vision process by injecting *new information* (not present in the initial message) on behalf of higher levels: the initial message is modified on the basis of this auto-referential process (bootstrap). Furthermore, (Restak 1979) has convincingly argued that these auto-referential loops govern the whole nervous system.

This mechanism is basically unstable (possibly explaining creativity). This means that brain zones are not seen as relays where data are stored, but as zones where the cortex produces sketches, modifies them and reinterprets them. This variability is a necessary feature to allow for adaptation in a modifying environment, as argued by Edelman who suggests a Darwinist interpretation of the evolution of the brain (1989, 1992). (Harth 1993) takes the same kind of approach and shows how various kinds of feedback allow to account for normal vision processes as well as for optical illusions.

A priori, blackboards (initially presented in Hearsay II (Erman *et al.* 1980) and extended later (e.g., Hayes-Roth 1985; Nii 1986) seem interesting as a way of modeling this kind of process. They allow for an efficient solution to the control problem and for an efficient dynamic ordering of the processes to be triggered. However, this opportunistic behavior does not allow higher modules (e.g. semantics, pragmatics) to feedback information to lower level modules (e.g. perception): within blackboards, as soon as a process has been triggered, it writes its results onto the blackboard, and further processes have no influence on these results. Feedback from higher levels truly exists in blackboards, but it rather concerns the strategy of choosing among several possible solutions: inferences from higher levels allow to choose the most relevant process in a given context, but cannot directly interfere with the behavior of this process. This kind of feedback does not model the situation where a given process is capable of adapting its own behavior in order to produce an improved result, better adapted to higher level knowledge.

Even with a sophisticated system such as BB1 (Hayes-Roth 1985), which is an excellent result of a declarative conception of control (a blackboard is devoted to the control itself), some problems are still present (without mentioning the complexity problem). In particular, declarativity of knowledge, necessary for a flexible and opportunistic control, does not guarantee that the constructed objects be coherent, nor does it guarantee that they converge towards elements useful for the final solution. This problem is linked to the notion of *meaning*: constructed objects have meaning only from an external point of view, not for the program itself, and therefore cannot be the basis for a semantic control. Furthermore, the fact that the more recent GUARDIAN (Hayes-Roth *et al.* 1992) claims to be based on a rational use of knowledge does not seem to me to be a significant response to the problem raised here.

Thus, in the blackboard model, incoherence can result from conflicts between resources or from duplicate actions (Davis and Smith 1983; Lesser and Corkill 1983). Hewitt (1986) discusses the complexity of this problem: agents must reason about the intentions and knowledge states of other agents that use heterogeneous representations. This requirement implies the use of sophisticated communication means such as blackboards or message passing. Blackboard models have no local memory linked to their agents. Hence, the agents are not able to reason about their behavior, neither are they able to communicate directly with other agents: control is centralized and completely independent from the other agents of the system. On the other hand, in the actor model, as shown in Hewitt's early (Hewitt 1977) and subsequent work (Agha and Hewitt 1986; Hewitt and Jong 1984), there is no public zone for communication. Hence, control remains implicit, making it difficult to maintain global coherence while reasoning.

> In the Sketchboard model, the higher levels of knowledge can provide feedback to lower levels so that the latter can adjust their behavior in order to be as coherent as possible with the results of the former.

Moreover, the same data can be considered under various points of view, implying different processings and producing different results. For example, concerning natural language understanding, several contexts may be relevant (e.g., perceptive context, conceptual thought context, intention context, communicative context), and the various interpretations with respect to these heterogeneous contexts may bear contradictions. Even in such situations these different contexts should interact and give rise to a global interpretation.

> We show below how the Sketchboard allows these kinds of interactions to be modelled, in a possibly parallel way.

Another blackboard characteristic implies that all data is handled the same way as others. In other words, as soon as a process has discovered data that may act as an input, it is triggered. Since the global solution is not yet available, no control — be it as sophisticated as possible — will be able to evaluate the importance of partial results towards the final solution.

> Our Sketchboard contains a mechanism for reactive feedback loops, generalized across all the modules that interact when solving a problem. As higher and higher level modules are triggered, the initial sketches become more and more precise, taking into account the whole knowledge of the system.

2.2. *The Sketchboard Model*

Our Sketchboard (strongly inspired from Harth's work) is an extended blackboard: in addition to being a general input-output zone, responsible for triggering modules and managing their communication, it sets up specific relations between a specialized process and the processes that use its results. These relations allow for feedback from the latter and explain how the former adapts its behavior according to these interactions.

Our extension allows the processes not only to be triggered in an opportunistic way (as in usual blackboards) but also to be considered from two different points of view: (a) either they build a given kind of result (a *sketch*, possibly rough and vague) or (b) they return a simple response that indicates the degree of confidence concerning their results.

2.2.1. *Interactions between modules*

In Figure 2, agent A produces an input for agent B. In addition to its own result, B sends $\mathcal{R}(\mathcal{S})$ to A — a response from B (not its result!) — in order to indicate to A how pleased B is with what it has done with $\mathcal{S}$, — A's result. This allows the Sketchboard to set up original relations between active processes. While higher and higher-level processes are triggered, the first sketch will become more and more precise, for these processes will give feedback to the first module, indicating the relevance of what is computed with regard to their own knowledge. At the end of the process, though the system remains truly modular, its entire knowledge is taken into account. In the example illustrated in Figure 2, the two modules are in a loop: the first one (A) modifies its result in order to optimize the response from the second one (B). In other words, a first sketch $\mathcal{S}$ is computed by A, and further refined, such that $\mathcal{R}(\mathcal{S})$ be *optimal* according to a given criterion. A's behavior is purely reactive:[3] if its previous modification produces a better response from B, it continues modifying its sketch in the same way, otherwise it performs modification the other way round.

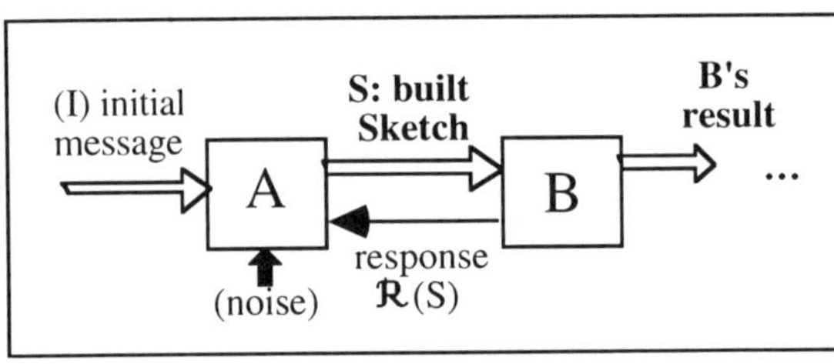

Figure 2. Retroactions between Processes

This mechanism continues until some stability is reached: *this amounts to modifying the signal until B's interpretation is as relevant as possible, given the context and the available knowledge.* (*Context* is represented here by active modules and the sketches they have written onto the Sketchboard. *Available knowledge* is the knowledge stored in the long term memory and used by these modules. The measure of *relevance* depends on the kind of the module itself — we shall see some examples below.) This loop and the noise generator allow for small variations for a given solution and may amplify such variations and produce features, absent in the initial message (bootstrapping). This results in a non linear process: high level agents act not only as filters for perceptions, but as a device adding features, thus modifying the characteristics of the input.

In its turn (and while the previous loop is still running), B too will receive feedback from higher level processes. As shown in Figure 3, this will influence its own sketch and, consequently, the strength of its response to A.

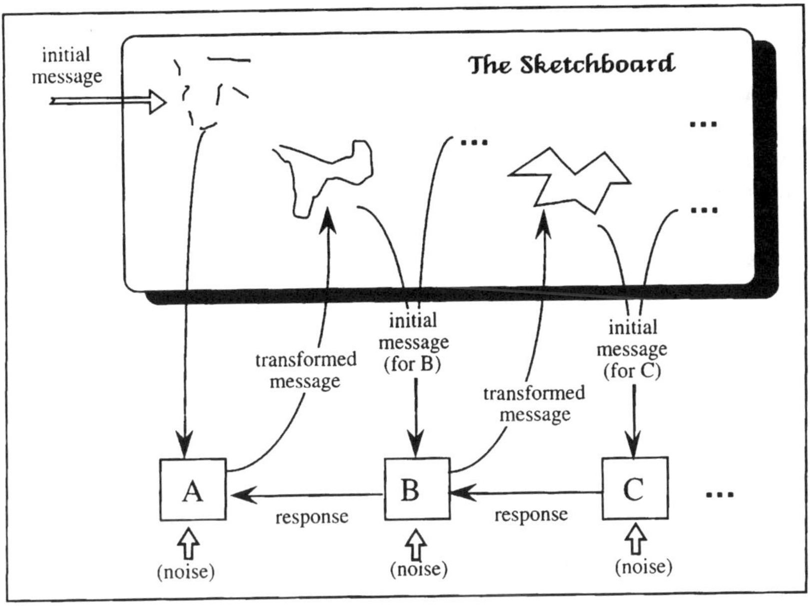

Figure 3. Relations between Processes inside the Sketchboard

Progressively, the various sketches will *simultaneously* become more and more refined. As higher-level modules provide feedback, the relevance of the results produced by a module with regard to higher-level knowledge is established. The Sketchboard then has to fulfil three main tasks: to detect which module is working on what data, to connect the modules that will give rise to elementary connections, and to handle their resulting exchanges.

One important feature is that there is no "output" from the system: the Sketchboard is read by the modules that are working on it, and the *feeling of understanding* results from some stability of the whole system. It should be noted that some modules may be specialized in computing expectations. Therefore, some results written into the Sketchboard account for the fact that the system is waiting for a given kind of data. Hence, a special process, related to consciousness, may be aware of a difference between what is computed and what is expected. This feature allows us to model the process of unconscious attention (cf. § 4.4).

The model has been implemented and tested on simple examples that show its feasibility and usefulness (Sabah 1996). Currently, we are in the process of applying the Sketchboard model to natural language understanding within the

CARAMEL project. The precise definition of feedback measures for the different processes is still under study.

2.2.2. *Multiple interactions and relations to memory*

Similarly to blackboards, the Sketchboard is divided into several layers in order to simplify the control process: when data appear at a given level, only a limited number of modules (a priori known) may be triggered. Figure 3 shows a simple example where three modules interact at three different levels.

But things may be more complex, for at least two reasons: at a given level, (a) previous data may stay for a variable length of time, and (b) several modules may act in parallel on the same data. To allow for such behavior, a given amount of memory is allocated to each level of the Sketchboard. This allows the system to maintain the data active for a certain amount of time, and to store alternative results from the active modules.

In relation to the memory model presented in paragraph 1.3, the *working memory* is the set of memories linked to these various processing levels. Data that are still in such a memory may receive feedback from higher levels and may be compared with competing data of the same level. Those receiving the highest amount of feedback are most likely to lead to a correct solution.

This allows us to model a *forgetting* mechanism as well. Another important role of the Sketchboard is to keep track of the various relations among the active modules. For this to be possible, it manages the limited amount of memory at every level and keeps a hierarchical representation of the various solutions currently under development. This allows the system to give preference to data receiving positive feedback from higher level, and lower the expectations for those receiving negative feedback or more radically, space becoming short, to forget them.

Further experimentation is necessary in order to determine the exact amount of memory necessary for each level. This may imply quite different behaviors of the system, since, if something has disappeared from a working memory, and if the chosen solution appears to be poor, it may be necessary to re-analyze the data from scratch. Had the size of the working memory been different, feedback from higher levels would have performed the task in a different way. However, reasonably-sized experimentation is needed to choose the optimal size of the working memories (one hypothesis I wish to test is that there is a strong relationship between those various working memory sizes and the *cognitive style* of entity modelled).

Garden-path sentences are a good example of such a situation.[4] Our model predicts three possible behaviors of the system, depending on the accessibility of the element that allows to solve the ambiguity — which seems to be in agreement with psychological experiments (Ferreira and Henderson 1991). If the various working memories are such that the different solutions are still present when the element that solves the ambiguity appears, feedback will only favor the correct meaning (since the other ones will receive negative feedback from the higher levels of interpretation). But, if the memory sizes are such that feedback from the resolving level comes only after feedback from previous levels (which *naturally* lead to a wrong interpretation — this is the characteristic of this kind of sentence), the wrong interpretation may receive positive feedback and can be developed simultaneously to the correct one. Lastly, if the element that allows to solve the ambiguity arrives too late, the correct interpretation is no more in the working memory, and understanding will result in a dead-lock.

3. Controlled Processes (the Old Caramel)

As I have already said, blackboards are very useful when various processes interact in different styles according to a goal or a situation, but at a certain cost. Since different modules contribute in unpredictable order to the final solution, there is a problem of control. Studies have already show how to solve this problem efficiently, in particular in the domain of natural language (Sabah *et al.* 1991; Bachimont 1992). We also showed (Sabah 1990a; Sabah and Briffault 1993) that (a) modularity is a practical need, (b) an independent control is needed in order to choose the most useful agent to be used in a given situation and (c) in order to adapt the agent behavior to the situation, this control should be distributed among the agents, giving them the capability to represent themselves, as well as what they are doing. This allows to dynamically choose the processes that will solve a problem, and to dynamically compute the order in which they should be triggered, depending on the global task (understanding a text, abstracting it or managing a dialogue) as well as on the specific current context.

3.1. *Explicit Control and Reflectivity*

Such a control is easier to implement within distributed artificial intelligence; it requires that each agent be a *reflective system*, i.e., a system that uses a representation of itself to reason about its own behavior.

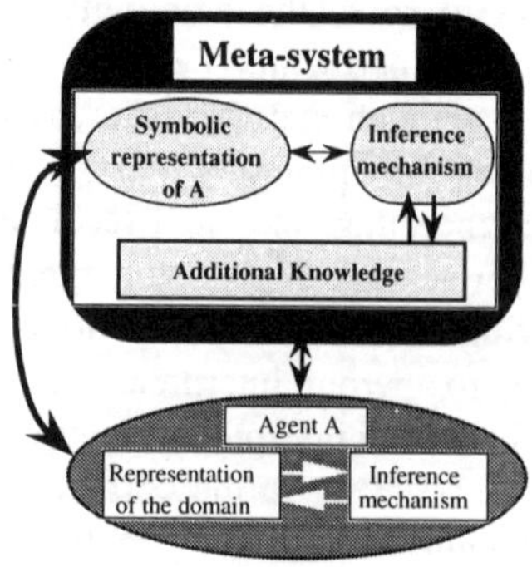

Figure 4. A Reflective System

Therefore, for every agent, we need to clearly distinguish the knowledge about the domain as well as the knowledge about the agent behavior and its interaction with the other agents (cf. Figure 4). If one accepts an analogy between the behavior of a complex system and mental activity, having knowledge about its own activity may correspond to an implementation of a partial consciousness (Minsky 1985).

The object (agent A in Figure 4) and its representation are two different systems, each being a classical system. Thus, meta-reasoning is performed just as any other kind of reasoning.

Nevertheless, for the meta-level to represent the initial system, a causal link between the system and its representation must exist (the representation is always *truthful*) (Maes 1987). In the same vein, Smith (1986) introduces the notions of *introspective integrity* (significant properties of the meta-representation are related to the object-level) and of *introspective force* (how the goals of the object level of a reflective system are performed through the top-down causal connection).

Our CARAMEL architecture is designed to integrate the representations and expertise needed to perform various natural language processing tasks. The representations are stored in a memory containing the knowledge and intermediate structures of the system (cf. Figure 6). CARAMEL is a multi-expert system with a continuous control and a dynamic management, based on blackboards mixed with message passing (this allows a first static plan, depending only on the task, to be dynamically adapted to the current data to be understood). It is a recursive implementation of reflective models. Two kinds of extensions (developed in the next paragraph) have been implemented to allow for a more efficient use of reflective models:

- a metasystem is a system that may have control over *several* agents, and
- metasystems themselves are considered *usual agents* and are represented at a meta-level.

3.2. *Extensions to Classical Reflectivity*

First, rather than having a flat set of agents, each of them being linked to its meta-system, we propose a more general kind of meta-system controlling several agents. This allows us to take advantage of the fact that, when several agents act in the same sub-domain, the additional knowledge used to control them (the heuristics that allow for a choice among them) may be the same.

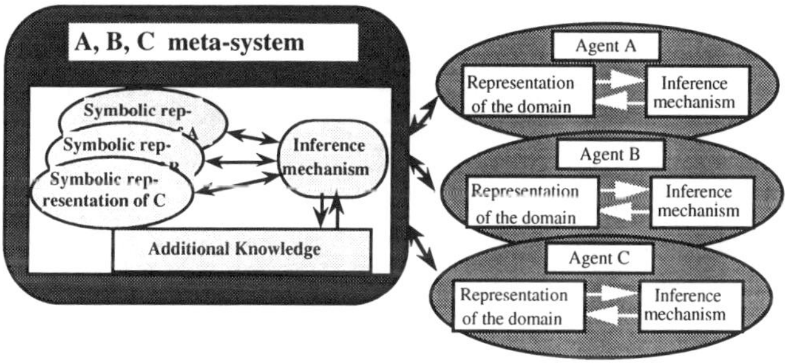

Figure 5. A Compound Meta-System Controlling Three Agents

Another extension consists in considering meta-systems themselves as usual agents (the set composed of a meta-system and the agents under its control is itself considered an agent — called *compound*). This set plays the same role as a single agent, and can thus be represented in, and associated with, its metalevel, which makes it reflective as well. The reflective mechanism applies recursively at various levels, resulting in a hierarchical organization of the agents: to reach a given goal, an agent has at its disposal several means (the agents it controls and may trigger).

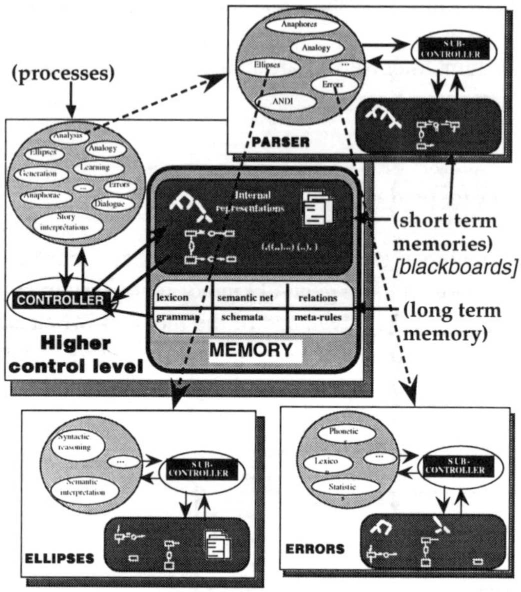

Figure 6. An example of a hierarchy of active processes. The goal of the higher control level is to understand a story. This level triggers the parser for every sentence of the text. The parser encounters an unforeseen problem and, to solve it, triggers both the ellipsis interpreter and the error corrector.

Meta-systems may themselves be reflective, controlling other agents, which may also be reflective and so on. The structure of the system is fully recursive, but each level manages its own blackboard, in which only the relevant piece of data is written. Figure 6 shows a hierarchy of compound agents: the higher level of control whose task is to understand a text, triggered the sentence parser. The parser, in its turn, encountering an unforeseen problem, triggered simultaneously the ellipsis solver and the error manager.

The embedding of meta-agents does not make the system overly complex nor inefficient. On the contrary, since it implies an *a priori* decomposition of the problems into sub-problems, it allows for a simpler planning process, which is the main part of our control mechanism. This “reflection” applies recursively, which corresponds to a hierarchical organization of the agents, and its behavior implies a continuous interaction between the usual reasoning of an agent and the reflective reasoning of its controller.

It is important to note that regarding the main controller and the sub-controllers, there is a continuous exchange between usual reasoning of an agent and reflective reasoning of its controller: the task for a complex agent M being to perform a goal G, M computes a static plan that consists of a sequence of agents it controls $\{A_i\}$.[5] Such reasoning is top–down, recursive and does not depend on the specific data to be processed.

The plan is then executed within M's common blackboard: the A_i produce their results in this memory or send a message to M to point out an unforeseen problem. Such a message, as well as any modification of the memory, triggers a reflective activity: a dynamic planning process tries to find what agents may solve the problem raised (when no solution is found, M itself sends a help message to S (M)), to check consistency, and to handle contingent aspects (figure 7). These reasoning processes are close to hierarchical or meta-level planning (Stefik 1981; Wilkins 1984).

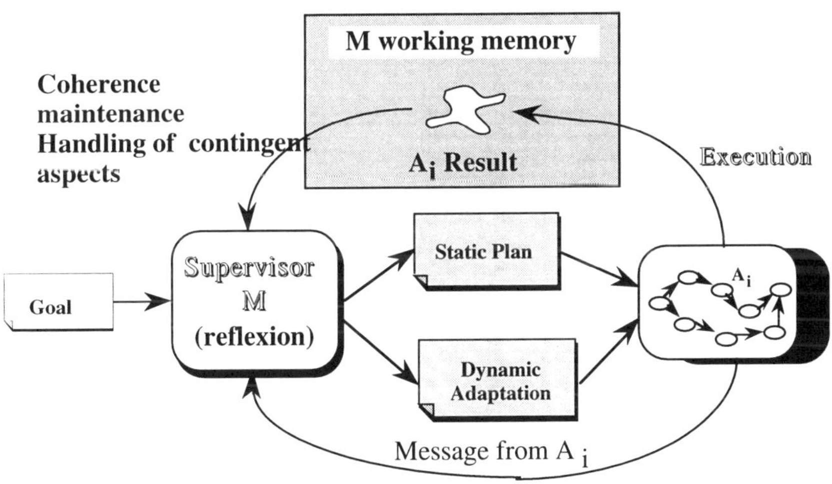

Figure 7. Continuous Exchange between Usual Reasoning and Reflective Reasoning

The controller also builds in its working memory a representation of the dynamic reasoning process, that is the sequence of agents that has allowed it to perform its task. This gives the system the ability to explain its actions and the chosen strategy. Memorizing what has been done and reasoning about it is, from our point of view, a useful part of consciousness. Nevertheless there are cases where more than a dynamic control is necessary. Our goal in the next paragraph is to go closer to a true notion of the role of consciousness …

4. A Bridge between the Sketchboard and the Controlled Processes: The Role of Consciousness

We will give here some precise ideas regarding the implementation of the various notions presented. This will result in a "revisited CARAMEL," used not only to understand and generate natural language, but also as a general model for intelligent behavior, since, as we will argue in the conclusion, language understanding should be a basis for intelligence in general.

As in Figure 8, consciousness is modelled as a controlled process able to trigger various sub-processes. Its data are stored in a blackboard. These data are permanent data (linked to the self) as well as results coming from the Sketchboard (candidates to become conscious). The various sub-processes are in charge of managing and evaluating the permanent goals, of evaluating the relevance of the candidates from the Sketchboard, of maintaining the self representation and so on.

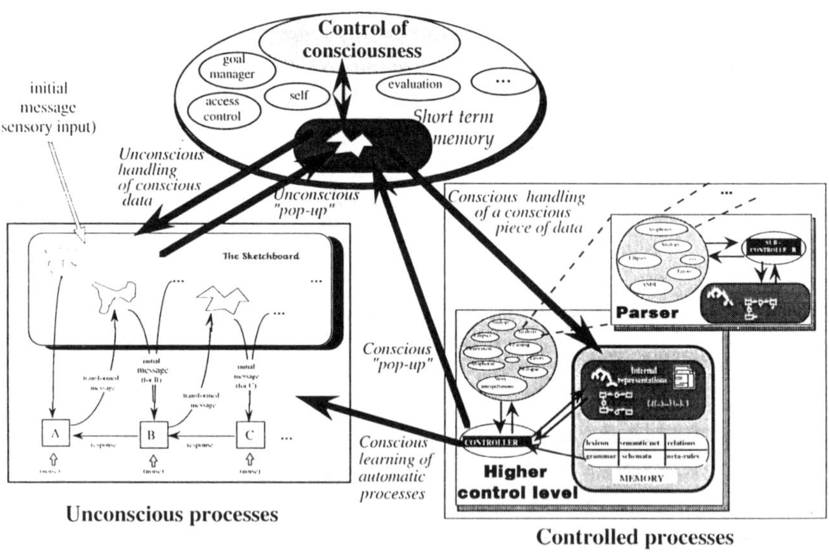

Figure 8. Consciousness as a bridge between subliminal and controlled processes. A piece of data in the Sketchboard is evaluated as deserving to become conscious (unconscious "pop-up") and later processed by controlled (conscious) processes. Relevant results of this conscious processing may also be made explicitly conscious (conscious "pop-up"). Control of consciousness decides whenever a piece of data or a specific problem deserve to be consciously processed or not; it also evaluates conscious processes, which, when encountered several times, give rise to unconscious processes (compilation).

When a piece of data deserves to become conscious (see § 4.1J, it is written in the short term memory which acts as a blackboard for consciousness. Therefore, our short term memory is very analogous to Baars' global conscious workspace (Baars 1988J which, unconscious processes are competing to take control of. This global blackboard is managed through a stack mechanism: any conscious event that becomes redundant may be replaced by a more informative subsequent event.

An important role of the interaction between the Sketchboard and the controlled processes through consciousness is to unify disparate results into a coherent whole. Therefore, it has a constructive function that neither unconscious processes, nor controlled processes, are able to perform on their own.

4.1. *Criteria*

We give here some criteria that may be used to decide whether a piece of data should be made conscious or not.

As mentioned above, different sequences of unconscious processes may act in a parallel way within the Sketchboard, and these sequences are analogous to Baars' competing contexts. Therefore, consciousness may be seen as permanent ly reading the Sketchboard and managing the writing within the short term memory. Only stable results — and *deserving* it — are written into this blackboard.

The notion of "*deserving*" relates to three other notions: the "*feeling of understanding*," the "*feeling of ambiguity*," and the "*feeling of contradiction*."

The "*feeling of understanding*" leads to some results becoming conscious. As we saw in section 2.2, this feeling of understanding is represented through some sort of stability within the Sketchboard. This stability is recognized when no significant shift appears, after several iterations. When such a situation is detected, the final result is transmitted to the short term memory, with the goal of integrating this new information within the currently active knowledge, in order to produce locally coherent data.

Another important reason for a result to become conscious is the "*feeling of ambiguity*." Cases where several solutions are relevant correspond to situations where data in the Sketchboard waver between several configurations (a "lasting instability"J. This will give rise, not to a conscious result, but to a conscious problem to be consciously solved: when no clear decision is suitable, the choice will be left to a controlled, conscious process. Therefore, partial results obtained from the Sketchboard are written into the higher control level blackboard, with the goal of resolving the remaining ambiguity.

Finally, there is the "*feeling of contradiction*." Such a situation is detected when a stable result in the Sketchboard is in contradiction with a goal managed within consciousness. In the same way as above, these results are transmitted to the conscious level in order for the problem to be solved by conscious means.

These criteria imply that only the results obtained through conscious perceptions are made available to other sources of knowledge, even though unconscious perceptions may influence further processing. This is explained in our model by the fact that fleeting results in the Sketchboard may imply some reordering of the goals of consciousness, even if those are not made conscious thereafter.

4.2. *Hierarchical Representation and Effects of Goals*

The process that models consciousness keeps a representation of various relevant objects (including a representation of the self) that represent the goals of the individual. In order to direct the behavior of the individual these goals are organized as a hierarchy where the higher goals are the most permanent ones (cf. Baars' goals and conceptual contexts). Goals are represented in consciousness as sets of properties to be maintained or to be made true about these representations. The range of these more or less implicit goals is very large, since they include very basic goals (e.g., to survive, to eat, to drink) as well as much more immediate goals (e.g., *to get a given cake*).

The very basic goals (analogous to Edelman's basic values) may be seen as innate compiled processes or processes internalized to a point that their declarative aspects have disappeared. On the other hand, more immediate goals are written into the short term memory. Intermediate implicit goals account for pragmatic constraints (reasons for speaking), discourse constraints (turn-taking, sticking to the topic, etc.), lexical and syntactic constraints, and so on. Reaching simultaneously all these goals will, for example, produce the sentence *I want this cake*.

The priorities of these goals (their relative importance) are constantly and dynamically re-evaluated. The result of this evaluation has significant influence on the evaluation of the candidates coming from the Sketchboard in order to become conscious (a piece of data that will be made conscious is one that favors the most important goals). In other words, these goals represent constraints that a piece of data should respect in order to become conscious.

Since the behavior of controlled processes is based on a double planning process (computing in a static way how the current goal can be reached and how this first plan should be dynamically adapted (Sabah 1990)), the deep goals may also have a significant influence on the conscious processes, as they can be introduced into the plans that these processes try to consciously complete.

These pre-defined goals are activated in an unconscious way, depending on the situation; they drive processes in the Sketchboard (various competing goals will result in sequences of processes acting in a parallel way within the Sketchboard). When there is no competition or when a solution is reached without any control, the process remains unconscious. On the other hand, when a goal or a sub-goal cannot be reached in that way, it becomes conscious and heavier planning processes will be used to reach it consciously.

4.3. *Reaching Goals*

When a goal has been made explicit within consciousness (as a representation to be built or as a given property that a representation should hold), the system first seeks for an automatic means allowing to reach it. If it can be found (i.e., an unconscious process acting within the Sketchboard is a priori valuated as a good candidate for achieving the task), it is triggered and the hierarchy of goals is re-evaluated, depending on its result.

If such a process cannot be found, the corresponding goal is given to the first level of controlled processes, which triggers conscious reasoning and planning. How the repetition of such process may lead to learning will be evoked in paragraph 4.5.

We are envisaging the extension of the conscious reasoning processes so that a closer intrication with unconscious processes could be achieved: when a controller must reach a given goal, it can act in a similar way as consciousness by first trying to find an unconscious process that may reach the goal (before trying to reach its goal with complex planning processes). An open question here is to decide whether such a situation implies that processes need to be triggered in the Sketchboard as presented above, or if the Sketchboard itself is distributed as well (with several Sketchboards respectively linked to various conscious controllers). Such an implementation would produce an entanglement of conscious and unconscious processes closer to what we can suspect about what happens in our brains.

4.4. *Relation with Attention*

As shown in Figure 8, attention is the process that makes a result or a running process conscious. Since consciousness is being modelled as a reflective process (as described in section 4), its meta-level may be seen as the process that models attention. In other words, attention partially controls consciousness. Then, we can distinguish between:

- unintentional attention (automatic) triggered by surprise (a difference between a computed result and some expectation). This will produce the conscious emergence of the corresponding results; and
- wilful attention triggered by a conscious (controlled) process. This will produce the conscious writing of a goal (or sub-goal) to be reached. Processes used to reach this goal are consciously triggered.

The more a mental representation is predictable, the more rapidly it tends to weaken; the more it is new and informative, the more it is likely to be conscious. This accounts for the fact that novelty or distance from expectations are the main reasons for a stimulus to become conscious. But to emerge, a conscious experience needs a degree of internal coherence, otherwise its accessibility will rapidly decrease due to the competition among alternative solutions.

Figure 8 shows the relations between the three essential levels evoked here (*unconscious* processes, *controlled* processes and *consciousness*).

4.5. *Learning*

Our model also accounts for some aspects of learning. Entities that stay for a while in the short term memory, and therefore that have been consciously perceived, are not completely obliterated hen there are no longer useful. They keep a residual accessibility in the long term memory, accessibility which is proportional to their persistence and their frequency of use. This transfer of data from short term memory (very volatile) to long term memory (much more lasting) is a first kind of acquisition (precise criteria to decide when this should occur remain to be exhibited).

Secondly, we consider that some processes may exist in the system under two different forms: a) when it uses declarative rules, it is a controlled, conscious process, b) but if it has been compiled, when the resulting code is triggered, it acts as an unconscious process within the Sketchboard.

Questions still remain regarding the situation of this distinction with respect to learning: how declarative rules are first built through various experiments, how some stability among these rules is detected, and how this results in a compiled version of the process. When a new problem appears, it is handled by controlled processes.[6] At each step, these processes dynamically compute the relevant plans and rules to be used. If the same kind of problem is encountered several times, the same rules could be found. Since this kind of process has the ability to represent its behavior and to reason about it, it will be able to detect such a situation. Then, the corresponding rules would be compiled, and the characteristics of the result of this compilation given to the Sketchboard manager. Subsequently, the Sketchboard will be able to trigger this new process if it is needed. Thus, the whole system has "learnt" how to solve this new kind of problem, and will make use of his previous experience at the time of its current behavior.

5. Evaluation of the Model

5.1. *Four Essential Properties of Consciousness*

We examine here four attributes of consciousness that Harth (1993) considers essential: selectivity, exclusivity, linking and unity (*his* definitions). We then try to evaluate our model through these definitions.

- *Selectivity*. Not all neural activities enter consciousness (perceptions, sensations, feelings). Thus, an important role of consciousness (linked to free will) is to select, among random events, those possibly leading to "interesting" thoughts.
- *Exclusivity*. Being conscious of something prevents us from thinking of something else at the same time (cf. the classical Necker cube). Consciousness has therefore a sequentialising effect (e.g., we can consciously explain what we would do as a classical von Neumann machine, but not as a parallel machine). The fact that the brain can simultaneously carry out hundred of tasks reinforces the first point that all processings cannot be conscious.
- *Chaining*. Items in consciousness are chained together, linked by association and reasoning. This implies the serial character of consciousness as well as its constructive role (putting together disparate results of unconscious processes).

- *Unitarity*. Consciousness unifies both the person who possesses it and the content of his or her conscious mind. It provides the continuity of our selfhood throughout our life span. To dc that, consciousness may re-create or modify data as well as results of the perceptive processes.

Taking into account the last two elements, Dennett (1993) opposes a *stalinian* (sic) vision of consciousness (non suitable events are deleted or modified) to an *orwellian* image (sic-again) (representations of never-occurring events are artificially built).

5.2. *The New Caramel and the Above Properties*

- *Selectivity*. In the paragraph 4.1, we have presented the criteria that are used in order to determine when a piece of data deserves to become conscious. Therefore, the system selects cases where it has the feeling of understanding or when a difficult problem arises, in order to make these situations conscious, since they may lead to interesting situations.
- *Exclusivity*. When the model is implemented on a single computer, processes within the Sketchboard may act in a parallel way, while controlled processes (following static and dynamic planning) act sequentially. However the machine does not have to follow this constraint: it may be implemented on a net of machines. We have shown elsewhere (Fournier *et al.* 1988; Fournier *et al.* 1990) that even with controlled processes a parallel implementation could be a significant improvement.
- *Linking*. This is typically the role of the process that simulates consciousness in our model: as it is partially based on a coherence maintenance, various significant results coming from the Sketchboard are transmitted to the controlled processes in order to build a coherent whole with them. Therefore, consciousness has a constructive role too in our model.
- *Unity*. The Sketchboard is based on the idea of modifying the very first results of perceptions so that they become coherent with higher level previous knowledge. So, as unifying external world events, our model of consciousness is in accord with this point. Many questions remain to be answered so that it may be considered as unifying a mind.

5.3. *Language and Consciousness*

Bresnan and Kaplan (1981), following Chomsky (1966, 1968), discuss various constraints on syntactic mapping:

- *creativity* (possibility of constructing means for characterizing, any of the infinite number of grammatical sentences);
- *finite capacity* (the infinite number of grammatical sentences results from finite set of words and relations and finite storage ability);
- *reliability* (a syntactic construction can be judged reliably, independently of context and meaning);
- *order-free composition* (local grammatical relations can be constructed for arbitrary fragments of sentences);
- *universality* (any string of words is related to some mental structure by a universal procedure).

Edelman (1989) and Rosenfield (1992) underline that there exist strong relationships between these characteristics and some aspects of consciousness.

- the *creativity* constraint is provided by the linkage to cortical conceptual systems that allow enormous combinatorial power due to a special memory for new conceptual combinations (this means is supplied by reentrant memory systems [Broca's and Wernicke's areas, for example] that interact with already existing conceptual systems);
- the *finite capacity* constraint is explained by considering memory as a form of recategorization with strong capabilities of generalization;
- the *reliability* constraint is related to the fact that the already existing conceptual system can treat verbal productions as a set of objects to be classified;
- the *order-free* constraint is due to the fact that syntax emerges from semantics and phonology;
- the *universality* constraint is consistent with the categorization of productions by the conceptual system, performed by the same process used for learning in general.

The importance of inner speech. Language also has an extremely important role: the inner voice maintains a running commentary about our experiences and feelings, it helps to relate past events to plans for the future. Habit and learning imply that this inner speech becomes less and less conscious.

However, it is necessary to clearly distinguish between such an inner speech and a hypothetical "language of thought" since there can be no identity at any level.

With a given point of view, "to speak" is equivalent to "make conscious," and from this, language gives us extraordinary possibilities to extend our memory: our short term memory, first, with the inner speech, our long term memory afterwards, thanks to communication with other people, and lastly with written language and books Therefore, language is the greatest means of storing knowledge and plays an essential role within any *conscious* episode that makes past, present or future events explicit. This leads Edelman to a remark that undermines the main role of consciousness for natural language understanding:

> The main reason why computers are unable to tackle the problem of semantics becomes clear: the implementation cannot be correct since it does not lead to consciousness. (Edelman 1992)

6. Conclusion

It may be decades before we have the talking computers popularized in science fiction. Moreover, we cannot guarantee that we will even be capable of building a computer program that understands human language perfectly. We have elaborated in this paper two propositions concerning conscious and unconscious processes. The first is a conceptual model that allows for a flexible and efficient implementation of intelligent systems. We have proposed extensions of reflective systems to the domain of DAI. In our implementation, control is explicit at various levels, thus making possible to use strategic knowledge at varying degrees of generality. The second proposition concerns a new data structure, the *Sketchboard*, an extension of Blackboards, allowing different modules to collaborate while solving a problem. This model allows feedback from higher levels to lower levels of processing, without any explicit control.

While the ideas presented in the two first sections have been implemented (a running version exists in LISP for the first one and is currently being rewritten in Smalltalk, a Smalltalk version exists for the second), it is not the case for the ideas presented in section 4, which remain to be fully tested. Several aspects remain to be implemented and tested through reasonably-sized experimentations before a number of parameters of the model may be efficiently chosen. However, it seemed very important to show that consciousness is not as far from Artificial Intelligence modeling capabilities as some would claim.

I have argued for reflectivity to be a central point for intelligent processing and for natural language understanding. Semantics is nowadays the bottleneck for real size implementations that are to remain psychologically relevant. I think semantics should be better grounded and take perceptive aspects into account, and I showed here that our Sketchboard is a possible way of doing so.

Then, I presented some ideas in order to build a model of consciousness that is a bridge between these two models. Our proposal is primarily conceived as being implementable. But, we believe it to have true psychological relevance, since it takes into account the fact that neither interactive nor modular language-processing systems emerge. Instead, some components of language-processing should be modular and others interactive.

To go further I would assert that language understanding *must* provide the basic process for almost every other intelligent mechanism, since for us human beings, language is the essential basis that allows for many other cognitive abilities. Even if this should be a very long term research project, I hope I have been convincing enough for this alternative be considered a possible and promising way, and may be the real future of AI.

Notes

1. In this chapter, we will use indifferently the words *processes*, *agent* or *module*, to designate an entity of the system that uses knowledge considered as independent from the other agent knowledge. Thus, our point is that our proposition is general enough to apply to various entities, whether they are typical agents of distributed artificial intelligence, procedural modules, or even connectionist modules.
2. Several elements of this paragraph and the following one are inspired from (Bassi Acuña 1995).
3. Here, it may appear that this mechanism brings us out of the blackboard paradigm, where processes are not conceived as directly exchanging messages. This is not the case: *by definition of modularity* A does not have the necessary knowledge to interpret a sophisticated message from B; the only thing it can do is to compare the values of two successive responses from B, and act with the goal of making this response optimal.
4. *"The horse raced past the barn fell"* is the classical example introduced in (Bever 1971).
5. We note $M = S(A_i)$ to indicate that M controls the agent A_i (i.e., the meta-system that possesses a representation of A_i).
6. Here *new* means that there exist no automatic process to solve the problem, nor can any existing plan be used to solve the problem.

References

Agha, Gul and Carl Hewitt. 1986, *Actors: A Model of Concurrent Computation in Distributed Systems*. Cambridge, MA: MIT Press.

Baars, Bernard. 1988, *A Cognitive Theory of Consciousness*. Cambridge, UK: Cambridge University Press.

Bachimont, Bruno. 1992, *Le contrôle dans les systèmes à base de connaissances*. Paris: Hermès.

Bassi Acuña, Alejandro. 1995. *Un modèle dynamique de la compréhension de texte intégrant l'acquisition des connaissances*. Paris XI: thèse d'université.

Bever, T.G. 1971. Integrated study of linguistic behavior. In *Biological and Social Factors in Psycholinguistics*. London: Logos Press.

Bresnan, Joan and Ronald Kaplan. 1981. Lexical functional grammars: A formal system for grammatical representation. In *The Mental Representation of Grammatical Relations*. Cambridge, MA: MIT Press.

Chomsky, Noam. 1966. *Cartesian Linguistics*. New York: Harpert & Row.

Chomsky, Noam. 1968. *Language and Mind*. New York: Harcourt.

Davis, Randall and Reid Smith. 1983. Negotiation as a metaphor for distributed problem solving. *Artificial Intelligence* 20(1), 69–109.

Dennett, Daniel. 1993. *La conscience expliquée*. Paris: Odile Jacob (*Consciousness Explained*, Little, Brown and Company, 1991).

Eccles, John. 1992. *Évolution du cerveau et création de la conscience*. Paris: Fayard. (*Evolution of the brain: Creation of the self*, New York: Routledge, 1989).

Edelman, Gerald. 1989. *The Remembered Present: A Biological Theory of Consciousness*. Paris: Basic Books.

Edelman, Gerald. 1992. *Biologie de la conscience*. Paris: Editions Odile Jacob.

Erman, L.D., F. Hayes-Roth, Victor Lesser and D. Raj Reddy. 1980. The HERSAY-II speech understanding system: Integrating knowledge to resolve uncertainty. *Computing surveys* 12(2), 213–253.

Ferreira, Fernanda and John M. Henderson. 1991. Recovery from misanalyses of garden-path sentences. *Journal of Memory and Language* 30, 725–745.

Fournier, Jean-Pierre, Peter Herman, Gérard Sabah, Anne Vilnat, Nathalie Burgaud and Michel Gilloux. 1988. Processing of unknown words in a natural language processing system. *Computational intelligence* 4(2), 205–211.

Fournier, Jean-Pierre, Gérard Sabah and Caroline Sauvage. 1990. A parallel architecture for natural understanding systems. *Proceedings PRICAI'90, Nagoya* 787–792.

Harth, Erich. 1993. *The creative loop; how the brain makes a mind*. New York, Addison-Wesley.

Hayes-Roth, Barbara. 1985. A blackboard architecture for control. *Artificial intelligence* 26, 252–321.

Hayes-Roth, Barbara, R. Washington, D. Ash, R. Hewett, A. Collinot, A. Vina and A. Seiver. 1992. Guardian: A prototype intelligent agent for intensive-care monitoring. *Artificial intelligence in Medicine* 4, 165–185.

Hewitt, Carl. 1977. Viewing control structure as patterns of passing messages. *Artificial Intelligence* 8(3), 3243–364.

Hewitt, Carl. 1986. Offices are open systems. *ACM Transactions* 4(3), 271–287.

Hewitt, Carl and Peter de Jong. 1984. Open systems. In *On Conceptual Modeling*, 147–164. New York: Springer Verlag.

Kandel, Eric R. and Robert D. Hawkins. 1992. The biological basis of learning and individuality. *Scientific American* 2(3), 52–61.

Kosslyn, Stephen. 1980. *Image and Mind.* Harvard, MA: Harvard University Press.

Kosslyn, Stephen and Olivier Kœnig. 1992. *Wet Mind, The New Cognitive Neuroscience.* New York: The Free Press.

Lesser, Victor and Daniel Corkill. 1983. The distributed vehicle monitoring testbed: A tool for investigating distributed problem solving networks. *AI Magazine* Fall, 15–33.

Maes, P. 1987. Computational reflection. *AI laboratory,* 87–2. Bruxelles: Vrije Universteit.

Meyer, D.E. and R.W. Schvaneveldt. 1971. Facilitation in recognizing pairs of words: Evidence of a dependence between retrieval operations. *Journal of Experimental Psychology* 90, 227–234.

Miller, George. 1956. The magical number seven, plus or minus two: Some limits on our capacity for processing information. *Psychological Review* 63(2), 81–97.

Minsky, Marvin. 1985. *The Society of Mind.* New York: Simon and Schuster.

Newell, Allen. 1990. *Unified Theories of Cognition.* Cambridge, MA: Harvard University Press.

Nii, Penny. 1986. Blackboard systems: The blackboard model of problem solving and the evolution of blackboard architectures. *The AI magazine* August, 82–106.

Pitrat, Jacques. 1990. *Métaconnaissance, futur de l'intelligence artificielle.* Paris: Hermès.

Ramón y Cajal, Santiago. 1933. Neuronismo o reticularismo? La prueba objectivas de la unitad anatomica de la celulad nervioses. *Archos. Neurobiologicas* 13, 217–291.

Restak, Richard. 1979. *The Brain: The Last Frontier.* New York: Doubleday, Garden City.

Rosenfield Israel. 1992. *The Stange, Familiar and Forgotten: An Anatomy of Consciousness.* New York: Alfred A. Knopf.

Sabah, Gérard. 1990a. CARAMEL: A computational model of natural language understanding using a parallel implementation. *Proceedings ECAI, Stockholm,* 563–565.

Sabah, Gérard. 1990b. CARAMEL: A flexible model for interaction between the cognitive processes underlying natural language understanding. *Proceedings Coling, Helsinki.*

Sabah, Gérard. 1990c. CARAMEL: Un système multi-experts pour le traitement automatique des langues. *Modèles Linguistiques* 12, Fasc. 1, 95–118.

Sabah, Gérard. 1993. Vers une conscience artificielle? In *Modèles et concepts pour la science cognitive: hommage à Jean-François Le Ny,* 207–222. Grenoble: PUG.

Sabah, Gérard. 1996. Le «carnet d'esquisses»: une mémoire interprétative dynamique. *Proceedings RF-IA*. Rennes.

Sabah, Gérard and Xavier Briffault. 1993. CARAMEL: A step towards reflexion in natural language understanding systems. *Proceedings IEEE International Conference on Tools with Artificial Intelligence*, 258–265. Boston.

Sabah, Gérard, Christophe Godin and Anne Derain. 1991. Proposition for the Control Architecture of PLUS. Internal PLUS Paper in Deliverable 1.2: FUNCTIONAL DESIGN (Bill Black ed) ESPRIT CEE.

Sabah, Gérard, Lydia Nicaud and Françoise Forest. 1988. A modular system for natural language understanding. *Proceedings European Workshop ESSCS.* Varenna.

Sauvage, Caroline, Lydia Nicaud and Gérard Sabah. 1989. CARAMEL: A flexible model for natural language understanding. *Proceedings Colloque IA*, 283–292. Tel Aviv.

Smith, Brian. 1986. Varieties of self-reference. In *Proceedings Theoretical Aspects of Reasoning about Knowledge*, Morgan Kaufman (ed), 19–43. Los Altos, Monterey, CA.

Stefik, M. 1981. Planning with constraints, *Artificial Intelligence* 16, 111–170.

Swinney, D.A. and D.T. Hakes. 1976. Effects of prior context upon lexical access during sentence comprehension. *Journal of Verbal Learning and Verbal Behavior* 15(6), 681–689.

Vignaux, Georges. 1992. *Les sciences cognitives: une introduction.* Paris, La Découverte.

Wilkins, David. 1984. Domain-independent planning: Representation and plan generation. *Artificial Intelligence* 22, 269–301.

A Neurocognitive Model for Consciousness and Attention

James Newman[1]
Colorado Neurological Institute
Denver, Colorado

Bernard J. Baars
The Wright Institute
Berkeley, California

Sung-Bae Cho
Department of Computer Science
Yonsei University, Seoul

It is widely argued that consciousness is an intractable, or irrelevant, problem for cognitive science (for recent examples, see Harnad 1994, Penrose 1994). We argue that the problems it presents are not inherently more intractable than, for example, those presented by language. In this chapter, we briefly review a growing body of evidence that some of the basic neural mechanisms that underlie consciousness are actually becoming better articulated and understood than those of other complex cognitive functions, such as language. Note that we do not claim to "explain" consciousness, only to clarify certain relevant issues surrounding it. Clarity on this controversial subject can be furthered, we believe, by carefully articulated definitions. Our study of the literature in cognitive psychology and neuroscience bearing on these issues (Baars 1988; Baars and Newman 1994; Newman and Baars 1993; Newman 1995a) has convinced us that consciousness can be operationally defined as follows:

> Consciousness reflects the operations of a global integration and dissemination system, nested in a large-scale, distributed array of specialized bioprocessors; among the various functions of this system are the allocation of processing resources based, first, upon biological contingencies of novelty, need or potential threat and, secondly, cognitive schemes, purposes and plans.

In a series of books and papers Baars has developed a set of "Global Workspace Models," all based on a single pattern, which addresses a substantial domain of evidence that is explicitly related to conscious experience (Baars 1983, 1988, 1992, 1996). These models explicate an architecture in which many parallel, non-conscious experts interact via a serial, conscious and internally consistent Global Workspace (GW), or its functional equivalent. GW, or "blackboard," architectures have been developed by cognitive scientists since the 1970's (Reddy, Erman, Fennell and Neely 1973), and this framework is closely related to the Unified Theories of Cognition of Simon, Newell and Anderson (see Newell 1992). Practical AI applications of this class of models, discussed in a concluding section, are numerous.

The notion of a global workspace was initially inspired by the HEARSAY model of speech understanding (Reddy, Erman, Fennell and Neely 1973), one of the earliest attempts to simulate a massively parallel/interactive computing architecture. This architecture consisted of a large number of knowledge modules, or "local experts," all connected to a single "blackboard," or problem solving space. Activated experts could post "messages" (or hypotheses) on the blackboard for all the other experts to read. Incompatible messages would tend to inhibit each other, while the output of cooperating experts could gain increasing access to the blackboard until a global solution emerged out of this "jostling for attention." This sort of architecture is slow, cumbersome and error-prone, but can produce solutions to problems which are too novel or complex to be solved by any extant modular knowledge source (once over-learned, even the most complex problem may be allocated to non-conscious solution).

The significance of this set of models to subsequent developments in cognitive science is attested to by McClelland (1986) who describes HEARSAY not only as "a precursor of the interactive activation model," but "of the approach that underlies the whole field of parallel distributed processing." (p. 123). McClelland's (1986) own "Programmable Blackboard Model of Reading," discussed in a concluding section, is a connectionist instantiation of the global workspace metaphor.

As will become clear as we describe the neural architecture of our model, its architecture is also quite compatible with Arbib's (1989) characterization of "the brain [as] a layered somatotopic computer … holding multiple maps, each connected to one another." In his model, schemas are modular entities whose instances can become activated in response to certain patterns of input from sensory stimuli or other schema instances that are already active (Arbib, Conklin and Hill 1987). Schema theory differs from GW theory in that it describe units or modules in a network of active processors communicating with each other, rather than through a unitary blackboard. In connectionist terms, schemata are not so much "entities" as relatively stable sets of connection strengths which become active under predictable circumstances. This would suggest that schema theory is concerned more with implicit knowledge sources than with conscious information processing. Yet these differing perspectives result in a dilemma.

> On the one had, schemata are the structure of the mind. On the other hand schemata must be sufficiently malleable to fit around almost anything. None of the versions of schemata proposed to date have really had these properties. (Rumelhart, Smolensky, McClelland and Hinton 1986: 20J

We would propose that a solution to this dilemma is to allow schema instances to compete and cooperate for access to a global workspace, where modifications and novel combinations can serve to fit existing schemata "around" a broad range of contingencies.

As a final example of compatible architectures, this class of models has much in common with the branch of artificial intelligence called "distributed AI" (DAIJ, which studies "how intelligent agents coordinate their activities to collectively solve problems that are beyond their individual capabilities" (Durfee 1993: 84J. Examples of DAI applications are such things as generic conflict resolution, unified negotiation protocols, and search-based models of coordination/cooperation. DAI approximates human interpersonal behavior to a greater degree than purely logic-driven AI, in that agents must learn to be "knowledgeable and skilled in interacting with others" (*ibid.*, p. 86J. An intelligent balance between competitive self-interest and cooperative problem solving is essential to optimizing overall outcomes.

Such cooperative, globally-integrative strategies appear to characterize the optimal processing of conscious information. A conscious visual percept, for example, represents a unified gestalt of shape, texture, color, location and movement, despite the fact that these various qualia are initially processed in separate brain regions in each cerebral hemisphere. Likewise, a conscious intention is generally single-minded and goal-directed. Of course, conflicts can and do arise, but a central purpose of consciousness is the resolving of such conflicts (sometimes by inhibiting relevant, but incompatible inputsJ.

While such unified brain states are highly adaptive, they are not essential to many human activities. Most of our thoughts and actions are automatic, requiring only minimal conscious monitoring. The key strokes I am typing to produce this manuscript, the saccades my eyes make in reading it, the choice of individual words to type, all these cognitive acts proceed with a minimum of awareness on my part. Paradoxically, the nucleus of my awareness in this activity seems to be a question: "Am I writing what I intended?" To the extent that the answer is "Yes," my awareness tends to proceed fairly seamlessly along the lines of my current intention. Should I become conscious of the need to make corrections (spelling, replace words, move phrases around, etc.), my attention shifts appropriately. My intention, however, remains unaltered. And thus, the conscious stream of my thinking proceeds until I decide I have reached my intended goal, or something (hunger, the telephone ringing, remembering an appointment in 15 minutes) shifts my awareness to a more pressing concern.

This phenomenological description of my present awareness is entirely consistent with GW theory, and what it predicts about brain processes. The vast majority of cognitive tasks preformed by the human brain are automatic and largely non-conscious. Conscious processing only comes to the fore in instances where stimuli are assessed to be novel (anomalous), imminent, or momentarily relevant to active schemas or intentions. The defining properties of stimuli which activate conscious attention (i.e. the global allocation of processing resources) are that they: (1) vary in some significant degree from current expectations; or (2) further the current, predominant intent/goal of the organism. In contrast, the processing of stimuli which are predictable, routine or over-learned is automatically allocated to non-conscious, highly modularized cognitive systems (see Kihlstrom 1987; Baars 1983, 1988, 1995; Baars and Newman 1994).

In the most general sense, we are centrally conscious of what has the highest relevance to us in the moment, whether that be a momentary threat, a biological need, or the broadest insight into a previously unimagined reality. While the range of our awareness is only limited by our most developed cognitive capacities, the basic mechanism for the allocation of these processing resources remains constant under virtually all contingencies. And, we will argue, the basic neural circuitry of this resource-allocation system is becoming increasingly well understood, to the extent that key aspects of it can be modeled to a first approximation employing neural network principles (see below).

The applications of GW theory go beyond neural network modeling, however. As an example of how it can be fruitfully applied to central philosophical problems in consciousness studies, take the perennial conundrums of the homunculus and Cartesian theater. The two are, of course, related. The metaphor these images conjure up is of a "little man in our head" observing and manipulating the play of conscious images passing across the theater of our mind. That image, while beguiling, is absurd: for who is this strange being lodged in my (or your) mind? And who controls him?

Global workspace theory suggests a more complex and dynamic scenario: the single homunculus is replaced by a large audience. The theater becomes a workspace to which the entire audience of "experts" has potential access, both to "look at" others's inputs and contribute their own. Awareness, at any moment, corresponds to the pattern of activity produced by the then most active coalition of experts, or modular processors. Thus, there is not some fixed, superordinate observer. Individual modules can pay as much or as little attention as suits them, based upon their particular expertise and proclivities. At any one moment, some may be dozing in their seats, others busy on stage. Thus, a crucial difference between these local experts and a theater audience is that each can potentially contribute to the direction the play takes. In this sense the global workspace resembles more a deliberative body than an audience. Each expert has a "vote," and by forming coalitions with other experts can contribute to deciding which inputs receive immediate attention and which are "sent back to committee." Most of the work of this deliberative body is done outside the workspace (i.e., non-consciously). Only matters of central import gain access to center stage.

The beauty of this sort of system is that while it is a multiplicity, its behavior is largely coherent and adaptive. This later characteristic derives from the fact that the workspace serves as a "global integration and dissemination system." It is out of the operation of this system that conscious awareness arises (and, over time, our sense of being a coherent "I"). And it is this unitary awareness — not any agent or homunculus — that is globally superordinate, not in any hierarchical sense, but in the sense of the GW being accessible for representing and potentiating the activities of any of the competing arrays of "specialized bioprocessors."

Of course, such a system is prone to inefficiencies and pathological perturbations, but these seem consistent with the scientific literature concerning human consciousness (see Baars 1988, 1996), as well as the branch of AI research which studies cooperative distributed problem solving (Durfee 1993). A logical question, given that these specialized bioprocessors number in the hundreds, if not hundreds of thousands, is how does the CNS intelligently allocate such a vast array of processing resources?

1. Unity from Diversity: The Global Allocation of Processing Resources

We have sketched the outlines of the architecture and dynamics of our Global Workspace. Newman and Baars (1993) reviews anatomical and physiological evidence for a "tangential intracortical network" which Newman hypothesizes as the essential substrate for the representation of conscious percepts. As our discussion up to this point should have made clear, however, the workspace is a dynamic, interactive system, not a fixed structure; and its activities are by no means confined to the cerebral cortex. The aspect of the GW system we wish to focus on here is the means by which this highly distributed system allocates processing resources. The neural bases of resource allocation, or attention, have been extensively explored in recent years (see Mesulam 1985; Posner and Rothbart 1991; LaBerge 1995 for reviews), but generally without the distinctions made in GW theory between the conscious and non-conscious processing of information. In most studies, the focus is upon attentional mechanisms in a single sensory modality, typically the visual system (see, e.g., Crick and Koch 1990: LaBerge 1995).

But, as we have argued, conscious awareness is rarely unimodal and many fairly complex forms of attention can, and usually do, operate non-consciously. As an example from connectionist research on reading, it is common knowledge that in computer simulations, activated modules must be "sticky," that is "interactive activation processes [must] continue in older parts of the programmable blackboard while they are being set up in newer parts as the eye moves along…" (McClelland 1986: 150–151). This "stickiness" clearly entails a type of attention, but it normally proceeds quite automatically, both in a reading machine and a literate human being. It is only when the process is disrupted by, say, a never-before-encountered word, that the individual word becomes the focus of our conscious awareness. What we are normally conscious of is the overall meaning of the passage of text, not individual words — yet we must clearly attend to each word if we are to glean the overall meaning of a passage. It may be a great challenge for cognitive science to explain how "overall meaning" arises in the mind, but we do not believe problems such as this preclude the modeling of attentional processes associated with consciousness (any more than the considerable problems in producing a machine that can read fluently are deterring AI researchers from making progress on that problem).

Our point, however, is that conscious awareness involves a particular form of resource allocation which Newman and Baars (1993) terms "global attention." Consistent with the GW model outlined above, this concept is neither vague nor tautological. Rather, "it refers to a level of cognitive processing at which a single, coherent stream of information emerges out of the diverse activities of the CNS" (Newman 1993: 258). The focus of that stream could (under atypical circumstances) be an individual word or even a single quale, but this "global integration and dissemination system" seldom concerns itself with the processing of such rudimentary representations. Rather it seems to be biased towards increasingly multifaceted, yet unified images.

As a purely visual-spatial example, we are able to perceive a Necker Cube as projecting out of a two-dimensional page, alternately to the left, then to the right; but we are curiously incapable of perceiving these two perspectives simultaneously. Despite its capacity to represent multiple qualia (shape, size, orientation, depth perspective, in this case), conscious attention seems to have a "one track mind." The non-conscious allocation of processing resources does not operate under such a constraint. For example, neuroscience has shown that widely separated modular units in the cortex routinely process, in parallel, the contours, color and spatial location of a visual stimulus. Yet, our consciousness perceives all of these qualia as a single object of perception. And this is as true for multimodal as unimodal perceptions.

What might be the neural mechanism of this global form of attention? One would expect it to be fairly intricate and extensive, which we believe it is. But, surprisingly, its core element is an array of gating circuits contained in two nuclei with a combined surface area of less than 9 square centimeters. These mirror-image nuclei would be unremarkable in themselves, were it not for the patterns of connectivities they share with the cortex and brain stem reticular formation. Neuroanatomists refer to them as the nuclei reticularis thalami (nRt), or reticular nuclei. They cover the lateral surfaces of the left and right thalami, much like a shell covers an egg (Figure 1). Through them pass nearly all of the pathways coursing between the thalamus and cerebral hemispheres. Most of these pathways give off collateral axons to nRt, while nRt neurons themselves project mainly to cells of the particular thalamic nucleus lying directly beneath them. There is a quite orderly topography to this array of axon collaterals, "gatelets," and underlying thalamic nuclei (Scheibel and Scheibel 1966; Scheibel 1980: Mitrofranis and Guillery 1993). It essentially mirrors, in miniature, the modular architecture of the cortex (see Newman and Baars 1993; Newman 1995a for reviews).

A host of neuroscientists have contributed to the elucidation of this "central gating mechanism" in the thalamus (see Jasper 1960; Skinner and Yingling 1977; Jones 1985; Scheibel 1980; Steriade and Llinas 1988; Llinas and Pare 1991 for reviews), but it was Skinner and Yingling (1977) and Scheibel (1980) who first proposed a neural model for its putative role in selective attention. This model can be metaphorically represented as a "wagon wheel." The thalamus forms the hub. The reticular nucleus would correspond to the metal sleeve fitted around the hub. The upper rim of the wheel is the cortical mantle. The lower half of the wheel corresponds to the major sensory systems and subcortical nuclei whose projections converge upon the thalamus.

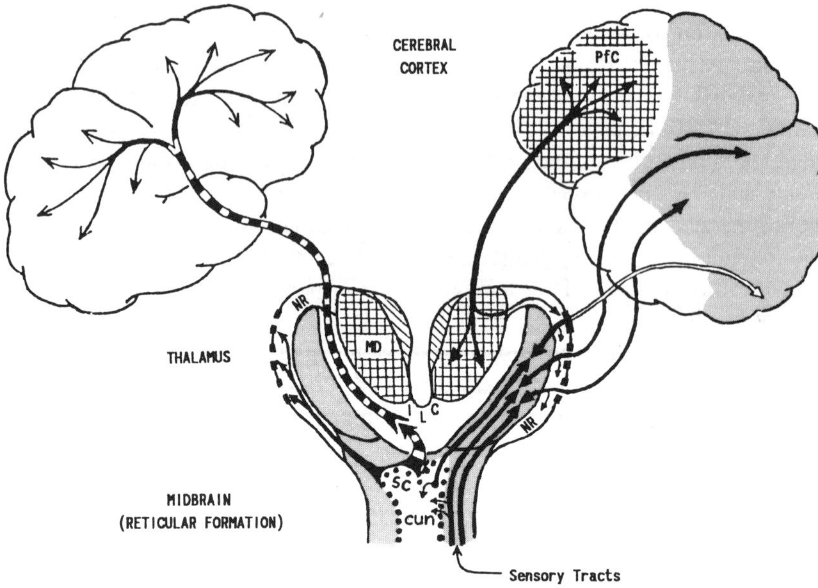

Figure 1. Model of a Neural Global Workspace System showing a schematic, coronal section through the midbrian *and* thalamus, *illustrating projections (arrows) between them, and with the* cerebral cortex. *The shaded areas represent: classical sensory pathways (in the midbrain); the ventral nuclei of the thalamus; and the areas of the cortex (right side) with which these nuclei share projections. The crosshatched areas designate the medial dorsal (MD) nucleus and pre-frontal (PfC). The unshaded areas in the* thalamus *and* midbrain *constitute the reticular core responible for the global activation of a Tangential Intra-cortical Network via projections (dashed arrows) from the Midbrain Reticular Formation (sc/un) and intralaminar complex (ILC). The heart of this extended activation system is the nucleus reticularis (NR, left and right), which both "gates" information flow to and from the cortex, and regulates rhythmic EEG patterns throughout the tangential network (from Newman and Baars 1993).*

Starting from "below," the first set of spokes are the major sensory pathways for vision, audition and the bodily senses. These project, in an orderly topography, to modality-specific nuclei in the thalamic "hub" (Th, Figure 2). As they ascend through the brain stem towards the thalamus, these sensory tracts give off collateral projections (bottom arrows) to the reticular core of the brain stem (Midbrain Reticular Formation, MRF, Figure 2). Scheibel (1980) reviews two decades of experiments indicating that these brain stem afferents serve as the

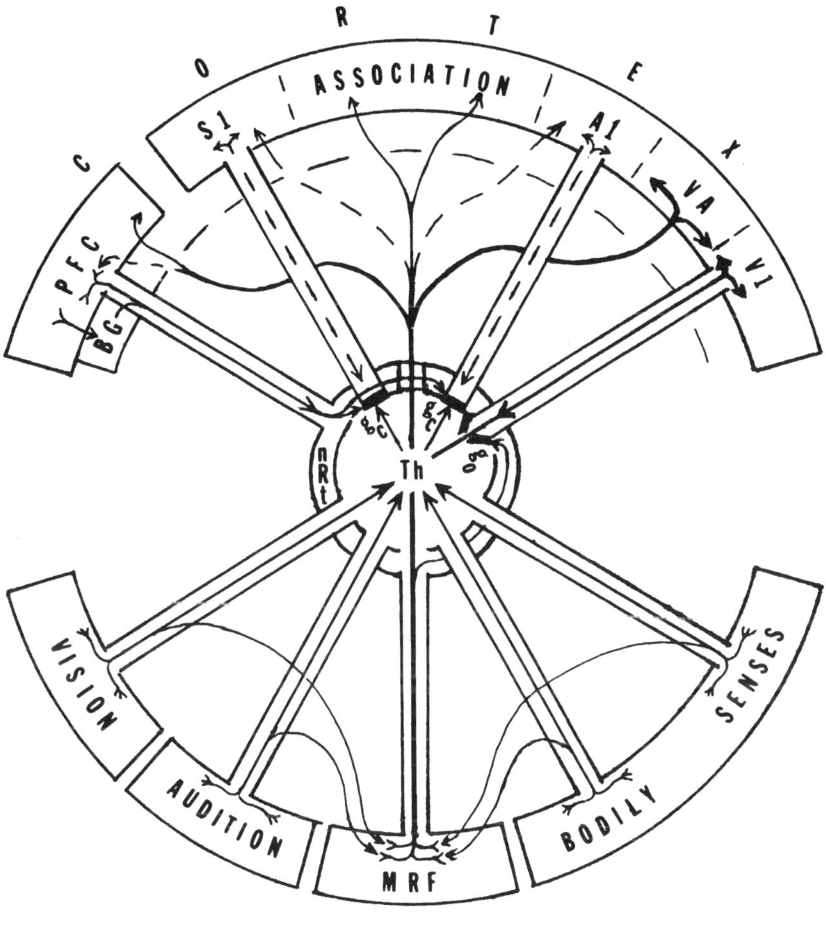

Figure 2. *"Wagon Wheel" model of CNS systems contributing to global attention and conscious perception. A1 — primary auditory area; BG — basal ganglia;* g_c *— "closed" nRt gate;* g_o *— "open" nRt gate; MRF — midbrain reticular formation; nRt — nucleus reticularis thalami; PFC — prefrontal cortex; S1 — primary somatosensory area; Th — ventral thalamus; V1 — primary visual cortex.*

basis for an initial "spatial envelope," or global map, of the sensory environment surrounding the animal.

> Most reticular neurons ... appear multimodal, responding to particular visual, somatic and auditory stimuli, with combinations of the last two stimuli most numerous. The common receptive fields of typical bimodal cells in this array show a significant degree of congruence. For instance a unit responding to stimulation of the hind limb will usually prove maximally sensitive to auditory stimuli originating to the rear of the organism. These twin somatic and auditory maps retain approximate register and overlap the visuotopic map laid down in the more peripheral layers of the superior colliculus ... These data might be interpreted to mean that each locus maps a point in the three-dimensional spatial envelope surrounding the organism. Further studies suggest the presence of a deep motor map closely matching and in apparent register with the sensory map. (p. 63)

The superior colliculus, which receives direct afferents from the retina, is known to play an important role in the coordination of eye movements as, for example, when an animal tracks a moving stimulus. Thus, it appears to serve as an early representational field for visuomotor perception. Crick and Koch (1990) refer to such representational fields as "saliency maps," and note that "a saliency map for eye movements is known to exist in the superior colliculus of mammals." (p. 957). They write:

> The basic idea of a saliency map (Koch and Ullman 1985) is that various visual areas send retinotopically mapped signals to a distinct area of the brain. This area does not code information on *what* it is that is salient but only *where* it is. (p. 957)

2. Topographic Maps, Gating Arrays and Global Attention

Scheibel's (1980) model suggests that Crick and Koch's (1990) visuotopic saliency map is actually an integral part the reticular core spatial envelope (MRF in Figure 2) globally orienting the sensorium. Of particular significance to the model for global attention we are presenting is the fact that this spatial envelope appears to project, with some precision, upon nRt. This topographic projection enables it to disinhibit particular arrays of nRt gatelets, selectively enhancing the flow of sensory information to the cortex. This enhancement of sensory transmission through the thalamus is most evident for stimuli that are novel, or potentially significant to the organism (e.g. a moving stimulus) (Scheibel 1980; Mesulam 1985).[2]

As one would expect, cortical projections upon the thalamus have more pervasive, and subtle, effects upon information flow. We can only describe these effects in the briefest terms here (see Newman and Baars 1993; Newman 1995a for more detailed accounts). These effects are largely inhibitory. While activation of the brain stem reticular core "opens" nRt gatelets, enhancing sensory transmission, the general effect of cortical projections is to "close" these same gatelets, blocking the flow of excitation to the cortex ("g_o" and "g_c" in nRt, Figure 2).

But as one would predict, given the modular architecture of the neocortex, cortical effects upon nRt's gating array are much more selective. Prefrontal projections were shown in animal experiments undertaken by Skinner and Yingling (1977) to be modality specific. Activity in one portion of a prefrontal-thalamic tract shut down sensory processing in visual, but not auditory, cortex. Activation of another portion of the tract shut down auditory processing, but allowed visual inputs to reach posterior cortex (see Figures 1 and 2). Within the posterior cortex itself, parallel cortico-thalamo-cortical loops would appear to serve to both amplify local inputs and (via nRt) inhibit the firing of adjacent, less active loops (Steriade, Curro-Dossi, Pare and Oakson 1991; LaBerge 1995). Using neural networks to model columns of thalamo-cortical loops, Taylor and Alavi (1993) concluded:

> symmetric inhibitory lateral connections between NRT cells, with excitatory NRT-thalamic feedback, sets up competition between thalamic inputs. There is enhancement of the direct input by the [cortico-thalamo-cortical] feedback, together with a reduced inhibitory contribution from nearby input. (p. 49)

The neural network Taylor and Alavi (1993) developed to simulate these nRt-mediated effects is shown in Figure 3 (p. 407). Its global properties will be discussed further below. Returning to Scheibel's (1980) model of converging influences upon the thalamic hub, he writes:

> From these data, the concept emerges of a reticularis complex [nRt] selectively gating interaction between specific thalamic nuclei and the cerebral cortex under the opposed but complementary control of the brain stem reticular core and the [pre]frontal granular cortex. In addition, the gate is highly selective; thus, depending on the nature of the alerting stimulus or locus of central excitation, only that portion of the nucleus reticularis will open which controls the appropriate subjacent thalamic sensory field. The reticularis gate [thus] becomes a mosaic of gatelets, each tied to some specific receptive zone or species of input. Each is under the delicate yet opposed control of: (a) the specifically signatured sensory input and its integrated feedback from [posterior] sensorimotor cortex; (b) the reticular core [MRF] with its concern more for novelty (danger?) than for specific

> details of experience; and (c) the [pre]frontal granular cortex-medial thalamic system [PFC] more attuned to upper level strategies of the organism, whether based on drive mechanisms (food, sex) or on still more complex derivative phenomenon (curiosity, altruism). Perhaps here resides the structuro-functional substrate for selective awareness and in the delicacy and complexity of its connections, our source of knowing, and of knowing that we know. (p. 63)

We can now say something more definite about the nature of these global attentional mechanisms, and suggest some ways to model them computationally. How would we succinctly describe the neural architecture suggested by the wagon wheel model? Beginning at the "top," it consists of a broad sheet of modular processing units with a generally columnar structure, but interconnected via an extensive network of cortico-cortical connections that allow the modules to exchange outputs. This is the cortical "rim." It is the most complex and densely interconnected assemblage of neural tissue in all of nature, and provides the substrate for nearly all of the higher cognitive functions of the brain. It is estimated that more than 90% of cortical connections are inter- and intra-cortical, while the thalamo-cortical tracts we have described above, make up less than 10%. This, of course, is consistent with the present model, since most non-conscious processing is assumed to be done automatically via densely interconnected cortical networks.

In terms of global attentional processes, however, Newman and Baars (1993) reviews a variety of evidence for cortico-thalamic selective gating. Such gating effects tend to fall into two anatomically-defined categories. The authors conclude that "prefrontal cortex acts as an executive attentional system by actively influencing information processing in the posterior cortex through its effects upon the nucleus reticularis. In this manner, the highly parallel [processing] functions of the posterior cortex are brought into accord with increasingly complex and intentional cognitive schemes generated within the prefrontal regions of the brain." (p. 281). In AI parlance, prefrontal cortex acts as a "control system" or "central knowledge store" (McClelland 1986). Posterior cortical areas act more like arrays of quasi-autonomous processing modules (or local experts) competing for access to the cortical GW. Similarities to Arbib *et al.*'s schemata theory (discussed above) are evident in this architecture.

Also involved in the global gating of cognitive resources are the basal ganglia (BG, Figure 2) which receive descending projections from virtually the entire cortical mantle, processing these internally along with limbic inputs, and finally relaying them to the thalamus and MRF. These BG circuits project to both specific and nonspecific (multimodal) nuclei, including nRt. They have been implicated in a number of functions, ranging from the suppression of unintended movements; to the initiation of voluntary movements; to selective attention/intention (see e.g. Edelman 1989; Miller and Wickens 1990). In a full-length review of the basal ganglia, focusing upon the "cortico-basal ganglia-thalamo-cortical loop," Parent and Hazrati (1995) noted:

> this projection allows pallidal [BG] neurons to influence vast collections of thalamocortical neurons indirectly via a relay in the reticular nucleus... complement[ing] the more massive and direct [pallidal] projection that appears to target specific subsets of thalamocortical neurons. (p. 116)

3. The Central Role of the Thalamus in Attention

Turning to the thalamus itself, its most obvious function is to relay the raw data of sensory experience out of which the cortex constructs its increasingly complex representations of reality. But these primary sensory inputs constitutes only a small fraction of the totality of thalamic outputs to the cortex. Moreover, virtually every area of the cerebral cortex projects back to the thalamic nucleus from which it receives inputs. These nuclei mirror the cortex in their modular organization (indeed, they have far fewer intrinsic connections). The overall picture, then, is of two assemblages of modular processors, reciprocally interconnected by "spokes" of segregated, parallel circuits (see Newman 1995a).

Superimposed upon this architecture is a much more compact sheet of neurons, the nucleus reticularis thalami. Its circuits can selectively block the transmission of thalamic outputs to the cortex. These gatelets can be activated by both direct and indirect (BG) projections from the cortex, allowing cortical modules to gate their own input. At the same time, projections upon nRt gatelets from MRF can selectively inactivate arrays of gatelets, enhancing the transmission of relevant sensory inputs to the cortex.

What is this wagon wheel architecture of reciprocal projections and gating circuits contributing to cognition? Based upon the model presented above, our answer should not be surprising: the selective allocation of the processing resources of the central nervous system.

We have already noted the multimodal representations generated by the reticular core "spatial envelope," serving to orient the animal in sensory space. Certainly the orienting response would qualify as global. It is a classic example of how rapidly our awareness can shift when we sense potential danger or unexpected stimulus. MRF is clearly central to orienting behavior, although in conjunction with a host of inputs from cortical, thalamic, limbic and BG tracts (see Newman 1995a, b).

What about schemes and intentions? The cortico-thalamic circuitry for these functions is even more complex and differentiated. But the basic principles remain the same: the nRt gating mechanism allows coalitions of modules to compete and cooperate in the selection of those processing resources which will be allocated to global contingencies of adaptation (i.e., schemes, drives, plans, etc.).

At this point, the perceptive reader may be thinking, "But the great bulk of the circuits you describe are parallel and modular; how do global forms of processing arise out of the operations of this "wagon wheel" model?" As mentioned above, Newman and Baars (1993) includes an extended discussion of a tangential intracortical network (TIN) hypothesized to spread waves of EEG oscillations across the cortex. This cortex-wide, self-organizing TIN is hypothesized to potentiate local cortical interactions via the excitatory spread of competitive and cooperative activation patterns, such that "global order can arise from local interactions ... ultimately leading to coherent behavior" (Von der Mahlsberg and Singer 1988: 71).

The reticular nucleus is integrated into this highly distributed activation system via the thalamo-cortical "spokes" of the CNS (Newman 1995a). In a mirror relation to the excitatory TIN, the reticular nucleus has its own inhibitory "intra-nuclear network" of tangential connections. The capacity of this network to globally inhibit cortical activation has been most clearly demonstrated for mechanisms involved in inducing sleep. LaBerge (1995) writes concerning this capacity:

> The RN cells are known to inhibit each other, and when inhibition hyperpolarizes an RN cell sufficiently, it produces a rebound burst. In this way a network of connected RN inhibitory cells can spread activity to every cell within the network, apparently without decrement in the intensity of the activity. (p. 184)

Taylor and Alavi (1993) have modeled this tangential nRt circuitry as part of a "competitive network for attention" (Figure 3), based upon the micro-architectural work of Steriade and his colleagues (Steriade, Domich and Oakson 1986; Steriade, Curro-Dossi and Oakson 1991). We have already noted Taylor *et al.* (and others) findings of the inhibitory effects of symmetric lateral connections upon nearby thalamo-cortical loops (see also LaBerge 1995). These connections are axonal-dendritic, typical of nearly all linear network models. Examples of such lateral inhibition effects, both in cortical and connectionist networks, are

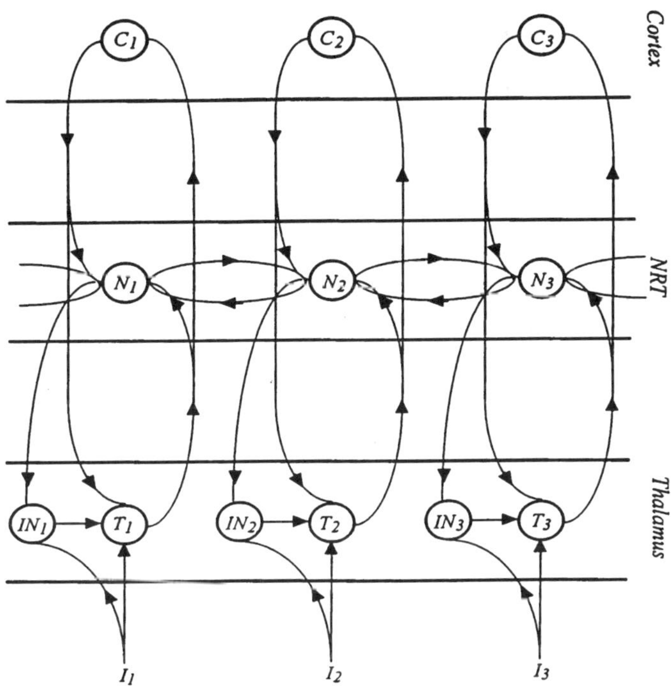

Figure 3. The wiring diagram of the main model of the thalamus- NRT-cortex complex. Input I_j is sent both to the thalamic relay cell T_j and the inhibitory interneuron IN_j, which latter cell also feeds to T_j. Output from T_j goes up to the corresponding cortical cell C_j, which returns its output to T_j. Both the axons T_jC_j and C_jT_j send axon collaterals to the corresponding NRT cell N_j. There is axonal output from N_j to IN_j, as well as collaterals to neighboring NRT cells. There are also dendro-dendritic synapses between the NRT cells (from Taylor and Alavi 1993).

numerous. They provide a basis for competitive networks such as Winner-Take-All (WTA) networks.

The WTA dynamic appears to be analogous to that posited by GW theory, in which cooperating experts are able increasingly to inhibit the inputs of competing, but incompatible ones, dominating the workspace until changes in input patterns allow some other coalition to win out. The problem with WTA-type networks is they are poorly suited to global forms of competition, where prohibitively long-range/geometrically-increasing numbers of connections would be required. Moreover, most long-range, reciprocal connections in the CNS are excitatory. In other words, inhibitory effects tend to be local, not global — except in the nucleus reticularis.

To model a global attentional network, Taylor and Alavi (1993) simulated the effects of dendro-dendritic connections in nRt. The dendrites of most nRt cells project out tangentially along the rostral-caudal axis of the reticular sheet, in both directions. The synapses of dendrites of adjacent cells are so extensive that "they allow the [nucleus] to be considered as a totally connected net even without the presence of axon collaterals…" (p. 351). It is this tangential feltwork of dendrito-dendritic synapses which "can spread activity to every cell within the network" (LaBerge 1995). Taylor and Alavi (1993) model this feltwork using non-linear equations which, "instantiate a form of competition in the spatial wavelength parameters of incoming inputs…" (p. 352). In their model, the entire nRt network oscillates with a wavelength, with the net strength given by the component of the input with the same wavelength.

> The way in which global control arises now becomes clear. Only those inputs which have special spatial wavelength oscillations are allowed through to the cortex, or are allowed to persist in those regions of the cortex strongly connected to the NRT: the thalamus-NRT system acts as a spatial Fourier filter. (p. 353)

Figure 4 shows the results of one of several simulation runs demonstrating the global, wave-like properties of Taylor *et al.*'s (1993) competitive model. Note that the overall pattern of activation in cortical units is exclusively dependent upon the wave pattern spanning across all of the NRT units. In a series of related papers, Taylor develops neural network models to include prefrontal and basal ganglia influences upon the nRt network (see, e.g. Taylor 1992; Taylor and Michalis 1995)

4. Parallels and Applications

Having presented the outlines of this neurocognitive model, we would like to conclude by describing some promising examples of current AI simulations which may relate to it. As noted above, a prototypical model for the GW system we posit has existed for some time in blackboard systems. The first blackboard system was the HEARSAY-II speech understanding system (Erman, Hayes-Roth, Lesser and Reddy 1980), which developed between 1971 and 1976. Subsequently, it has been exploited in a wide range of knowledge-based systems and psychological simulations. It has been used to structure cognitive models (e.g.

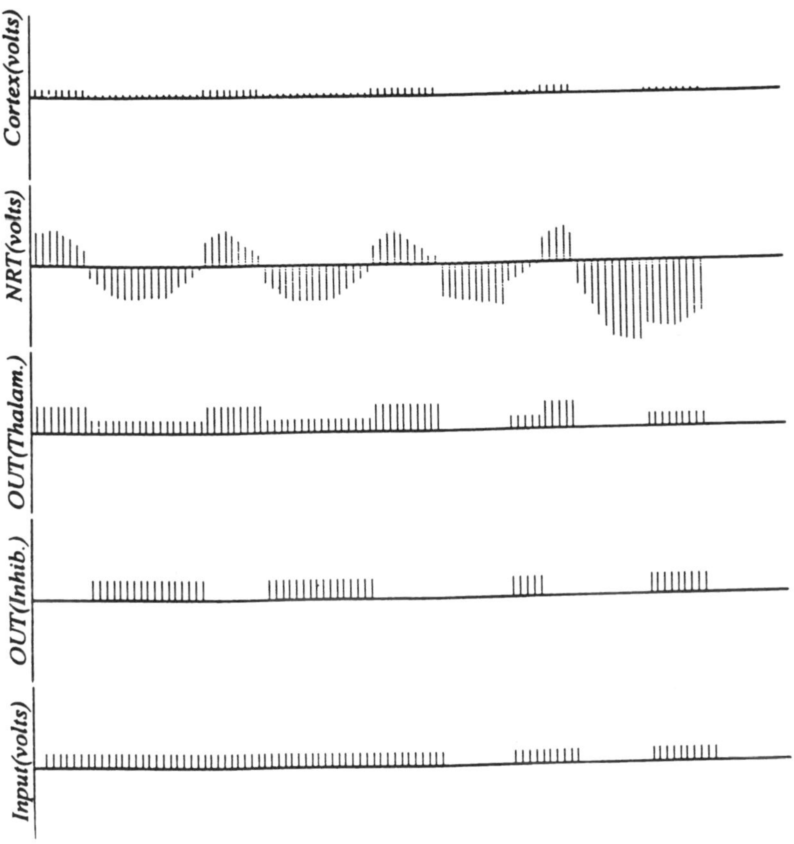

Figure 4. One of 15 simulation runs for thalamus-NRT-cortex model showing full global control with semi-constant spatial input. Note that cortex activity is influenced by the NRT alone (from Taylor and Alavi 1993).

the OPM system), which simulate the human planning process. Sometimes the blackboard model is used as an organizing principle for large, complex systems built by many programmers (Penny Nii 1986).

The blackboard architecture has three defining features: a global database called the blackboard, independent knowledge sources that generate solution elements on the blackboard, and a scheduler to control knowledge source activity. Blackboard systems construct solutions incrementally. On each problem-solving cycle some modules execute, generating or modifying a small number of solution elements in particular blackboard locations. Along the way some elements are assembled into growing partial solutions; others may be abandoned. Eventually, a satisfactory configuration of solution elements is assembled into a complete solution, and the problem is solved. Newman and Baars (1993) have suggested that analogous (although massively parallel) iterative processes are required to produce conscious representations: the series of "complete (or global) solutions" generated within our brains.

Among blackboard architectures, we have found McClelland's (1985, 1986) "Programmable Blackboard" a particularly promising model, not only because it employs a connectionist architecture, but because it is programmed via a gating array. This gating array is not modeled on the principles presented in the previous section, but it appears to perform some similar functions. The model is based upon multiplicative connections. Rumelhart, Hinton and McClelland (1986) write about such gates,

> Thus, if one unit of a multiplicative pair is zero, the other member of the pair can have no effect, no matter how strong its output. On the other hand, if one unit of a pair has value 1, the output of the other pass[es] unchanged to the receiving unit… In addition to their use as gates [such] units can be used to convert the output level of a unit into a signal that acts like a weight connecting two units. (p. 73)

McClelland's (1985, 1986) model starts with a typical set of connectionist networks, or modules, except that they are "programmable." This added feature is achieved by replacing all of the fixed connection strengths with multiplicative connections. The modules are programmed via a Connection Information Distributor (CID) which includes: (a) a central knowledge store, including the Central Module and Connection Activation System; (b) converging inputs to the central knowledge store from the Programmable Modules; and (c) diverging outputs from the central knowledge store back to the programmable modules. Figure 5 shows the basic elements of the CID system. Figure 6 illustrates the patterns of connections between the elements.

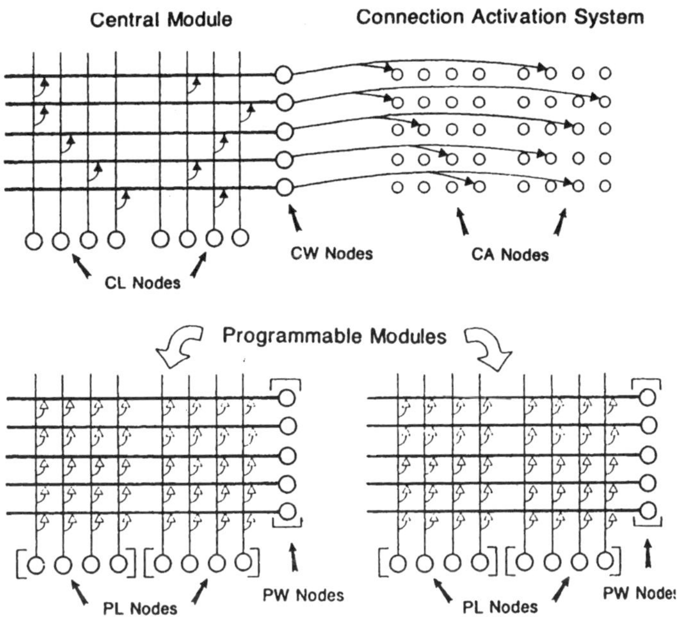

Figure 5. A simplified example of a Connection Information Distributor (CID), sufficient for simultaneous bottom-up processing of two two-letter words. The programmable modules consist of the programmable letter (PL) and programmable word (PW) nodes, and programmable connections between them (open triangles). The central module consists of a set of central letter (CL) nodes and a set of central word (CW) nodes, and hard-wired connections between them (filled triangles). The connection activation system includes the central word nodes, a set of connection activator (CA) nodes, and hard-wired connections between them. Connections between the central knowledge system (central module plus connection activation system) and the programmable blackboard are shown in Figure 6. (from McClelland 1985).

In the programmable blackboard, the array of time-varying gates allows the central knowledge store to momentarily set (and vary) the pattern of connection strengths of the programmable modules via a Connection Activation System of gating units. Such a system might simulate what Crick (1984) refers to as "transient cell assemblies" activated by a thalamic "attentional searchlight," momentarily binding perceptual features into a unified awareness of an object. Regardless of whether connectionist gating systems like McClelland's (1985,

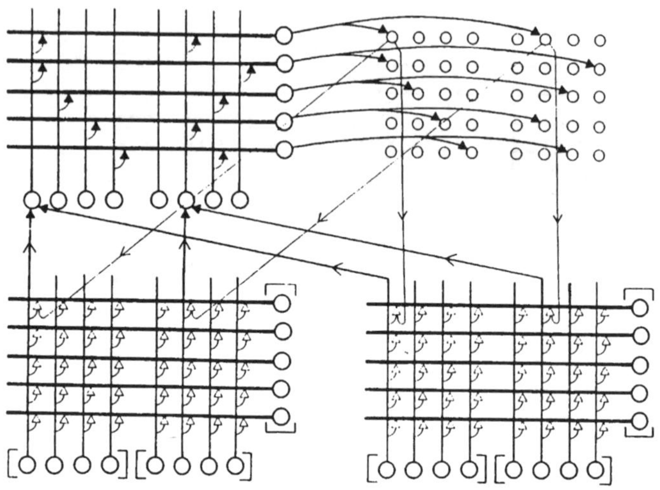

Figure 6. Each CA node projects to the corresponding connection in both programmable modules, and each central letter node receives projections from the corresponding programmable letter node in both programmable modules. The inputs to two central letter nodes, and the outputs from two CA nodes are shown (from McClelland 1985).

1986) approximate actual CNS functions, they have been shown to have a number of useful applications in the AI field.

For engineering problems like object recognition and robot motion control, the concept of combining modular networks using gating connections has been actively exploited to develop highly reliable systems (Jacobs, Jordan, Nowlan and Hinton 1991; Hampshire and Waibel 1992; Jacobs and Jordan 1993; Cho and Kim 1995). The key issue in this approach is how to combine the results of the individual networks to give the best estimate of the optimal overall result.

Architectures used in this approach consist of two types of networks: an expert and a gating network. Basically, the expert networks compete to learn the training instances, and the gating network facilitates cooperation by the overall mediation of this competition. The expert networks may be trained separately using their own preassigned sub-tasks and differing modalities (e.g., vision and touch), or the same modality at different times (e.g., the consecutive 2–D views of a rotating 3–D object). The gating network need only have as many output units as there are expert networks.

To train such a gating network, Hampshire and Waibel (1992) developed a new form of multiplicative connection, which they call the "Meta-Pi" connection. Its function is closely aligned with predecessors described in McClelland (1986). The final output of the overall system is a linear combination of the outputs of the expert networks, with the gating network determining the proportion of each local output in the linear combination. Figure 7 illustrates this architecture with three expert networks.

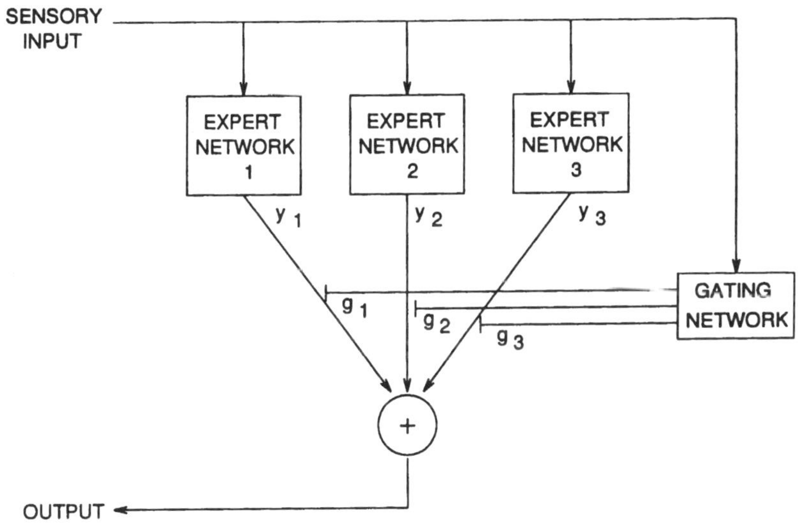

Figure 7. Schematic diagram of modular neural networks with three expert networks and a gating network. The output of the entire architecture, denoted y, is $y = g_1y_1 + g_2y_2 + g_3y_3$ *where* y_i *denotes the output of iTH expert network.*

The final output of the overall system is a linear combination of the outputs of the expert networks, with the gating network determining the proportion of each local output in the linear combination. The Meta-Pi gating network allocates appropriate combinations of the expert networks when stimuli are assessed to be novel, while an automatic ("non-conscious"), stochastic decision process operates in instances where a single expert can execute the task. This coupling of modular, expert networks and gating controls produces new levels of cooperative behavior. The expert networks are local in the sense that the weights in one network are decoupled from the weights in other expert networks. However, there is still some indirect coupling because if some other network changes its weights, it may cause the gating network to alter the responsibilities that get assigned to the expert network.

Such modular architectures may contain a variety of types of expert networks (e.g., networks with different topologies) that are more or less appropriate for particular tasks. By matching the modular networks to tasks, the system can produce superior performance to what can be achieved with a single, multipurpose network. Designers of neural networks often have some knowledge of the problem at hand; thus it may be feasible to choose particular classes of expert networks that are appropriate for particular tasks. Also, by partitioning a complex mapping, modular architectures tend to find representations that are more easily interpretable than those in fully connected networks. This is helpful for analysis and can be useful in incremental design procedures.

In summary, we believe the programmable blackboard model, and related modular architectures with gating networks, represent fair, first approximations to the much more complex reciprocal interactions occurring between prefrontal cortex, thalamus and posterior cortex in our wagon wheel model. Such "intermediate" models, bridging the gap between the neuronal and symbolic levels of computation (Smolensky 1988), hold promise for understanding conscious processes in computational terms (much as the processes of memory, categorization, and language are gradually yielding to AI/connectionist modeling).

Notes

1. All correspondence for reprints should be addressed to the principal author at 740 Clarkson, Denver, CO 80218. Email: newmanjb@aol.com
2. We would note that numerous other thalamic, cortical and striatal (basal ganglia) projections exist from and to the superior colliculi. Interestingly SC has been implicated in the phenomenon of "blindsight," which may be a primitive example of orienting behavior. See LaBerge (1995) for an in depth review of these subcortical attentional mechanisms.

References

Arbib, M.A. 1989. *The Metaphorical Brain 2*. New York: John Wiley and Sons.

Arbib, M.A., Conklin, E.J. and Hill, J.C. 1987. *From Schema Theory to Language*. Oxford: Oxford University Press.

Arbib, M.A. and Newell, A. 1993. Unified theories of cognition. *Journal of Artificial Intelligence* 59.

Baars, B.J. 1983. How does a serial, integrated and very limited stream of consciousness emerge out of a nervous system that is mostly unconscious, distributed, and of enormous capacity? In *CIBA Symposium on Experimental and Theoretical Studies of Consciousness*, G.R. Brock and J. Marsh (eds), 282–290. London: John Wiley and Sons.

Baars, B.J. 1988. *A Cognitive Theory of Consciousness*. Cambridge: Cambridge University Press.

Baars, B.J. 1992. *Experimental Slips and Human Error: Exploring the Architecture of Volition*. New York: Plenum Press.

Baars, B.J. and Newman, J. 1994. A neurobiological interpretation of a Global Workspace theory of consciousness (Ch. 4). In *Consciousness in Philosophy and Cognitive Neuroscience*, A. Revonsuo and M. Kamppinen (eds), 211–226. Hillsdale, NJ: Erlbaum.

Baars, B.J. 1996. *In the Theater of Consciousness: The Workspace of the Mind.* New York: Oxford University Press.

Cho, S.-B. and Kim J.H. 1995. Multiple network fusion using fuzzy logic. *IEEE Trans. Neural Networks* 6, 497–501.

Crick, F. 1984. Function of the thalamic reticular complex: The searchlight hypothesis. *Proceedings of the National Academy of Sciences, USA* 81, 4586–4590.

Crick, F. and Koch, C. 1990. Some reflections on visual awareness. *Cold Spring Harbor Symposium on Quantitative Biology* 15, 953–962.

Durfee, E.H. 1993. Cooperative distributed problem solving between (and within) intelligent agents. In *Neuroscience: From Neural Networks to Artificial Intelligence*, P. Rudomin *et al.* (eds), 84–98. Berlin: Springer-Verlag.

Edelman, G.M. 1989. *The Remembered Present*. New York: Basic Books.

Erman, L.D., Hayes-Roth, F., Lesser, V.R. and Reddy, D.R. 1980. The Hearsay-II speech-understanding system: Integrating knowledge to resolve uncertainty. *ACM Computing Survey* 12, 213–253.

Hampshire II, J.B. and Waibel, A. 1992. The Meta-Pi network: Building distributed knowledge representations for robust multisource pattern recognition. *IEEE Trans. Pattern Analysis and Machine Intelligence* 14, 751–769.

Harnad, S. 1994. Guest editorial — why and how we are not zombies. *J. of Consciousness Studies* 1(2), 164–167.

Jacobs, R.A., Jordan, M.I., Nowlan, S.J. and Hinton, G.E. 1991. Adaptive mixtures of local experts. *Neural Computation* 3, 79–87.

Jacobs, R.A. and Jordan, M.I. 1993. Learning piecewise control strategies in a modular neural network architecture. *IEEE Trans. Systems, Man, and Cybernetics* 23, 337–345.

Jasper, H.H. 1960. Unspecific thalamocortical relations. In *Handbook of Neurophysiology,* J. Field, H.W. Magoun and V.E. Hall (eds), 1307–1322. I, Washington, DC: American Physiological Society.

Jones, E.G. 1985. *The Thalamus.* New York: Plenum Press.

Kihlstrom, J.F. 1987. The cognitive unconscious. *Science* 237, 285–292.

LaBerge, D.L. 1995. *Attentional Processing: The Brain's Art of Mindfulness.* Cambridge, MA: Harvard University Press.

Llinas, R.R. and Pare, D. 1991. Commentary: Of dreaming and wakefulness. *Neuroscience* 44(3), 521–535.

McClelland, J.L. 1985. Putting knowledge in its place: A scheme for programming parallel processing structures on the fly. *Cognitive Science* 9, 113–146.

McClelland, J.L. 1986. The programmable blackboard model of reading (Ch. 16). In *Parallel Distributed Processing*, Vol. 2, J.L. McClelland and D.E. Rumelhart (eds). Cambridge, MA: MIT Press.

Mesulam, M. 1985. *Principles of Behavioral Neurology.* Philadelphia: F.A. Davis.

Miller, R. and Wickens, J.R. 1990. Corticostriatal cell assemblies in selective attention and in representation of predictable and controllable events. *Concepts in Neuroscience* 2(1), 65–95.

Mitrofanis, J. and Guillery, R.W. 1993. New views of the thalamic reticular nucleus in the adult and developing brain. *Trends in Neuroscience* 16(6), 240–245.

Newell, A. 1992. SOAR as a unified theory of cognition: Issues and explanations. *Behavioral and Brain Sciences* 15(3), 464–492.

Newman, J. 1995a. Review: Thalamic contributions to attention and consciousness. *Consciousness and Cognition* 4(2), 172–193.

Newman, J. 1995b. Reticular-thalamic activation of the cortex generates conscious contents. *Behavioral and Brain Sciences* 18(4), 691–692.

Newman, J. and Baars, B.J. 1993. A neural attentional model for access to consciousness: A Global Workspace perspective. *Concepts in Neuroscience* 4(2), 255–290.

Parent, A. and Hazrati, L-N. 1995. Functional anatomy of the basal ganglia. I. The cortico-basal ganglia-thalamo-cortical loop. *Brain Research Reviews* 20, 91–127.

Penny Nii, H. 1986. Blackboard systems: The blackboard model of problem solving and the evolution of blackboard architectures. *The AI Magazine* Summer, 35–53.

Penrose, R. 1994. *Shadows of the Mind — A Search for the Missing Science of Consciousness.* Oxford: Oxford University Press.

Posner, M.I. and Rothbart, M.K. 1991. Attentional mechanisms and conscious experience. In *The Neuropsychology of Consciousness*, A.D. Milner and M.D. Rugg (eds), 11–34. London: Academic Press.

Reddy, D.R., Erman, L.D., Fennell, R.D. and Neely, R.B. 1973. The Hearsay speech understanding system: An example of the recognition process. *Proceedings of the International Conference on Artificial Intelligence* 185–194.

Rumelhart, D.E., McClelland, J.L. and the PDP Group. 1986. *Parallel Distributed Processing.* Cambridge, MA: MIT Press.

Rumelhart, D.E., Hinton, G.E. and McClelland, J.L. 1986. A general framework for parallel distributed processing (Ch. 2). In *Parallel Distributed Processing*, Vol. 2, J.L. McClelland and D.E. Rumelhart (eds), 45–76. Cambridge, MA: MIT Press.

Rumelhart, D.E., Smolensky, P., McClelland, J.L. and Hinton, G.E. 1986. Schemata and Sequential Thought Processes in PDP Models (Ch. 14). In *Parallel Distributed Processing*, Vol. 2, J.L. McClelland and D.E. Rumelhart (eds), 7–57. Cambridge, MA: MIT Press.

Scheibel, M.E. and Scheibel, A.B. 1966. The organization of the nucleus reticularis: A Golgi study. *Brain Research* 1, 43–62.

Scheibel, A.B. 1980. Anatomical and physiological substrates of arousal: A view from the bridge. In *The Reticular Formation Revisited*, J.A. Hobson and M.A.B. Brazier (eds), 55–66. New York: Raven Press.

Skinner, J.E. and Yingling, C.D. 1977. Central gating mechanisms that regulate event-related potentials and behavior. In *Progress in Clinical Neurophysiology: Attention, Voluntary Contraction and Event-Related Cerebral Potentials*, Vol. 1, J.E. Desmedt (ed), 30–69. Basel: Karger.

Smolensky, P. 1988. On the proper treatment of connectionism. *The Behavioral and Brain Sciences* 11, 1–23.

Steriade, M., Domich, L. and Oakson, G. 1986. Reticularis thalami neurons revisited: Activity changes during shifts in states of vigilance. *Journal of Neuroscience* 6, 68–81.

Steriade, M. Curro-Dossi, R. and Oakson, G. 1991. Fast oscillations (20–40 Hz) in thalamocortical systems and their potentiation by mesopontine cholinergic nuclei in the cat. *Proceedings of the National Academy of Sciences* 88, 4396–4400.

Steriade, M. and Llinas, R.R. 1988. The functional states of the thalamus and the associated neuronal interplay. *Physiological Reviews* 68(3), 649–742.

Taylor, J.G. 1992. Towards a neural network model of the mind. *Neural Network World* 2, 797–812.

Taylor, J.G. and Alavi, F.N. 1993. Mathematical analysis of a competitive network for attention. In J.G. Taylor (ed), *Mathematical Approaches to Neural Networks*, 341–382. Amsterdam: Elsevier

Taylor, J.G. and Michalis, L. 1995. The functional role of the hippocampus in the organization of memory. (International Neural Network Society Annual Meeting, July 1995.) Washington, DC: International Neural Network Society Press.

Von der Malsburg, C. and Singer, W. 1988. Principles of cortical network organization. In *Neurobiology of the Cortex*, P. Rakic and W. Singer (eds), 69–99. Berlin: S. Bernhard, Dahlem Konferenzen, John Wiley.

Modeling Consciousness

J.G. Taylor
Centre for Neural Networks
Department of Mathematics
King's College, Londen

1. Introduction

There is currently considerable activity in trying to understand the nature of consciousness. Many disciplines have an interest, and an important contribution to make, in obtaining an acceptable solution: philosophy, psychology, neuroscience, pharmacology, physics, AI, engineering, computer science and mathematics. Whilst such interdisciplinarity makes the problem more exciting it also makes it more difficult. The languages of various scientific disciplines have to be used, and appeals to the knowledge bases in those disciplines also made. Yet progress in developing suitable interdisciplinarity appears to be occurring (Laukes 1994).

It is possible to gain a partial understanding of the mechanisms of the brain needed to produce consciousness by following one or other of the above disciplines on its own. In following such a less interdisciplinary path, important contributions have been made especially by neuroscience, psychology and pharmacology. Philosophy has also made key contributions, especially in elucidating some of the most important questions to be asked. Presently the deepest question raised by that discipline is that of determining "What it is like to be X," where X is any other living being other than oneself, which has the potentiality for consciousness (Nagel 1974). It is this question which has been claimed to be impossible to answer for science (Searle 1991), and has led to the useful distinction between 'hard' and 'easy' questions about consciousness (Chalmers 1994). The latter are those about the brain mechanisms which appear to support consciousness, and which are being ever more effectively discerned by the recent developments in psychology, neuroscience and pharmacology mentioned earlier. Non-invasive brain imaging, lesion studies and single and multi-electrode studies on animals and humans have given an enormous wealth of knowledge about such mechanisms. The harder sciences of engineering, physics and mathematics are now making important contributions to a theoretical framework for the various micro and ever more global information strategies used.

This progress in solving the easy problems of consciousness is to be contrasted with the clear lack of progress in the 'hard' problems of consciousness. The most important of these is the one raised above by Nagel (1974), where it was claimed that scientific method can never probe the subjective or phenomenal nature of consciousness. That impossibility arises due to the objective, 3rd person nature of science when opposed to the subjective, 1st person character of consciousness. There is a measure of irony in the situation as perceived by many today; consciousness is based solely on physical activity in the brain, but yet supposedly cannot be probed by science, which is still to be regarded as the ultimate analyzer of the physical world.

It is the purpose of this paper to present a possible physical model of consciousness which would allow for a solution to Nagels hard problem. We will detail how this could occur after a presentation of the physical basis of the model. That will initially be at a level of specifying a 'boxes'-type of information flow in the brain. Parts of the model will then be made more specific in terms of putative neural structures able to perform the functions specified in the boxes model. Some supporting evidence for these structures and their function, as well as for the more global program being carried out, will be brought forward from neurophysiology and psychology. On the basis of the model, and supporting evidence for it, a solution to the hard problem of describing 'what it is to be X' will then be outlined at the end of the paper. Before commencing what is initially a modeling exercise on the physical sub-structure for supporting consciousness, it is clearly necessary to justify, why an approach could ever work. How, using specific features of the physical world, could one ever attempt to answer what appears to be the purely philosophical question of Nagel (1974): how ever to discover, from analysis of the material world, what it is like to be an X? The initial justification is that if science is to build a bridge from the physical world to the apparently non-physical, mental one, it must do so starting from the physical structures available. Indeed there is nowhere else from where science can commence; the material world is the only possible domain of scientific analysis. This means that science must attempt to be able to discern how mental features of brain activity may arise from the latter. These mental features must possess seemingly non-physical attributes. It is through such 'mental' aspects that the beginnings of an answer to Nagel's problem may emerge.

The main thrust of the answer to be developed here is that through the *relations* between brain activities that mind might emerge. Such relations, to be specified in detail in the next sections, are not in themselves physical. In the same way, relations between numbers (such as less than or greater than), are not themselves numerical. It is this non-physical essence of relations which opens up the possibility of meeting the philosophers' difficulties over the mind-body problem.

2. The Relational Mind

Relational structures have long been recognized as an integral part of brain and mind. Aristotle proposed over two millennia ago that thinking proceeds by the basic relations of contiguity, similarity and opposites, an idea developed strongly by the associationist schools of psychology in the 18th and 19th centuries. The manner in which ideas led one into another was considered seriously by the empirical philosophers Locke, Berkeley and Hume in the 17th century. More fundamentally Hume stated in his treatise:

> Mind is nothing but a heap or collection of different perceptions, unified together by certain relations, and suppos'd tho' falsely to be endowed with a perfect simplicity and identity. (Hume 1896)

Later associationists developed this notion of relations between ideas to that of relations between stimuli and responses or rewards. It is the purpose of this paper to attempt to bring about rapprochement between the later and earlier forms of associationism. In so doing we will try to delineate the way in which the relations between ideas corresponds to those between neural representations corresponding to the ideas. Even more fundamentally we will try to put more detail into the phrase of Hume's "mind is nothing but a heap or collection of different perceptions, unified together by certain relations,… ." In particular we will try to specify those relations in ever greater detail, so as to move towards justifying the basic thesis of the Relational Mind model:

> The conscious content of a mental experience is determined by the evocation and intermingling of suitable past memories evoked (usually unconsciously) by the input giving rise to that experience. (Taylor 1973, 1991)

This model goes beyond that of Hume in that the Humean relations between perceptions are now extended to include a range of past experiences entering the relation with a present one. These past experiences need not necessarily have been conscious when they were originally experienced or as they are evoked to enter the relation.

There is increasing support for such a relational model of mind from experiments by psychologists during the last decade. In general there is strong evidence for the thesis that past experiences, however stored, influence present behavior. Present behavior must influence, or be related to, present consciousness, even if the latter only has a veto effect on developing behavior (Libet 1982). Thus evidence for the influence of past experiences on present responses is supportive of the relational mind model.

An important theory in social psychology is that of Norm Theory (Kahneman and Miller 1986) in which the norms of social behavior were shown to be manipulable by modifying past experiences. Thus Kahneman and Miller concluded from their experiments that "specific objects or events generate their own norms by retrieval of similar experiences stored in memory." The norms used by people in determining their social responses are assumed to modify the conscious experience of those people during the specific responses; that would not be true only if these responses were made in automatic states, which would be expected to be rather infrequent. It is not being claimed that the norms themselves become conscious, but that they determine some of the contents of consciousness.

Another area in which past experiences are used to color consciousness is that of categorization, which appears to be heavily sensitive to the past (Barsalou 1987). Thus if a person is asked to name a bird, if they have been in an urban landscape they may say 'robin,' whilst if they have just been on a hunting trip they might well say 'hawk.' Context dependence of prototypes of categories was found quite strong from the studies of Barsalou, and he concluded "The concepts that people use are constructed in working memory from knowledge in long-term memory by a process sensitive to content and recent experience" (Barsalou 1987).

The manner in which past experience alters response in numbers of other situations has also been explored recently. One fertile area has been that of memory illusions. Thus Witherspoon and Allan (1985) had subjects first read a list of words on a computer screen, and later judge the duration of presentation of words presented individually on the screen. Subjects judged the exposure duration as longer for old words (read on the list in the first phase) than new words, although the actual duration was identical. They misattributed their fluent perception of the old words to a difference in duration of presentation. Their conscious experience of the old words had thus been altered by the prior experience of exposure to those words.

A number of similar features, including the well-known 'false-fame' effect are recounted in a more recent paper (Kelley and Jacoby 1993). The 'false-fame' effect itself (Jacoby and Whitehouse 1989) involves two phases. In the first people read a list of non-famous names; in the second these old names were mixed with new famous and non-famous names in a test of fame judgements. Names that were read earlier were more likely to be judged as famous than were new names; this was especially so if subjects were tested in the second phase under a divided attention condition, preventing the use of conscious recall of the earlier list. Kelley and Jacoby (1993) concluded that "Past experiences affect the perception and interpretation of later events even when a person does not or cannot consciously recollect the relevant experience."

The above are only a brief selection of the many experimental results which support the basic thesis of the Relational Mind model stated earlier in this section. The model itself was developed in a formal manner initially in Taylor (1973), then Taylor (1991), and more explicit neural underpinning has been given to it in more recent papers (Taylor 1992a, 1992b, 1993a and 1993b). The initial, formal extension was by means of the notion of a relational automation, for which the usual quintuplet $<I, O, S, f, g>$ of input, output, states, next state function and next output were augmented by a relation R on the states. This binary relation leads, for a given state, to the set $R(s)$ of states in S in relation R to s. The state space was assumed to have a measure of size $\|s\|$ of each state s, so that the 'meaning' attached to a state was defined as the size:

$$\|\boldsymbol{R}(s)\| = \max_{s_1 \in \boldsymbol{R}(s)} \|s_1\|$$

More especially an overlap theory of meaning was introduced, in which the meaning of the sequence of two states s_1, s_2 was given by the 'size' of the overlap of $R(s_1)$ with $R(s_2)$: $\|R(s_1) \cap R(s_2)\|$.

This relational automata approach was used as a framework in Taylor (1973, 1991) to develop the relational mind approach to consciousness. An input is encoded by a semantic net W (such as in words in Wernicke's area or parts of objects in IT). The output s of W is sent to an episodic memory store E, so as to evoke suitable activity of related past experiences stored in E. The output of E and the direct one of W are combined in some manner in a comparison or decision net D. That module functions so as both to delimit the range of output of E and to combine it with the output of W. The resultant combination is supposed to have conscious content. In particular the mental state of the system is defined as:

Mental state = set of memories m suitably similar to s. (1)

(where similar denotes that set of memories determined by the decision net D). This allows an estimate to be given of the change of the 'level of consciousness' over experience, where the level of consciousness is defined as an average, say over a waking day. of the size of the mental state. An overlap theory of the meaning of inputs was also described, using that defined for a relational automaton, the relation R being that defined by the set of memories m equal to the set $R(s')$ defined by (1), where s' is the next state of E produced by the input s to E. This allows meaning to be given to linguistic input. Thus the sentence "This piece of cake is ill" has little meaning, since there is rather small overlap between the set of memories evoked by the phases "This piece of cake" and "is ill."

In order to develop the Relational Mind model further it is necessary to consider in more detail the control mechanisms involved in the comparison or decision system. Moreover extensions must be made to the model to take account of the present understanding of the global information processing strategies in the brain, as well as to include the different modalities of consciousness and the actual dynamics of the development of awareness. An even greater problem arises when one faces up to the question of the near-uniqueness of consciousness. How can this unique stream of consciousness be achieved in spite of there constantly being new objects of awareness. William Jame's description of consciousness as "a stream flowing smoothly along, now eddying around, then flowing smoothly again" (James 1950) must somehow be incorporated in any model, or arise naturally from it. In the next section a neuroscientifically-based control system will be considered which will allow such features, and other related ones, to be seen as natural properties of the stream of consciousness discernible in the model.

3. The Global Gate

It is difficult to consider awareness as an object of psychophysical study, but ultimately it must be considered so. A set of high level analyzers must exist which are working on inputs in various modalities from early processing modules (speech, 'where' vision, 'what' vision, motion detection, somatosensory, etc.). These can directly access that network or set of networks, denoted the awareness network, whose activation is necessary and sufficient for consciousness. The high level modules, as well as the awareness network, can directly access the motor response system. Thus any high level analyzer might give a motor response independently of activating consciousness, so leading to automatic activity which may be equated with the well-known forms of it in the case of humans. Note that no (ideal) observer has been used or seems needed, the homunculus has disappeared. One might thus regard the model as an attempt to model the observer as well.

It seems difficult to consider the access to awareness by the high level analyzers other than by a winner-take-all process. Such access would guarantee uniqueness to conscious content at any one time. It would also allow for analysis of the dynamics of conscious access over time if implemented by one of many equivalent forms of the winner-take-all competitive process. It would also explain the 'all-or-none' aspect of consciousness: either one is, or is not, conscious of a given object. There is no half-way house. There are suggested divisions of consciousness, such as into the 'nucleus' and 'fringe' (James 1950; Mangan 1993). However we are here confining discussion solely to 'core' consciousness, or to the nucleus; the fringe will be considered later. Other possible combinations of the outputs of the higher level analyzers, such as summation, would seem to be difficult to reconcile with this unique, all-or-none feature of consciousness. Thus we will only consider the competitive combination of the outputs of the higher level modules here, leaving to a separate account other possibilities.

So far a 'boxes-systems' approach has been adopted, in which the boxes' functionality and connectivity have only weakly been constrained by evidence from psychology. In order to make progress, especially to develop further the above psychophysical model, it appears helpful to incorporate neuroanatomical and neurophysiological constraints. These arise from the possible neural modules which might be able to support global competition between the various modality-specific analyzers. It is also necessary to be able to include mechanisms for competition between inputs in a given modality.

One might suppose that an effective global competition between distant modules in cortex can be achieved by suitable inhibition between these modules carried by distant excitation of local inhibitory neurons. There is however, a certain number of arguments against such a mechanism being actually present in higher vertebrates. Inhibition is known to be necessary for orientation selectivity in striate cortex, but only appears to occur over a range of about half a hypercolumn laterally [Worgörtter, Niebur and Koch]. However the main determinant of orientation selectivity in *V*1 seems to be oriented feedforward excitation from LGN. Long-range feed-forward activity also appears to be mainly excitatory to the temporal or parietal lobes, as single cell recordings would seem to indicate. Thus there is little evidence for long-range lateral inhibition in cortex, and other mechanisms have been proposed, such as conservation of weights on a given neuron, to achieve winner-take-all competition. However none of these other suggestions appear to have any experimental support.

All inputs to cortex (other than olfaction) pass through one or other thalamic nuclei. It might be possible that there is sufficient lateral inhibition in these nuclei to support 'winner-take-all' processing. However there is little neuroanatomical evidence for this, since the existent inhibitory interneurons there are only of the local-circuit variety. Past ideas of lateral connections between different nuclei in thalamus to explain coherent cortical activity in sleep have not been supported by further investigation. The thalamus does not seem able to provide a mechanism for winner-take-all competition either.

One region, often described as a thalamic nucleus, is the nucleus reticularis thalami (NRT for short). It consists of a thin sheet of mutually inhibitory neurons draped over the thalamus. Any axon from cortex to thalamus which passes through NRT gives off collateral excitation to NRT, as does any cortico-thalamic axon. NRT occupies such a strategic position in thalamo-cortical processing that it has been implicated in numerous aspects of brain activity from a local to a global level. It has long been termed a gateway to cortex, and more specifically "The reticularis gate becomes a mosaic of gatelets, each tied to some specific receptive field zone or species of input" (Schiebel 1980). He goes on to suggest "Perhaps here resides the structurofunctional substrate for selective awareness and in the delicacy and complexity of its connections, our source of knowing, and of knowing that we know."

It is posited here that NRT functions, through its intrinsic global connectivity, to allow a long-range winner-take-all competition to be run between various specialized cortical modules. Such competition arises from the combination of activity in various thalamic NRT and cortical regions, so form what can be called the TH- NRT-C complex. Loss of NRT action would lead to loss of the global features of this competition; if NRT were dissected, so only be locally connected, then the system would function in the manner suggested by Schiebel above as a set of local gatelets. In its waking mode of action it is posited that NRT functions as a global gate, supporting a global competitive method of cortical processing. The support from this thesis comes from a variety of sources, which are now discussed.

The first is neuroanatomical. The important position NRT occupies vis à vis cortical input and outputs has already been noted. In terms of the gating analogy, if the cortex is the storehouse of the inputs, both past and present, then the thalamus comprises a set of nuclei identifiable as the entrances for different modalities, whilst the NRT acts at least as a set of gates at these entrances

(Schiebel 1980). There are also other thalamic nuclei which do not act as relay centers for external inputs. One such nucleus is the pulvinar which does, however, have important cortical inputs and outputs organized in a reasonably well-defined manner. Thus pulvinar and its appropriate region of NRT will also be expected to play a role in attentional processing. We should also add that every thalamic nucleus does receive input from a brainstem or other cortical source (Jones 1983) so every such nucleus can ultimately be regarded as a relay nucleus.

The second line of support for the thesis is from lesion studies. These show clear deficits in reaction time in the re-engage process (Posner and Petersen 1990) when thalamic lesions occur. This is true especially for lesions in pulvinar. It is not clear how extensively the related region of NRT is damaged in this process, although the closeness of NRT to thalamus would lead one to expect important effects to occur there.

A third line of support comes from direct measurements of enhanced pulvinar activity in monkeys during visual attentive tasks (Posner and Petersen 1990) and of similar results in humans by PET studies (LaBerge 1990). Moreover injections of GABA-related drugs into pulvinar in the monkey altered animals' performance on an attentional task. Again it is not possible to conclude how strongly NRT is involved or affected in the various conditions used during the performance of these tasks. However, the well- established strong excitatory connectivity between pulvinar and adjacent NRT (Jones 1975) would strongly suggest the concomitant involvement of the nucleus as part of the processing.

Fourthly, there is important evidence accumulating on a possible global mode of processing by means of MEG measurements of brain activity during attentive processing using multi-SQUID detection systems (Llinas and Ribary 1992). This has shown the presence of a rostro-causal sweep of brain activity at about 40 Hz, that of cortex being led by 3 msec by that in thalamus, during a human subject's listening to a tone. This thalamo-cortical wave has been posited as being guided by NRT, this being the only appropriate connected structure at mid-brain level. The 40 Hz nature of this activity has also been supported by the discovery of thalamo- cortical cells with 20–40 Hz intrinsic membrane oscillations by Steriade *et al.* (1991). Such oscillatory activity may be related to that originally observed in cat visual cortex by Singer and colleagues (Gray and Singer 1987), although the state dependence of these oscillations is not clear.

A final approach, which will be discussed more fully in the next section, is based on the results of modeling. A suitably accurate mathematical model of NRT and adjacent cortical areas and corresponding thalamic nuclei must be created. It must then be shown either analytically or by simulation, that the system of equations transforms inputs in a way corresponding to a competitive net. The levels of resulting competitive control of inputs may be dependent on some of the details assumed in the model, such as the nature of the neurons, of the connections present and their strengths, and so on. We turn to discuss that in more detail in the next section.

4. The 'Conscious I' Gating Model

A certain amount of spade-work has already been done to tackle the problem of constructing a viable model for NRT as a competitive net. That lateral competition could be effective in enhancing the maximum of a set of local inputs has been shown by simulation of a simple model in La Berge *et al.* (1992) based on the mode of NRT inhibitory action suggested in Steriade *et al.* (1986), and mentioned in an earlier section. This explicitly used axon collateral feedback from different cortical areas to achieve a competitive relationship between the different cortical activities. An independent approach (Taylor 1992a, 1992b) was based more on the existence of local inhibitory connectivity on the NRT sheet. This allowed the use of an analogy with the outer-plexiform layer (OPL) of the retina, which has dendro-dendritic synapses which are gap junctions between the horizontal cells (Dowling 1987). Hence the epithet 'Conscious I' for the model.

A mathematical model of the OPL was developed (Taylor 1990). The model was analyzed particularly in the continuum limit, where analytic expressions can be obtained for the OPL response. The OPL itself emerges in this limit as a Laplacian net, so-called since the leaky integrator neuron equations for the membrane potential have a two-dimensional Laplacian operator giving the gap-junction lateral contributions. A similar model for the NRT sheet has the same Laplacian contribution in the continuum limit, but now with a negative coefficient multiplying the operator (arising from the inhibitory action of NRT cells on each other). The resulting negative Laplacian net has an analogy with quantum mechanical scattering, in which oscillatory waves occur in a potential well. Negative Laplacian nets also arise in the analysis of pattern formation, where heterogeneous wave patterns occur (Cohen and Murray 1981) as also in the modeling of hallucinatory effects in cortex (Ermentrout and Cowan 1978). It is these waves which give a strong hint of control. However the NRT equations, when coupling with thalamus (TH) and cortex (C) is included, have a contribution non-linear in the input and membrane potential which appears crucial. This term leads to the possibility of global control of cortical activity by that incoming over a limited region of thalamus or NRT. This opens the way to the possibility of top-down control in attentional tasks.

These results are developed in more detail elsewhere (Taylor and Alavi 1993a, 1993b, 1994). The initial model uses the simplest possible neurons in which the output of each neuron at any time is a sharply thresholded function of the membrane potential. Dendro-dendritic synapses are assumed on NRT. The NRT, in the continuum limit, becomes a negative Laplacian net. In the hard-limiter case (infinitely sharp threshold) an explicit analytic form can be obtained for the solutions for the activity in TH and NRT (C activity is dropped in the simplest case), as described in Taylor and Alavi (1993a, 1993b). The solutions are sine or cosine waves in space, with regions of NRT activity corresponding to cortical input and neighboring NRT inactivity to lack of cortical input. The distance between 'pockets' of thalamic and cortical activity is the wavelength of these harmonic waves, and is determined by the parameters of the NRT sheet (see also Taylor 1990) for a discussion of the parameters in the retinal OPL model). A softer gating model was also analyzed (Taylor and Alavi 1993a, 1993b). The conclusion of this is that a similar global wave structure of NRT activity arises, the latter acting as a controller of thalamic throughput and cortical input, but with the wavelength acquiring input dependence. The model without dendro-dendritic synapses was also analyzed in Taylor and Alavi (1993a, 1993b), where it was concluded that, when a certain class of feedback controls on NRT is assumed, the global control features appear to be lost, and localized states of activity can persist which are independently sited, provided they are distant enough from each other. The implications of this for a rat without dendro-dendritic synapses are interesting: can it be said to possess a single consciousness in this model or a set of them?

To fit with the MEG data reported earlier and discussed elsewhere (Llinas and Ribary 1992) it is to be expected that the whole anterior pole of the NRT exercises its authority over more posterior activity. How could this be achieved? To help answer this, known neuroanatomy (Lopes da Silva *et al.* 1990) indicates the following relevant facts. Firstly a number of regions in the limbic complex have reciprocal connections with anterior thalamic nuclei. However, these latter have no connections with NRT, as mentioned earlier (except in rat). The only clear connection with NRT appears to be from the subicular complex, with a one- way projection to the mediodorsal thalamus. This latter is known to have connections, specifically with the anterior pole of NRT. Thus it would appear that guidance from neuroanatomy indicates that the solution to the control problem we raised above — as to what regions exert control over the anterior part of NRT — has been achieved by using external input from limbic cortex, which is part of the Papez limbic circuit, itself being outside direct NRT control. The limbic cortex involves nervous tissue relevant to emotional drives as well as to stored memories so that the general nature of the control on NRT activity will be from these sources. We note that rat is not expected to have such control input in which limbic activity is down loaded on to NRT, but has feedback from the anterior thalamus to the limbic circuit and NRT, so its emotions will be expected to be inextricably linked with its consciousness.

The discussion of this section leads us to posit a more developed flow chart compared to that of section 2. Neural net inputs may activate several stored memories simultaneously (Amit 1990), and in that manner it may be possible to implement the relational structure of the model of section 2. Thus we posit that the working memory W, the episodic memory E and the decision unit D are identifiable with:

$$W = V1 + V2 + V4 + IT,\ E = HC,\ D = NRT + LIMBIC$$

Thus the NRT analogue of the comparison unit D seems thereby to be used in making its decision between the input, with its elaboration, along the whole range of striate and extra-striate cortex and IT, and the episodic memory (and goal activity) arising from earlier limbic activity (in LC and HC). It could be that competition is occurring all along the NRT sheet. That might be helpful in solving the binding problem of properly uniting the various features of an object, such as line segments, motion and color properties, which had been processed separately in extra-striate areas. It might also play a role in achieving competition between goals and drives.

If NRT does exercise global control (as we have claimed so far) then it would be correct to conclude that attention (and even consciousness) arises due to such global correlation of cortical activity. The interaction between the limbic control pattern impressed on the anterior NRT pole (and related limbic cortical activity) and that coming posteriorly from external (or bodily) input produces endogenous attentional control and a concomitant conscious state. Attentional search, for example, would cease after a good match had occurred between some anterior template set up, say in frontal lobe by the search process and the posterior input. On the other hand, the level of exogenous attention control exerted is also to be regarded as determined by the degree of mismatch between these two forms of activity, the bottom-up and the top-down, as in the case, for example, of finding one's 'sea-legs' when first going on a boat. Consciousness will have its unity because of the existence of only a single winner on the NRT sheet at any time, but the extended nature of the activity being correlated will allow for the rich inner content of consciousness.

Finally, let us consider exogenous attention. It is known that the shift to the waking state from sleep arises from ascending modulatory neurotransmitter systems, of which the cholinergic and noradrenergic components play a key role. A possible source of exogenous attentional control is by the midbrain reticular formation (MRF) cholinergic input on thalamic and NRT cells. From the previous references, and in particular Steriade *et al.* (1980), the action of such activity is excitatory on thalamic cells and inhibitory on NRT and inhibitory interneurons. The direct MRF inputs to NRT (Steriade *et al.* 1980) allow the expectation that an increase in MRF activation will reduce, if not totally destroy, the ongoing globally controlled activity on the NRT sheet. The direct input of MRF on NRT appears to be a very efficient method of deleting any ongoing NRT activity. Such a mode of action will be effective in allowing new competition to proceed on NRT so as to take account of any strong inputs, coded by striate cortical input, which may have caused the MRF activity in the first place. Thus we posit MRF as one of the sources of exogenous attention. Its specific mode of action can be modelled by adding dynamics for the threshold for the NRT cells, which causes them to be increased considerably by a suitable weighted sum of inputs, as supposedly assessed by MRF.

5. Winning Global Control

Interesting results, now thirty years old (Libet *et al.* 1964), appear relevant to the nature of global control, and give support to the model being developed. The experiments of Libet were to determine the threshold current for conscious experience when a 1mm diameter stimulating electrode was placed on the post central gyrus and the just conscious experience of what seemed like a localized skin stimulus reported by the patient. The stimulus was delivered as a series of short (of around 0.1 or so msec duration) pulses. There are two features of the data which stand out, which can be summarized as two quantitative laws (Taylor 1994a):

a. for threshold current to be consciously experienced over a short (<0.5 sec) duration, the applied electrical energy (frequency times duration times square of current) must be greater than a critical value;
b. for a duration longer than about 0.5 sec the applied electrical power must be large enough to allow the conscious experience to continue.

The requirement of enough applied electrical energy to capture, or turn on, conscious awareness in the short term would seem to fit well with the NRT control structure model above. For that is functioning essentially as a resistive circuit, with some non-linearity to provide stability, and such a circuit would be expected to function in terms of electrical energy requirements for capture of the dominant mode. The second result (b) above leads to a need for enough injected power to keep the control system going; there will be a certain amount dissipated, and so power above that critical level will have to be injected to hold the control of consciousness achieved by the earlier injected electrical energy.

It is possible that other approaches may be derived to explain the above results. Thus, they appear similar to many psychophysical results on threshold discrimination levels, such as Bloch's law in vision (Barlow and Mollon 1982). This law states that for visual input patterns with persistence less than some 100 msec the discrimination level satisfies a law similar to law 1, where electrical in that law is now replaced by light, primary cortical region by retina and electrode by light surface. For times beyond 100 msec, law 2 applies (again with the same replacements). The simplest explanation of this (and similar) laws is in terms of peripheral summation. Thus rods and cones have a temporal integration period of 200 and 30msec respectively (Barlow and Mollon 1982). Such an explanation cannot be easily used in the case of direct cortical excitation, since the temporal summation time in cortical cells has been found to be about 50–100 msec (Libet *et al.* 1964); this value is very different from the 500 msec or so value of the minimum duration for a pulse at liminal current intensity needed to cause the occurrence of conscious awareness.

There are further reasons why a traditional psychophysical approach is unsatisfactory:

1. A strong pulse, 40 times the threshold value, applied to thalamus, was unable to cause awareness in spite of the strength of the input (Libet *et al.* 1964). At least four times more summated energy was applied than needed at threshold. Hence a simple summation model cannot apply.
2. The back-dating of the time of onset of the awareness experience itself to close to the start time of the input pulse (Libet *et al.* 1979). This backward referral of the experience cannot be explained in any simple psychophysical terms, and has led to considerable discussion (Dennett and Kinsbourne 1992).
3. The phenomenon of the existence of a threshold of awareness itself is difficult to incorporate into standard psychophysical models. This led to the competitive model for access to awareness, in which the nature of the competition determines the nature of the 'entry' into consciousness. Thus subjective reports from patients indicate a sudden experience of awareness after the neuronal adequacy time (Libet 1994). It does not seem possible to understand such a feature in traditional psychophysical terms, but simula-

tions (Alavi and Taylor 1994) of the Th-NRT-C model produced results which support a minimum time of 18 units to win the competition. The results (Alavi and Taylor 1994; Taylor 1994a) are very close to that of the experiment results in figure 1 of Libet (1982). Interestingly there was a very short period taken for the smaller input actually to be annihilated by the larger one, of the order of one-tenth of the total time taken for the competition to be run. This corresponds closely to the results reported by Libet.

Let us now turn to the temporal ordering of signals for awareness. The phenomenon of "backward referral in time" has caused a great deal of excitement, as evidenced by the discussion and references in Dennett and Kinsbourne (1992). We can summarize the results of Libet *et al.* (1979) as to the nature of the stimulus needed to achieve conscious awareness (a continued train of pulses verses a single pulse) and the presence or not of backdating of the first experience of the stimulus. A peripheral stimulus to the skin only requires a single pulse, which, however gets backdated, which also happens to a pulse train of "neuronal adequacy" duration applied thalamically to VPL or to LM only. The cortical stimulus neither gets backdated nor can be achieved by a single pulse but only by a pulse stream of temporal duration equal to the neuronal adequacy time.

We wish to propose an explanation of these features which is consistent with, and supports, the competitive model of consciousness presented in the previous section. Let us suppose that the negative after-potentials noted in response to a single pulse skin stimulus (Libet 1982) corresponds to that input having rapidly gained access to the appropriate working memory, and continuing to be active there. That continuation of activity is, indeed, our definition of a WM, and agrees with increasing numbers of observations of such continued activity in various brain sites under trial conditions in monkeys (Anderson 1993). The evidence indicates that more artificial sources of input from VPL, LM or C, are not able to gain direct access to the WM that the single skin pulse did. Instead the former inputs have to be injected by creating their own working memory, in other words as a train of pulses, not a single pulse. The mechanism by which the skin pulse can activate the *WM* rapidly (in, say, 50 msec), whilst the other inputs cannot, may be in terms of the mesencephalic reticular activating system (MRAS). This is thought to be excitatory to NRT, and as such may convey input directly to the appropriate part of NRT so that the skin input, processed through somato-sensory cortex, easily activates the appropriate WM. This activity then proceeds to attempt to win the competition on NRT, as discussed in the previous section.

We note, that from simulations of the competitive model, however strong an artificial input is to cortex, it still takes a minimal but non-zero time to win the competition. This agrees with the experimental results reported in Libet (1982). Moreover a strong single pulse to VPL was reported there as not causing awareness, due to being below the requisite level of neuronal adequacy of the input being needed to cause such awareness. The same happens in the simulations, when large new input is rapidly switched off almost immediately after being turned on — there is little change, and little chance to win the competition.

The backdating that occurs has a different pattern of distribution from that of the nature of the requisite stimulus (single pulse versus pulse train). A direct sub-cortical input, most likely at thalamic level, appears necessary to achieve backward referral in time. One conjecture as to how that might occur, consistent with the presence of WM structures, is that a thalamic-level timing mechanism is turned on by the peripheral or thalamic input on its first appearance. This is then updated constantly, essentially as a counter, until the neuronal adequacy is reached, when the counter is reset to zero. The contents of the counter are used to tag the input time and so allows for the backward temporal referral. If this occurs for two competing inputs, then the first counter has a larger or stronger output, so could give the conscious experience of having occurred before the second, later input. Such a timing mechanism is expected to be hard-wired, since there would appear to be little value in having it adaptive. However the size of the timing interval may be modified by neuromodulatory effects, so leading to the well-known variations of subjective time in, say, emergency situations.

A timer of the above sort has to be able to count up to about 500 msec, and do so with an accuracy of, say, 25 msec. (although other units of time, such as 100 msec, would also do, related to 10Hz oscillations, as compared to 40Hz giving 25 msec). Thus it must be able to count at least 20 different states, each leading to the next as succeeding intervals of 25 msecs pass. That could be achieved by a recurrent network with recurrence time t_o, equal to some fraction of 25 msec. Each recurrence would lead to recruitment of an identical number m of neurons, so that after time nt_o there would be mn neurons active. Different inputs would recruit from different areas of the timing network, and would then compete against each other in their contributions to inputs winning the consciousness competition. Undoubtedly other neuronal models of timing circuits exist, although the one proposed above appears to be one of the simplest. We note that this timed competitive approach to consciousness by WMs can also give an explanation of the color-phi and the cutaneous rabbit phenomena described in Dennett and Kinsbourne (1992), although there is not space to enlarge on that here.

One of the clear predictions of the above discussions is the existence of timing circuitry, possibly in thalamic regions, although there could be other sites, such as in cerebellum or basal ganglia. This timing mechanism would be activated by a peripheral, LM or VPL stimulus, and would be predicted to give an increasing input until neuronal adequacy occurred. Such a linear ramp function would have a clear signature, although it might only be seen, for example, by non-invasive techniques, such as by multi-channel MEG measurements, at the short time scales involved. A further prediction is that a direct cortical input would not be able to activate the timing circuitry, so not allowing backward referral in time. At the same time there are further predictions as to the activation of suitable WM circuitry by a peripheral stimulus, whilst the absence of such activity would be clear if VPL, LM or C stimulation were used. Measurement of mid-brain reticular activity system (MRAS) activity during such tasks would be important here to probe the activity further.

6. Return to Consciousness

The discussions so far can be summarized under the two themes:

a. The Relational Theory of Mind
in which semantically coded input, and its related episodic memories are compared in some decision unit, as proposed in section 2, and
b. The 'Conscious I' NRT,
in which global competition is carried out through activity on NRT, and more specifically by means of the feedback control system, so that the flow chart, proposed in section 4, occurs.

It is necessary to try to combine the above features (a) and (b) to obtain consciousness. More specifically, we have to attempt to understand how the TH-NRT-C system of section 4, which produces the comparator/decisions unit of section 2, and other more local control activity leads to consciousness as experienced as the 'private' or subjective world. To do that the manner in which various forms of memory seem to be involved will be considered.

From the discussion in section 2 it is clear that the re-excitation of episodic memories needs to be of a whole host of earlier memories, which can give the consciousness 'color' to experience. That might not best be achieved by a pattern completion or attactor network (Amit 1990), since only a single pattern would result at one time. The most suitable memory structure is a matrix memory for a one-layer feedforward net in which the connection weight matrix A_{ij} has the form:

$$A_i j = \sum a_i^{(m)} a_j^{(m)} .$$

The response $\underline{y} = A \underline{x}$ will then have the form:

$$\underline{y} = \sum \underline{a}^{(m)} (\underline{a}^{(m)} \cdot \underline{x})$$

being a weighted sum of projections of $\underline{x}$ along the various memories. Assuming that attention has been caught by $\underline{x}$, it would seem reasonable to assume that memories $\underline{a}^{(m)}$ suitably close to $\underline{x}$ would therefore be allowed to be reactivated in earlier cortex by the TH-NRT-C complex. The details of how close such memories should be to $\underline{x}$ (the 'metric' of the comparator) depend on the nature of the TH-NRT-C complex, and are presently being analyzed by extensive simulations (Alavi and Taylor 1994). However, it is reasonable to assume that such a metric exists. Thus the 'Conscious I' leads to a specific class of comparators in the 'Relational Mind' theory of section 2.

On the other hand, there have been strong arguments proposed for the action of the hippocampus as a pattern completer (McNaughton and Morris 1989; Marr 1971; Rolls 1989; Reiss and Taylor 1991). In addition recent experimental results from McNaughton and colleagues (Wilson and McNaughton 1994) support the suggestion that random pattern completions may be occurring in SW sleep (in monkeys). It is possible to consider episodic memory being activated by pattern completion controlled by the hippocampus provided that (i) some sort of temporal sequence storage had occurred there, and that (ii) sufficient time has occurred for completion of a relevant sequence of patterns. If these criteria were met then it is possible to consider as an alternative to (i) the hippocampus as playing a controlling role in pattern completion of representations stored in various parts of cortex.

Feedback to associative and even primary cortex may need to be present to allow re-excitation of features of the appropriate memories $\underline{a}^{(m)}$. This is necessary in order that the detailed content of these memories be 'unscrambled' from the high level coding they have been represented as in medial temporal lobe areas. Indeed the important feedback connections in cortex, and also those through the TH-NRT-C complex, may have exactly the form required to achieve this amplification of the memories. In this approach, part of consciousness arises from the 'imagery' caused by feedback of excited memories allowed to persist with the input by the action of the TH-NRT-C competitive complex. In order for such feedback to be effective, some form of priming or temporally extended input activity would seem essential.

However consciousness can exist in people with temporal lobe loss (who have a normal short-term memory (STM) and normal priming). It would seem that either STM and /or priming is essential for consciousness; as is said in Crick and Koch (1990) 'no case of a person who is conscious but has lost all forms of STM has been reported.' Thus the content of consciousness must have another component apart from that suggested to arise from episodic memory; this further part must involve STM.

We have already suggested (Taylor 1992, 1993) that the crucial component needed to support the persistent input activity to allow return to consciousness is STM, or more precisely working memory (WM) of Baddeley and Hitch (1974; Baddeley 1986). This involves continued activity in a 'store' or 'scratch-pad' over a period of about 2 seconds before decay occurs without rehearsals.

More can be said, however, on the need for persistent activity support by working memory. The simulation results of the preceding section, together with the experimental results of Libet and his colleagues (Libet *et al.* 1964), indicate the need for input activity persisting over a period of at least 0.5 sec (for minimal current to achieve awareness). The need for this continued activity in an electrode placed on somatosensory cortex or in ventrobasal thalamus can be interpreted as a method of creating an artificial working memory, which has thereby activity sustained over a long enough time to win the competition for awareness. Peripheral stimulation is expected to achieve activation of an actual working site, as in the phonological store activation by speech input, the visiospatial sketch-pad by visual pattern input, etc (Baddeley 1986). Thus, in normal processing, for consciousness to be achievable it would appear essential to have cortical working memory modules to allow persistent activity to have enough time to (a) win the consciousness competition (b) achieve activation of suitable stored memory in short-term memory representations, to help in the consciousness competition. We note that the ability store 7±2 objects may be related to the need for WM activity to persist for 300–500 msec; if such activity could last for 2 seconds then 4–7 objects could be stored in WM.

This latter activation may not have to be as extensive as suggested earlier, i.e., to recreate at near-conscious level the lower-level features contained in the high-level memory encoding. Such detail may be unnecessary unless conscious recognition/recall or imagery processing were occurring. The situations mentioned briefly in section 2 (the 'false fame' and other illusory effects, and other similar content-dependent effects) do not seem to involve such a conscious level of involvement of past memories in consciousness. Nor, in cases mentioned in

Jacoby and Kelley (1993), could such conscious recall occur in the case of amnesics (such as in the famous Claparede case) who still, however, have their conscious experience colored by past implicit, non-declarative memories.

It is now possible to extend earlier ideas to indicate some of the possible circuitry after the high-level analysis, now regarded as a semantic encoder, has been activated. A semantic module S_A receives input IN_A (say in the visual modality), and responds rapidly to give output going elsewhere or going to its dedicated cortical working memory WM_A. This has persistent activity, over a suitable length of time, as discussed above, which activates and receives feedback from an associated memory store E. This latter store may be common to a number of working memories, such as another one WM_B. WM_A and WM_B are coupled through the TH-NRT-C complex, as well as possibly directly cortico- cortically. A competition is run between WM_A and WM_B till one or other wins. The winning WM then destroys the activity on the other WM, and is expected to lay down its content in hippocampus. This is regarded as conscious memorization, with stored representatives being later available by reactivation of the relevant working memory state at time of memory deposition.

Subliminal effects of polysemous words (Marcel 1980) allow a detailed quantative explanation in terms of the above model of coupled WM/semantic memory systems (Taylor 1994b). The model assumes (Taylor 1994b) that each semantic memory transfers its input to the associated WM in an excitatory fashion, with little feedback. There is excitatory support between semantically similar nodes on WM, inhibition between those involving semantic dissimilarity of polysemous word nodes (supported by lateral cortico-cortical connections and possibly by the TH-NRT-C complex).

In the experiments of Marcel (1980) there is inhibition between the 'hand' and 'tree' interpretations of the polysemous word 'palm.' It is possible to develop a simple perturbation theory for reaction times, in the Marcel (1980) paradigm in which a subject observes three words in succession, the middle one being polysemous and possibly masked, and has their reaction time to the last word timed. This perturbation is about the direct encoding from SM to WM, in terms of the effects of the lateral connections on WM and those between nodes on SM to different nodes on WM. This allows the 18 different reaction times (for a given SOA) to be understood in terms of 6 free parameters (in the simplest model) (Taylor 1994b). Much more detailed predictions can be made using more complete models, in which the nature of the semantic memory (to be discussed in section 7) and working memory (Hastings and Taylor 1994) are incorporated. It might even be possible to use this paradigm to probe the structure of SM and WM and their interaction more closely.

It is important to note that the experimental results of Marcel (1980) already indicate that 'both *p* and not *p*' might be considered valid at semantic level (semantically contradictory nodes being activated subliminally). This may go some way towards explaining creativity, in which the working memory template (or that arising from the anterior attentional system) picks out the most appropriate of *p* and not *p* to help develop a reasoning process. Such a feature also indicates the danger of neglecting such unconscious processes in any functional approach to the mind. In particular it seems to negate the use of the Gödel sentence in arguing about the nature of mind (Lucas 1960; Penrose 1989).

It is possible to extend this model to include, as output from the winning S_A/WM_A, the process of learning a sequence of actions. A system to achieve this could have the output from the winning WM_A/S_A system feed directly to the frontal lobe/basal ganglia/thalamic feedback system, and also to it through hippocampus. The feedback system allows for context-dependent sequence learning at various levels, which context would be provided by the S_A/WM_A to FL feedforward activation on adapting synapses, guided by hippocampal activation as well. After a suitable amount of learning it is then to be expected that the feedforward activation by S_A of the supplementary motor area, say, for motor action sequences, may be strong enough to bring about the requisite motor activity without S_A needed to become conscious nor concomitant activation of hippocampus occur. This gives an explanation of the changeover from explicit, conscious motor control of action sequences to the implicit, automotive form of their utilization. Other modules, such as cerebellum, are also involved, but their detailed manner of inclusion still needs exploring.

The principles of the competitive model of consciousness may be summarized as:

1. For each input, with specialized semantic coding, S_A, there is an associated working memory WM_A in which activity persists for up to 2 seconds (or longer).
2. S_A/WM_A feedforward and feedback activation to memory representations E lead to augmentation of the WM_A activity in a relational manner.
3. Competition occurs between the various working memories WM_A,WM_B, etc, over a critical 'neuronal adequacy' time (Libet *et al.* 1964), in which one WM which we denote, WM_{win} finally wins, and vetoes the losers' activities in their WMs.
4. Competition also occurs on each WM between possible 'interpretations' of input sequences arising from the SM_A, and still active on the WM_A (so helping solve the binding problem).

5. The activity in WM_{win} is stored in hippocampus and nearby sites as a buffer, to be usable (possibly non-consciously) at later times.
6. WM_{win}/S_{win} activity guides thinking and planning sequences.
7. The activities in the non-winning S'_As can still be used for automatic, non-conscious level processing, so this non-winning activity itself is not vetoed completely, though that in non-winning WM's may be temporarily.
8. Upgrading of memory, either of declarative form in the hippocampal complex or in semantic memories or non- declarative form in frontal-lobe-basal ganglia thalamic feedback circuits, may have to occur off-line.

The upgrading of memory under point 8 above would especially appear to be necessary for the unloading of hippocampal memories to prevent overload, and also to develop semantic categories. Such off-line upgrading may occur in sleep, a feature we will return to later.

We note that the details of the competition in point 3 above do not need to be specified in general, although it was already noted that finer features of the competitive process cannot be explained without a specific model. However the details of that model need not be restricted, at this stage, only to the TH-NRT-C complex. There may be other candidates, involving inhibitory interneurons in some as yet to be understood manner, or other even less neurobiologically likely choices. Finally the manner in which memory representations E are activated in a relational manner, as part of the Relational Mind model described in section 2, have still to be explored. We turn to that in the next section.

It is to be noted that recent experimental results on intrusive thoughts (Baddeley 1993) provide direct psychological evidence for competition between working memories. The experiments involved investigating the sorts of disruption of intrusive thoughts (a problem for depressed patients, often exacerbating depression) by activation of visual or auditory working memories. Both activation of the phonological loop and of the visuospatial sketchpad by suitably designed tasks, whilst subjects were seated in an isolated sound-isolated room, reduced the occurrence of stimulus- independent thoughts from about 80% to 30% of the times the subjects were probed. The speed of the requisite additional tasks turned out to be important; shadowing digits (and repeating them) at the rate of one every few seconds caused little disruption on the intrusive thoughts in comparison to an 80% to 30% reduction for shadowing every second. The disruption was effective on coherent sequential intrusions, but had almost no effect on the relatively infrequent fragmentary independent thoughts. Moreover competition was observed between intrusions and the generation of random sequences of letters.

These data fit the competitive relational model of consciousness very well in a qualitative fashion. There is seen to be competition between sequential intrusive thoughts, random generation of letter, phonological tasks and those using the visuospatial sketch-pad. The interpretation in Baddeley (1993) that it is the central executive (Shallice 1988) that governs the occurrence or suppression of intrusions is consistent with the competitive relational mind model if the central executive is equated with the TH-NRT-C complex (or more generally the competitive system, whatever it turns out to be). Is that identification viable? We propose to consider that elsewhere.

7. Building Relations

So far the most important structures in the relational mind approach have been outlined, these being the competitive TH-NRT-C complex and the WM/semantic modules between which the competition for consciousness is supposed to be run. Other structures have also been introduced, such as the hippocampus for declarative memory and frontal lobe/basal ganglia for, among other things, implicit or skill memory. However the relations being used in constructing the content of consciousness, following the model of section 2 and 6, have not been clarified. What is their specific form? How are they achieved — are they prewired or are they learnt? Some form of learning will be essential to ensure flexibility. Furthermore these relations depend on preprocessing through semantic modules. The coding in these latter modules themselves have to be learnt, at least in part. How is this achieved to be consistent with the competitive structure of the TH-NRT-C complex? There are various deep problems here, especially associated with the nature of categorization, and its development from infant to adult. This is an aspect of development which has not yet been understood, especially associated with language and the meaning of words and concepts. It is only possible to scratch the surface of these difficult problems, but we will attempts to give preliminary answers to some of the questions raised above to allow the relational structure of mind to be clarified a little. It is also hoped that some guidance is given as to the nature of solutions to the problems in terms of the structures so far introduced into the relational mind model.

There are a variety of approaches possible to relations themselves. One such is in purely mathematical terms, as a binary (or higher order) graph or function on symbols. These latter would correspond, say, to states of the neural modules in the brain, following the relational automaton approach to the relational mind model in section 2. However this appears to be a static approach, with relations pre-wired in the automaton. We wish to have a flexible relational structure to allow for the development of new concepts and the learning of language and other symbol manipulation.

A second approach is to follow the developments in relational semantics in which a basic set of relations are determined by clustering analysis of language (Chaffin and Hermann 1987). Even deeper features arise from the analysis of these semantic relations into what are called semantic primitives (Chaffin and Hermann 1987). Interestingly enough a basic feature of these primitives is that they all involve some sort of action in order to achieve a comparison. This brings an important feature into the arena, that of activity. We will return to that action-based interpretation shortly.

A third method of analyzing relations is in terms of the various features into which the objects (between the relations are supposed to occur) can be analyzed. Thus the relationship between an apple and an orange can be given in terms of those features which they possess in common (round, edible, colored, etc) and those which are different (red or green v orange in color, etc). The objects themselves have been collected into categories (apple, orange), and they and others into superordinate categories (fruit). However it has been noted for at least a decade that which especial features are to be chosen to achieve the definition of an object category in the first place is somewhat arbitrary. The notion of "similarity" to help pick out "similar" features has been strongly criticized by a number of developmental psychologists (Fivush 1987), and support has developed for a script- based development of categories (Schank and Abelson 1977). Thus the 'eating-routine script' will allow the welding together of the actions (get in high chair, put on bib, eat, drink, etc) to help build the functional categories of food (cereal, bread, banana), drink (milk, juice) or utensil (cup, glass, bottle). At the same time thematic categories can be created, involving temporal sequences of action → object → action → object →, as in: put on bib → have bowl → eat cereal → have cup → drink juice, etc. It is known that quite young infants will sort objects in thematic categories (Fivush 1987) and moreover that adults use both functional and thematic categories as well as those defined by prototypes and classical rules. The schematic features of functional and thematic category formation by scripts may then follow (Taylor 1994c).

These results indicate that actions are important in developing categories, especially in the infant. However there would appear to be category information, even in the 3–month old, in which little action takes place (Quinn *et al.* (1993). The infant is supine, and is presented with visual patterns of rows or columns of alternating light and dark squares. After six 15 second exposure sessions the infant appears to have formed a representation of the pattern which helped it, in a Gestalt manner, to process similar patterns which it was presented later. Such category learning does not appear to use any movement sequence. That is quite contrary to the observation (Rovee-Collier *et al.* 1993) of category inclusion of an object, such as a butterfly, hung from a mobile (and rocked back and forth) by the experimenter which a 3–month old infant had earlier been able to move by kicking when yellow blocks with alphanumeric characters printed on them were attached to the mobile. This is clearly action-based categorization, apparently very different from the earlier visual pattern categorization. Yet there is expected to be considerable eye movement associated with the former categorization, as is found in the viewing of pictures by adults (Yarbus 1967). Moreover frontal eye fields (FEF) and supplementary eye field (SEF) modules are in reasonable proximity to supplementary motor cortex, which is known to be necessary for learning visually- initiated motor sequences. Thus we might suspect that FEF and/or SEF modules are involved in learning processes when visual eye movements are made over a scene. There is some evidence for this (Noton and Stark 1971), and it is proving effective in artificial vision systems (Ryback *et al.* 1994; Zrehen and Gaussier 1994).

A possible structure for modeling the learning of action related inputs may now be developed. A visual input, which may have been preprocessed by motion detectors (in MT) or by positional coding (in parietal cortex), etc, is fed to a WM involved in the competition for consciousness. If the WM wins, its activity is fed to an LTM net for (semi) permanent storage, where concomitant action input from the motor cortex is also used in the LTM representation. The suggestion of Noton and Stark (1971) based on experimental evidence they gathered and, implemented in Rybak *et al.* (1994) is that an object is encoded by means of an alternating sequence of visual feature memory and eye movement memory. A similar use of eye movements for feature binding to make object categories was proposed in Taylor (1994c). If there are many movements associated with a given input then one may suppose that ultimately the encoded memory in LTM may appear to be independent of the action inputs (although could still be activated by it). This was the mechanism for developing semantic memory in Taylor (1994c): simulations (Taylor 1994c) showed how features f, f^1 sequently viewed by the action *A*, so by the sequence $f \rightarrow A \rightarrow f^1$ can be used to build the category (ff^1) composed of the joint features f, f^1. This was achieved by the input f leading to the action (eye movement) *A* by the associative action memory $f \rightarrow A$ activated by the visual input f, which itself leads to the new input f^1. The initial input f to the semantic memory *SM* activates the associated working memory WM; the continued activity of f and the new input f^1 on WM brings about learning of the lateral connections $f \leftarrow f^1$ on SM. Later activation of f of f^1 on SM then brings about lateral activation of the other feature, so of the composite object (ff^1).

In general the encoding of different inputs will be guided by the action sequences involved, so that functional and thematic encoding is expected to occur most naturally in LTM. Thus categories and superordinate categories are expected to be constructed thereby. There is also the fact that in REM sleep (in which eye movements may be used to reactivate visual memories with the associated actions) posterior associative cortex is mainly active along with frontal eye fields. This suggests that the LTM being constructed during REM is the semantic memory associated with the relevant WM under consideration. We note that in REM sleep the TH-NRT-C complex is still expected to be functional, so that different WMs will be winners at different times. However there will be no external input to drive the competition, and the internal activity playing such a driving role may well be mainly from the frontal and supplementary eye fields and similar motor areas (SMA, PMA). Thus action-based replays are conjectured as occurring, but those are only based on memories stored up in the near past. The bizarre character of REM-based dreams may be explained in terms of the different nature of the source of activity during the competition on the TH- NRT-C complex. Thus the above discussion leads to the conjecture that in REM sleep semantic memory is augmented. There are still the action-based relations between the concepts stored in semantic memories, embedded in the script-based action-object sequences. That is furthermore suggested as the origin of semantics.

In non REM slow-wave sleep, on the other hand, the thalamus is hyperpolarized, and motor output is blocked in brain stem neurons. Consciousness is regarded as dull or absent, when people are awoken in SWS. This is explicable in terms of the lack of the TH-NRT-C system to handle any competition between working memories. On the other hand some frontal lobe circuits are active, as is hippocampus. The latter even appears to be involved in consolidating memory representations (Wilson and McNaughton 1994). It is therefore natural to suggest that during SWS episodic memories may be being laid down, and in particular incorporated into frontal lobe contents, such as determining norms of social behavior, etc. The context-dependent models of frontal lobe of Cohen and Servan-Schreiber (1991) would fit into such processing very effectively. These episodic structures do not appear to be action-based from the absence of REM. However more careful analysis of brain activity (say using non- invasive techniques) would seem to be required before any more detailed modeling can be developed.

One of the important criteria of a modeling exercise such as this is to attempt to explain as many of the phenomena relevant to consciousness as possible. Having earlier considered normal states of consciousness and (albeit briefly) both REM and SW sleep, it would therefore be appropriate to turn to altered states of consciousness. It is possible to do that for deficits brought about by brain damage or malfunction, and for states of consciousness brought above by modified processing due to drug ingestion. Some progress on this can be made in terms of the Relational Mind model (Taylor (1994a), to which the interested reader is referred.

8. What is it like to Be

In order finally to approach the very difficult problem of the inner point of view, it might first be useful to consider other features of consciousness which may be understood in terms of the relational model. This will help give confidence in the model, as well as adumbrate further certain of its features. An interesting list of twelve aspects of consciousness has been presented in Searle (1991), the most relevant of these , which will be considered here, are: [1] finite modalities; [2] unity; [3] intentionality; [4] familiarity, and finally [5] subjective feeling. The other seven aspects are also of considerable interest, but would require far more space than is available here.

To begin with, modality. There are only five senses, sight, touch, smell, taste and hearing, together with internal body sensations, the sense of balance, and possibly those of self. Why not more? Let me suggest a reason from the competitive relational mind model which may appear extremely naive but may have a modicum of truth. The competition, on which the model is based, can only handle distinguishing between a maximum and a minimum of activity along any direction on the sheet. This leads to four possible ways of distributing the activity in two dimensions, with one winner out of four in a 2×2 square of cortex. This limit of four distinct possible winners holds for each cerebral hemisphere, so that a maximum of sixteen different winners could be allowed at different times. Cross-connections between the hemispheres would reduce this, but not completely since there is known lateralisation of function. The model may be naive in the extreme, but it does give a possible first approximation to a reason for the limited modality in which consciousness is expressed.

The unity of consciousness has already been built in as part of the relational mind model. There can only be one winner on the NRT sheet, so only one mode of consciousness, and accordingly consciousness content. There are two hemispheres, two NRT, two thalami, so that there are actually two winners. However these winners would not compete with each other, since the NRT sheet of each thalamus does not continue to the other sheet. Thus the two winners would co-exist side by side. Indeed these activities would be correlated so allowing the sixteen different modes of consciousness mentioned above. Only when the corpus callosum joining the two hemispheres is cut would there be a dissociation of these two sets of consciousness centers. The consciousness of the two halves would then be separate and reduced, even to the point of inability to respond linguistically to questions. However each side would still only have one winner at a time. It is important to add that in general lateralization of function is pronounced in the human brain, possibly related to the asymmetric position of the heart. Whatever the cause, usually the left hemisphere is the language-dominant one, whilst the right supports affect.

Intentionally is based on the fact one is conscious of something — an object, a feeling or whatever. As stated in Searle (1991), "... consciousness is indeed consciousness of something, and the 'of' in consciousness is the 'of' of intentionality." This feature may be seen to arise from the relational structure mentioned earlier in the paper and the nature of the working memory which wins the competition at any one time. The detailed content of consciousness can only be 'fleshed out' by the feedback support from the episodic and semantic memory. The specific nature of the object of consciousness will be that arising from the coding of the particular working memory involved. In the case of an auditory input, for example, an input to Wernicke's area would be encoded into phonemes, but at the same time activate the appropriate word centers. The most active of these would be representing the actual words heard. There would, however be words with lesser activation energized as part of the semantic coding giving meaning to words. These would be part of the relational structure at the semantic level. Beyond this would be activation of episodic memories appertaining to personal beliefs and to personal experiences. Feedback from these memories would add to the relational structure constraining further activity and aiding the competition which the auditory input could sustain on entering the phonological working memory. These feedback activations give intentionality to the words fighting their way to consciousness. In other words they give the perspectival character to the consciousness experience, and may be considered as giving the aspectual shape to the intentional state (Searle 1991).

The above account can also help explain the notion of familiarity which the objects of consciousness display so effortlessly. There will be initial familiarity brought about by the semantic coding, since the memories laid down earlier have the essence of familiarity. They present no novelty, no jarring from the expected into unfamiliar territory. Above and beyond this somewhat general source of familiarity is that arising from episodic memory. This gives additional contextual familiarity as a background to the ongoing words as they are processed in the phonological loop. Even though the present context of the words may be unfamiliar, the memory of past environments when the words were experienced is expected to ameliorate this sensation of novelty. The words are then helped to possess more familiarity than without such feedback.

Finally we come to subjective feeling, to the problem of the inner view. To echo Nagel (1974) again "...every subjective phenomenon is essentially connected with a single point of view and it seems inevitable that an objective, physical theory will abandon that point of view." How can the relational mind model approach this supposedly scientifically impossible 'inner point of view'? To answer this question, the possible nature of such a point of view must be described from within the model, to see how far the *singleness* of the point of view can be arrived at. In such a manner it might be possible to arrive at an answer to the problem raised by Davies and Humphrey (1993): "*how it is* that there is something that is like to have those processes going on in one's brain."

By the point of view of the sentient being X will be taken to be meant the detailed content of conscious awareness of X. This detail, according to the competitive relational model, is composed of the working memory activity winning the consciousness competition. To that is to be added all the fed-back and parallely activated semantic and episodic memory activities related to the input. Moreover there are also parallel activities in the other, coding, working memories, which may be involved in linguistic report (as in the case of the phonological loop). But most especially the 'point of view' is determined by the semantic and episodic memories related to the input in the appropriate working memory.

It is possible to explore this view of the subjective character of experience in more detail by considering separately the contributions made by the semantic and the episodic memory feedbacks. The former of these has been developed by interaction with the external world in a manner which might appear closer to that of the supervised training schemes of artificial neural networks than the unsupervised learning approach of Hebb and more recent neurophysiologists. For the supervision required in the former of these schemes can be seen to arise from the parents, siblings and school teachers of the developing child, or the corresponding peer group pressure in the growing animal in its natural home environment. Only certain responses are allowed to specific inputs, only specific behavior patterns permitted in given situations. Infringement of the rules of response are met with chastisement of physical or verbal form, whilst conforming to the norm is greeted by acceptance, praise and increased stature in the community. It may be more appropriate to regard the training schemes for semantic memory as of reinforcement rather than one or other of the extremes of complete lack of, or presence of, supervision. The reinforcement schedule has only a single (or few) signals from the environment to guide the learning of appropriate responses to gain such rewards from the learner. Underlying curiosity or exploration drives may be needed (with internally generated rewards) to achieve initial domain exploration before external rewards are encountered (Toates 1986). But for a broad spectrum of such learning styles, with varying degrees of internal and external reinforcement, the engraving of semantic memory structures is expected to occur in regions of cortex readily accessible to working memory structures. Then the short-term activity in such structure, arising from an input, can call on such memory to both constrain and amplify it.

Such semantic memory content of consciousness will in general have general culture and species-specific characteristics. Thus for words, there will be words of the natural language in which the human has been brought up, as well as words of other languages learnt by the person during its schooling or as part of its general life experiences. For animals such as dogs similar encoding of the few words with which they are familiar would be expected to occur, although there may be no corresponding phonological loop to give conscious experience to words but only direct output to spatial working memories allowing the conscious experience of such words to acquire a visual form. Similarly other modalities will have culture-specific semantic memories, such as for shapes in the visuospatial sketch pad (Baddeley 1992; Baddeley and Hitch 1974). There will be expected also to be more person-specific encodings in the semantic memories, such as dialects or particular shapes, although these will usually still be shared with others in a local area. Thus the semantic encodings are expected to have a certain degree of objectivity associated with them, although that will not be absolute.

It is in the episodic memory context that the inner point of view comes into its own. The personal 'coloring' added to consciousness, the inner 'feel' of the experience, is, it is claimed in the relational model, given by the parallel activation and feedback of memories associated with the present input. Particular people, buildings, or more generally sight, smells, touches and sounds from relevant past events are excited (usually subliminally) and help guide the competition for consciousness of the present input in its working memory. The laying down of these episodic memories only if they were consciously experienced is an indication of the filter character of the activity of the competition winner. The level of relevance and significance of these past experiences to ongoing activity will not be absolute but will depend on mood and emotion. These stored memories are energized both by inputs and basic drives, and guide the ongoing cortico-thalamic activity in a top-down manner from limbic structures (Taylor 1994a).

Most crucially the episodic memory is a private diary of the experiences, and responses to the experiences, of an individual. It is so private that it would be difficult to discern its meaning, since it is expressed in a coding that will not be easily accessible to an outsider. The engraved traces in the diary are trained connection weights, developed by using unsupervised and/or reinforcement training algorithms. But the knowledge of the values of these weights is not necessarily enough to know all that the inner point of view would receive for its specification. Mood, and the resulting neuromodulation of the relevant nerve cells by catecholamines and other chemicals, would be an essential boundary condition, as would the contextual inputs involved in the experience. In other words not only is the internal record of the connection weights of memory needed but also a possibly imperfect record of mood and place is needed to recreate the conscious experience of the sentient being X for any one period.

It is of interest to consider the distinction made by James (1950) and amplified by Mangan (1993) of the nucleus and the fringe of consciousness. We can consider the nuclear conscious state C_{nucl} as critically determined by winning (primed) semantic memory SM_{pr} and the associated working memory, which are denoted $(SM_{pr}\ uWM)_w$. Then we identify:

$$C_{nucl} = (SM_{pr}\ UWM)_w.$$

If $(SM_{pr}\ UWM)_{rest}$ denotes the other semantic and working memories, $(M_{tot})_{act}$ denotes the current activity in the episodic memory store (expected to be in various parts of cortex well connected to the hippocampus), and L_{act} denotes the active part of the limbic system, then we may identify:

$$C_{fringe} = C_{fringe1} + C_{fringe2} + C_{fringe3}$$
$$C_{fringe1} = (SM_{pr}\ UWM)_{rest}$$
$$C_{fringe2} = (M_{tot})_{act}$$
$$C_{fringe3} = L_{act}$$

We may further consider $C_{fringe1}$ as roughly the inattentive information in the fringe, which can be upgraded to be at the nucleus if so needed. $C_{fringe2}$ then would correspond to the experience of familiarity, involving tip-of- the-tongue and familiarity of knowledge type events (Mangan (1993). Finally $C_{fringe3}$ might be expected to consist of the feeling of rightness or wrongness of the environment, although this might also arise partly from $C_{fringe2}$. In total these identifications have begun to give more detailed suggestions for the neurophysiological underpinnings of the fringe, and could lead to non-invasive investigations of the activities of the corresponding regions.

It behoves us, at this juncture, to discuss an essential limitation to the nature of the explanation of inner experience put forward so far. It is not being claimed that we are thereby able to have that experience. If it was attempted to build, say, a virtual reality machine, on the basis of the competitive relational model, could it give one the feeling as to what it would be like to be X? One would first require X's input sensor transforms, such as achieved by the retina, the ear or by the nose. Then one would need built in to the virtual reality head-set the further transforms performed at primary cortical areas on that input. But why stop there? Why not build the associative cortexes and their actions-semantic memories, working memories, episodic memories and action plans. It would be necessary, for the latter, to build additional sub-cortical structures, such as basal ganglia and cerebellar cortices, for further control. At this juncture momentum has built up to include also the drive/goal and affect structures. One would thus either have built a silicon replica of X's brain, or if one wished to have the experience oneself, and after a suitable trepanning, a silicon version of X's brain in one's own head. But then one has again built a version of X, and one is no longer oneself. One might be able, by brain transplantation, to add cortical (and sub-cortical) circuits so that one can do the processing required to experience as X at one time and to revert to oneself at another. However such a possibility does not seem to add anything new to the previous possibilities, other than being able to keep one's personality more intact.

The above is based on the desire to have the inner experiences that X does. One could attempt to do so without going to such extremes, by developing and extending one's imagination of what it is like to be X, as suggested by Nagel (1974). However it would appear impossible to experience in that manner, color vision, say, if one was color blind from birth. Thus it is not being attempted to try to be or to experience X's 'inner view' from one's own necessarily limited brain circuitry. Suitable brain exercises could give one the increasing illusions that one was X, as might have been pursued by Kafka during his writing of the story of the man who became a cockroach in 'Metamorphoses.' What those exercises may be trying to do is indeed to change connection weights of suitable circuits so as to give the impression that one was a cockroach, or whatever. Yet the complete set of X's experiences would still not be able to be felt unless all the crucial brain circuitry was achieved by such a learning process. Since the possibility of achieving such a transformation without considerable structural change seems extremely slight, and in any case we would then be back at the trepanning scenario, it would appear that we have to accept the limitation noted above. It just does not seem possible to acquire X's own inner view without effectively becoming X. But we claim that an understanding has been gained of how it is that it is like to be X. The subjective, unique, multi- modal, intentional characteristics of that view have been given an explanation in terms of the competitive relational model of the mind. We expect that science can go no further. But then neither, we might remark, can X. Indeed if we know all of X's episodic and semantic memory relationship then we would know more than X might ever be able to experiencc in later encounters with the world. Yet we would still not have the resulting inner experiences, only understand how they could occur and the nature of any of X's description of them. We would know, as part of the relational structure of X, any report that X might make of his inner experiences.

9. Discussion

A model has been presented which is claimed to enable one to discover how it is that it is like to be the conscious being X. It is assumed that X has suitable neural network structure to enable us to recognize separate modules in X performing the activities of global competition, and episodic memory, and also sets of sensory modality-specific working and semantic memories. Then the inner view that X has of its experiences are given by the relations set up by the memory activities accompanying the winning working memory in its global competition with other such memories. The various features of X's consciousness, of finiteness of number of modalities, of uniqueness, intentionally, familiarity and subjective character are then derived from the model, as it is supposed other features of Searle's list (1991) would be. Moreover identification of neural structures involved with the nucleus and fringe of consciousness has been made (James 1950; Mangan 1993). What has been gaincd thereby compared to earlier models of the mind?

There are models, at least which deserve such comparison, since they have proved influential in the development of thinking on the subject very recently, and in particular in the development of the competitive relational model. The first is Dennett's "Pandemonium of specialists" (Dennett 1991). This could be identified with the set of competing working memories, the added value of the relational model being the manner in which semantic and episodic memories are crucial to the ongoing amplification of conscious experience by their activation and feedback. In the language of the Cartesian theater which is eloquently discussed, and rejected, by Dennett (1991), our model has a Cartesian theater peopled only by actors (the working memories), each of whom struts his or her time, only to be replaced by a new winner from amongst the competing actors clamoring in the stalls to be crowned. Each actor carries with them extra loud speakers which resonate especially with certain inputs, to allow the corresponding actor to win time in the limelight. The inner conscious experience is that shared by all the actors, although only being broadcast by one.

Such a picture is seen to be an extension of the "global blackboard" model of Baars (1988), in which input sensors compete to broadcast their activity to others and effectors by the medium of the global blackboard. The relational mind model goes beyond this, in a similar manner to Dennett's version, by emphasizing the need for relational memory structures to amplify and constrain incoming signals. It is these which are claimed to give the inner view, something absent from Baars' model. At the same time some resemblance (although still far from reality) to neurophysiological structures in the brain is required to effect the various models of operation of the competitive relational mind model. This is given in the relational mind model by the thalamo-NRT- cortex complex.

There is similarity also noticeable between the competitive relational model and that of Edelman (1989). His emphasis on reafference is very similar to the need for feedback from memory structures for guidance and prediction. These themes were not augmented in Edelman (1989) by the working memory structures discussed earlier, nor in the manner in which the crucial and unique experimental data of Libet and colleagues (Libet *et al.* 1964, 1979) could be explained and used to extend the model. Nor were the themes of global and local competition and the resulting features of consciousness considered in any manner in Dennett (1991).

Finally the supervisory attentional system (SAS) of Norman and Shallice (1986) has been very influential, especially in guiding experiments in working memory (Baddeley 1992; Baddeley and Hitch 1974). However it has been admitted by one of the proponents that the model has a problem of underpinning conscious experience (Shallice 1988), so it does not appear easy to discuss. In particular, as pointed out by Horne (1993), who does the contention scheduling for the SAS? It would seem the buck stops there, whilst in the competitive relational mind model the buck is constantly handed from winner to winner. It is true that the buck stops altogether if there are no activations of working memory-consciousness then is lost — a prediction which could ultimately be testable by real-time MEG as sleep takes over from wakefulness, and the fires of the working memories are dimmed (although possibly at different levels in REM, S1, S2, S3, S4 stages of sleep). Already the phenomenon of blind-sight can be explained by the lack of suitable activation of the appropriate visual working memory (and the need for suitable long-lasting pulses in artificial vision systems for blind-sight patients). The competitive relational model of mind is testable in a large number of ways. This was already indicated earlier, with observable thalamic timing systems and specific timings for the turn-on of appropriate working memories, differentially for peripheral, thalamic and cortical stimulation. It is also testable in the phenomena of neglect, as indicated above for blind-sight. Non- invasive MEG and rCBF measurements should allow details of the essential circuitry for consciousness ultimately to be ascertained.

The model presented also allows an understanding of the "slippages of consciousness" (Marcel 1993) in which the unity of consciousness is lost. In particular in multiple personality disorder (MPD) (Hilgard 1977) there is a relatively complete splitting of consciousness into several (serial) personalities. Slippage-type of phenomena can be understood in terms of lack of transmission of detailed (or any) information to some working memory sites, so that responses directed by then will be different from those for which good transmission has occurred. MPD appears more complex, possibly involving the manner in which the amygdala (involving reward and penalty memories) may inhibit full hippocampal contextual encoding of long-term memories, so that separate strands of personality develop. Further analysis requires careful modeling of affective memories and their relation to the development of epsidic memories and personality. That is beyond the scope of this article, but will be explored elsewhere.

What about the possibility of consciousness in a being without the requisite neural circuitry mentioned above? It would be necessary to consider each case on its merits, however. It may be necessary to discuss beings with different orders of consciousness , say with a three-dimensional neural or other form of network. However, it would be more useful to wait until such beings have been met and investigated before one should speculate further on this. All vertebrates on earth have nervous systems which allow their Inner Experience to come under the guise of the competitive relational model. At least we have given an answer as to how it is that it is like to be a sentient earthling.

Acknowledgments

The author would like to thank Dr D. Gorse, Prof. S. Grossberg and Prof. W. Freeman for trenchant comments, Dr P. Fenwick of the Institute of Psychiatry, London for a useful conversation, Prof. B. Libet for useful discussions on his work and Dr D. Watt on neuropsychological implications.

References

Amit, D. 1990. *Modeling Brain Function*. Cambridge: Cambridge University Press.

Baars, B. 1988. *A Cognitive Theory of Consciousness*. Cambridge: Cambridge University Press.

Baddeley, A. 1993. Working memory and conscious awareness. In *Theories of Memory*, A.F. Collins, S.E. Gathercole, M.A. Conway and P.E. Morris (eds), 11–28. Hillsdale NJ: Erlbaum Associates.

Baddeley, A. 1986. *Working Memory*. Oxford: Oxford University Press.

Baddeley, A. 1992. Is working memory working? *Quart. J. Exp. Psyc.* 44, 1–31.

Baddeley, A. and Hitch, G. 1974. In *The Psychology of Learning and Motivation*, G.A. Bower (ed). New York: Academic Press.

Barsalou, L. 1987. The instability of graded structure: Implications for the nature of concepts. In *Concepts and Conceptual Development*, U. Neisser (ed), 101–140. Cambridge University Press.

Chaffin, R. and Hermann, D.J. 1988. Effects of relation similarity on part-whole decisions. *J. Gen Psychology*. 115, 131–139.

Chalmers, D. 1994. Invited talk at the Arizona Conf. April 1994. To appear in *Towards a Scientific Basis For Consciousness*, S. Hameroff (ed). Boston: MIT Press.

Cohen, D.S. and Murray, J. 1981. *Math. Biol.* 12, 237–249.

Crick, F.H.C. and Koch, C. 1990. Towards a neurobiological theory of consciousness. *Seminars in the Neurosciences* 2, 237–249.

Dennett, D. 1991. *Consciousness Explained*. London: Allen and Lane.

Dowling, J. 1987. *The Retina*. Cambridge, Mass: Harvard University Press.

Edelman, G.J. 1989. *The Remembered Present*. New York: Basic Books.

Ermentrout, G.B. and Cowan, J.D. 1978. Studies in mathematics. *The Math Assoc. of America* 15, 67–117.

Fivush, R. 1987. Scripts and categories: Interrelationships in development. In *Concepts and Conceptual Development*, U. Neisser (ed), 234–254. Cambridge University Press.

Gray, C.M. and Singer, W. 1987. IBRO Abstr. *Neurosci Lett Suppl* 22, 1301.

Hastings, S. and Taylor, J.G. Modeling the articulatory loop. In *Proc. ICANN94*, Marinaro and Morasso, P. (eds). Berlin: Springer.

Hilgard, E.R. 1977. *Divided Consciousness*. New York: John Wiley.

Horne, P.V. 1993. The nature of imagery. *Consciousness and Cognition* 2, 58–82.

Hume, D. 1896. *A Treatise on Human Nature*, E.A. Selby-Briggs (ed). Oxford: Clarendon Press.

Jacoby, L.L. and Whitehouse, K. 1989. An illusion of memory: False recognition influenced by unconscious perception. *Journal of Exp. Psych. Gen.* 118, 126–35.

James, W. 1950. *The Principles of Psychology*. New York: Dover Books.

Jones, E.G. 1983. In *Chemical Neuroanatomy*, P.C. Emson (ed). New York: Raven Press.

Jones, E.G. 1975. *J. Comparative Neurobiology* 162, 285–308.

Kelly, C.M. and Jacoby, L.L. 1993. The construction of subjective experience: Memory attribution. In *Consciousness*, M. Davies and G.E. Humphries (eds), 74–89. Oxford: Basil Blackwell.

Kahneman and Miller. 1986. Norm theory: Comparing reality to its alternatives. *Psychological Review* 93, 136–153.

La Berge, D., Carter, M. and Brown, V. 1992. *Neural Comp.* 4, 318–333.

La Berge, D. 1990. *J. Cognitive Neuroscience* 2, 358–373.

Laukes, J. (ed). 1994. *Towards a Scientific Basis for Consciousness.* (To appear.)

Libet, B., Alberts, W. W., Wright Jr, E. W., Delattre, D.L., Levin, G. and Feinstein, B. 1964. Production of threshold levels of conscious sensation by electrical stimulation of human somato-sensory cortex. *J. Neurophysiol.* 27, 46–578.

Libet, B., Wright Jr, E.W., Feinstein, B. and Pearl, D.K. 1979. Subjective referral of the timing for a conscious experience. *Brain* 102, 193–224.

Libet, B. 1982. Brain Stimulation in the Study of Neuronal Functions for Conscious Sensory Experience. *Human Neurobiology* 1, 235–242.

Llinas, R. and Ribary, U. 1992. Chapter 7. In *Induced Rhythms in the Brain*, E. Basar and T. Bullock (eds). Boston: Birkhauser.

Lopes da Silva, F.H., Witter, M.P., Boeijinga, P.H. and Lohman, A.H.M. 1990. Anatomic organization and physiology of the limbic cortex. *Physiol. Rev.* 70, 453–511.

Lucas, J.R. 1961. Minds, machines and Gödel. *Philosophy* 36, 120–124.

Mangan, B. 1993. Taking phenomenology seriously. *Consciousness and Cognition* 2, 89–108.

Marcel, A.J. 1993. Slippage in the unity of consciousness. In *Experimental and Theoretical Studies in Consciousness*, Ciba Foundation Symposium no. 174. Chichester: John Wiley and Sons.

Marcel, A.J. 1980. Consciousness and preconscious recognition of polysemous words. In *Attention and Performance VIII,* R.S. Nickerson (ed). Hillsdale NJ: Erlbaum.

McNaughton, B.L. and Morris, R.G.M. 1989. Hippocampal synaptic enhancement and information storage with a distributed memory system. *Trends in Neurosciences* 10, 408–415.

Nagel, T. 1974. What is it like to be a bat? *Philosophical Reviews* 83, 435–450.

Noton, D. and Stark, L. 1971. Scanpaths in eye movements during pattern perception. *Science* 171, 308–311.

Penrose, R. 1989. *The Emperors New Mind.* Oxford: Oxford University Press.

Posner, M. and Petersen, S. 1990. *Ann. Rev. Neuroscience* 13, 25–42.

Quinn, P.C., Burke, S. and Rush, A. 1993. Part-Whole Perception in Early Infancy. *Infant Behav. and Development* 16, 19–42.

Reiss, M. and Taylor, J.G. 1991. Storing Temporal Sequences. *Neural Networks* 4, 773–788.

Rolls, E.T. 1989. Functions of Neuronal Networks in the Hippocampus and Neocortex. In *Neural Models of Pasticity*, Byrne, J.H. and Berry, W.O. (eds), 240–265. San Diego: Academic Press.

Rovee-Collier, C., Greco-Vigorito, C. and Hayne, H. 1993. The time-window hypothesis: Implications for categorization and memory modification. *Infant. Behav. and Development* 16, 149–176.

Ryback, I., Gusakova, V., Golavan, A., Shertsova, N. and Podlachikova, L. 1994. Modeling of a neural network system for active visual perception and recognition. 12th Conf. Pattern Recog., Jerusalem.

Schank, R. and Abelson, R. 1977. *Scripts Plans Goals and Understanding*. Hillsdale, NJ: Erlbaum.

Schiebel, A.B. 1980. In *The Reticular Formation Revisited*, J.A. Hobson and B.A. Brazier (eds). New York: Raven Press.

Searle, J. 1991. *The Rediscovery of Mind*. Cambridge, Mass.: MIT Press.

Shallice, T. 1988. *From Neuropsychology to Mental Structure*. Cambridge University Press.

Steriade, M., Domich, L. and Oakson, G. 1986. *J. Neurosci.* 6.

Steriade, M., Curro-Dossi, R. and Oakson, G. 1991. *Proc. Nat. Acad. Sci.* 88, 4396–4400.

Steriade, M., Ropet, N., Kitsikus, A. and Oakson, G. 1980. In *The Reticular Formation Revisited*, J.A. Hobson and M.B. Brazier (eds). New York: Raven Press.

Taylor, J.G. 1973. A model of thinking neural networks. Seminar, Inst. for Cybernetics, University of Tübingen (unpublished).

Taylor, J.G. 1990. A silicon model of the retina. *Neural Networks* 3, 171–178.

Taylor, J.G. 1991. Can neural networks ever be made to think? *Neural Network World* 1, 4–11.

Taylor, J.G. 1992a. Towards a neural network model of mind. *Neural Network World* 2, 797–812.

Taylor, J.G. 1992b. From single neuron to cognition. In *Artificial Neural Networks 2*, I. Aleksander and J.G. Taylor (eds). Amsterdam: North-Holland.

Taylor, J.G. 1992c. Temporal processing in brain activity. In *Complex Neurodynamics, Proc. 1991 Vietri Conference*, J.G. Taylor *et al.* (eds). Berlin: Springer-Verlag.

Taylor, J.G. 1993a. A global gating model of attention and consciousness. In *Neurodynamics and Psychology*, M. Oaksford and G. Brown (eds). New York: Academic Press.

Taylor, J.G. 1993b. Neuronal network models of the mind. *Verh. Dtsch. Zool. Ges.* 86(2), 159–163.

Taylor, J.G. 1994a. A competition for consciousness? *Neurocomputing* (In press).

Taylor, J.G. 1994b. Breakthrough to awareness, *Biol. Cyb.* (In press).

Taylor, J.G. 1994c. Relational Neurocomputing. Invited talk, Special Interest Group Meeting, WCNN 1994, San Diego CA, and talk at the IEE Conf on Symbolic Computing, IEE Savoy Place, London.

Taylor, J.G. and Alavi, F. 1993a. A global competitive network for attention. *Neural Network World* 5, 477–502.

Taylor, J.G. and Alavi, F. 1993b. Mathematical analysis of a competitive network for attention. In *Mathematical Approaches to Neural Networks*, J.G. Taylor (ed), 341–382. Amsterdam: Elsevier.

Toates, F. 1986. *Motivational Systems*. Cambridge, UK: Cambridge University Press.

Worgotter, F., Niebur, E. and Koch, C. In press. *J. Neurophysiol.*

Wilson, M.A. and McNaughton, B.L. 1994. Reactivation of hippocampal ensemble memories during sleep, Science. In press.

Witherspoon, D. and Allan, L.G. 1985. The effect of a prior presentation on temporal judgements in a perceptual identification task. *Memory and Cognition* 13, 101–111.

Yarbus, A.L. 1967. *Eye Movements and Vision*. New York: Plenum Press.

Zrehen, S. and Gaussier, P. 1994. Why topological maps are useful for learning in an autonomous agent. In *From Perception to Action*, P. Gaussier and J. Nicond (eds). Los Alamos: IEEE Computer Sci. Press.

Mind and the Geometry of Systems

William C. Hoffman
Professor Emeritus
Tucson, Arizona

"Geometry is a magic that works."
(René Thom)

1. Introduction

The conveners of the recent "*Foundations of Cognitive Science Workshop*" noted a crisis surrounding the present approaches to cognitive science. Two main fronts were identified: (i) are mental processes best studied in terms of information processing or as phenomena of neuroscience and consciousness; and (ii) the matter of unifying consciousness, affect, and social psychology with the science of mind. The answer, I suggest, lies in firmly grounding an approach in terms of the known structures of the brain.

The approach presented here involves both of the above fronts. In it there occur four main neuropsychological structures:

1. the Ascending Reticular Activating System (Hoffman 1990) and Baars' (1988) ERTAS (Extended Recticular Activating System) as a basis for the awareness fundamental to consciousness;
2. the limbic system to impart an emotional cast to percepts and cognitions;
3. the cortex of the brain as the seat of the higher mental faculties of perception and cognition;
4. the nuclei, gyri, and other subcortical centers in the midbrain and forebrain region, where the seat of consciousness lies.

All of these may be analyzed (Hoffman 1966, 1980b, 1981) in terms of the mathematical structures of the Geometry of Systems (Mayne and Brockett 1973). These structures encompass Lie transformation groups, fibre bundles,[1] fibrations and certain other structures of algebraic and differential topology (Hoffman and Dodwell 1985). Lie transformation groups and fibre bundles are fundamental to geometric structures and to the invariance of such transformations as those encountered in form memory and the psychological constancies. Many years ago, Pitts and McCulloch (1947) noted that the figures invariant under the transformations imposed by viewing conditions must be "the geometric objects of Cartan and Weyl, the Gestalten of Wertheimer and Kohler."

Extensive research on psychological categories (Rosch and Lloyd 1978; Smith and Medin 1981) has found that (1) the basic categorization consists of a division of the world into alternative categories in which within-category similarity dominates between-category similarity; and (2) categories not only can overlap but their members may be only probabilistically determined as well. But this is just the structure of the symmetric difference operation (XOR in computer parlance) of set-theoretic topology and its complement. It is therefore postulated that this two-stage process, the symmetric difference and its complement, constitute a basis for cognition and affect. It also happens that this combination provides an apt description of Klaus Riegel's (1973) dialectical psychology in addition to many features of what we call consciousness. Affect involves a subjective statistical decision process (Hoffman 1980b) as well. We are all statisticians, subjectively at least: "What's happening?" "What sense can we make of it?"

Although these, as mathematical structures, are amenable to digital computation, they can also be expressed in coordinate-free terms. This is the way the brain does it, by coordinate-free *flows* (in the technical, mathematical sense) along neuronal processes.[2] The two disparate approaches ("fronts") are thus readily unified by appeal to the Geometry of Systems. The latter can be computational, but it can also be "coordinate-free." In the present context, therefore, Rene Thom's dictum quoted above seems singularly apt.

The brain is *not* a digital computer. In the Central Nervous System, in contrast to muscle end-plate, the response is graded rather than all-or-none (Talbot and Marshall 1941). The "Binary Fallacy" has come under attack in a number of quarters (Selfridge 1990). As noted above, the brain does what it does by *flows* along neuronal processes, which appear in Golgi-Cox stains of brain tissue for all the world like the local phase portraits of dynamical systems (Hoffman 1994b). Learning and memory keep pace with the growth of neuronal processes all through life (Jacobson 1971). Memory *grows* (Hoffman 1971a), right along with the neuronal arborescence (Pribram 1971: Fig. 2–1). In post-traumatic amnesia or senile dementia, it is the most recent memories that fail the first (Ribot's law). Alzheimer's disease is characterized by the breakup of neuronal morphology and the formation of so-called "plaques" which replace the neuronal arborescences.

Yet the computational metaphor, that the brain actually does do digital computations, pervades cognitive science and motivates much of AI. If anything, however, the brain *simulates* such computations by *flows* which are expressible in *coordinate-free* terms. Since flows are present, a rich variety of mathematical structures cannot be far behind. Available then, along with Lie groups for analysis of neuropsychological phenomena, are continuous transformation groups ("continuous symmetry"J (Cassirer 1944; Hoffman 1966, 1968, 1970, 1977, 1978, 1984, 1989, 1990; Palmer 1983J, fibrations and fibre bundles (Hoffman 1985, 1986, 1989J, and category theory (Hoffman 1980a, bJ, and in particular the category of simplicial objects. In information processing psychology, concepts and percepts are regarded as nodes of a network and thoughts or trains of thought as the edges relating them. In other words, information processing has a simplicial structure consisting of points, edges, triangles, and polygons. The simplicial objects that are embodied in the "chunks" of information processing psychology are isomorphic to the elements of information processing psychology (Hoffman 1980a, 1985J. In the latter, the 0,1 basis for digital computation becomes such "chunks," i.e., simplicial chains embodied in prefrontal and inferotemporal cortex (Hoffman 1985J. The brain imaging technique of positron emission tomography (PETJ (Posner and Raichle 1994; Sereno *et al.* 1995J provides evidence for the presence of such a brain-wide array of localized simplicial complexes. The dual of such simplicial graphs is the area enclosed by them, and this is apparently what is seen in the areas of activation (Barinaga 1995J observed during PET and functional magnetic resonance imaging (fMRIJ. I therefore broach the following principle:

Principle 1: *To the extent that information processing psychology applies in cognitive phenomena, memories and processes are simplicial sets structured by the simplicial category.*

Simplicial complexes of this sort are also to be found in current research in synchronous multiprocessing in parallel computation (Herlihy and Shavit 1994, 1995J. Such an algebraic topology approach to distributed and concurrent computation offers an unexpected new avenue to understanding the problems of synchronization, decision, and consensus in parallel digital computations.

I also advance the following postulate:

To the extent that "computation" is involved in brain processes, it consists of the operation of Beck's finite information process. (Beck 1977)

The basic structure of Beck's finite information process consists of a *flow of information* in a composable sequence of information transfers

$$A_i \xrightarrow{f_i} A_{i+1} \ , \ (i=1,2,\ldots n-1)$$

that is embodied in an n-dimensional simplex whose vertices are the objects A_i and edges, the transfers f_i. When $n=3$, for example, the simplex is a tetrahedron $\Delta(3)$. In higher dimensions such flowcharts are simply simplicial sets built up from elementary simplices. Such a simplicial set can be constructed for any topological space and is called its *nerve.* The information processing in such a structure is both serial and parallel.

The *finite information process* itself consists of an *information complex* $\mathbb{I}$, which is a simplicial set of the kind in the example above, a *control complex* $\mathbb{C}$, and a vertex-preserving simplicial map from information complex to control complex $\mathbb{I} \to \mathbb{C}$. The map $\mathbb{I} \to \mathbb{C}$ is a fibration which possesses the homotopy lifting property in the category of simplicial sets. The homotopy lifting property admits the possibility of finite recursions upon $\Delta(n)$ (Beck 1977: 115). Figure 1 shows how the process would look in the simple planar case.

Here k_0, k_1, and k_2 denote the vertices of the control complex $\mathbb{C}$. The point x_0 lies on the initial fibre $\mathbb{I}(0)=\pi^{-1}(k_0)$ and represents an array of data and instructions at the initial state k_0 of the process. The path γ: $\Delta(2) \to \mathbb{C}$ with initial value $\gamma(0)=k_0$ describes how the operations in $\mathbb{C}$ are to be performed. γ is a homotopy of the projection of x_0 into $\mathbb{C}$. Hence by the homotopy lifting property there exists a lifting homotopy γ': $\Delta(2) \to \mathbb{I}$ with x_0 as the initial point. The path γ' moves through the information complex $\mathbb{I}$ in response to the path in the control complex $\mathbb{C}$. The point x_2 in the terminal fibre $\mathbb{I}(2)=\pi^{-1}(\gamma(2))$ is the final value of the path emanating from x_0. This process models what happens in the n-dimensional case when the point x_0 moves to the point x_n in the terminal fibre $\mathbb{I}(n)=\pi^{-1}(\gamma(n))$ of the n-dimensional information complex.

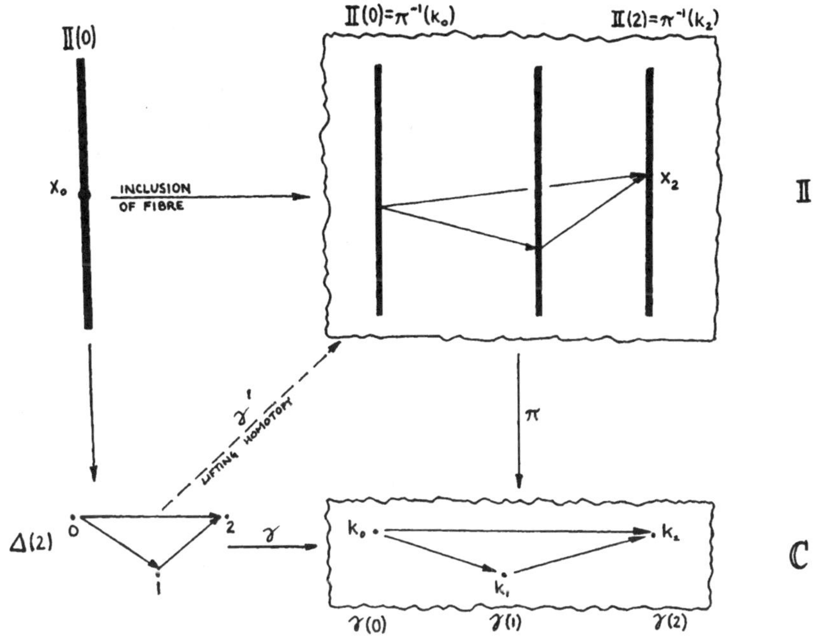

Figure 1. Beck's finite information process. $\mathbb{I}$ is a Kan fibration called the information complex whose fibres $\pi^{-1}(k)$ over the points k of the control complex $\mathbb{C}$ represent an array of data and instructions. The simplicial sets $\Delta(k)$ in the control complex generate a path γ that constitutes a homotopy. By the homotopy lifting property γ generates a lifting homotopy γ' that maps $\Delta(k)$ to $\mathbb{I}$. See text for details.

Such a structure is amenable to both computational (digital) and geometric (coordinate-free) treatment. The neuropsychological implications of Beck's finite information process have been explored at length in (Hoffman 1985).

Such simplicial sets can also be acted upon by the symmetric difference operation, thus giving rise to a structural model of Riegel's dialectical psychology (Riegel 1973). This theoretical structure consists of a two-phase cognitive process that first of all performs discrimination and classification and is then followed by synthesis *cum* context. This theory provides a ready explanation for such cognitive elements as learning, memory, intelligence, affect, and creativity (Hoffman 1995). The traditional "similarities-differences" paradigm is thus given a formal structure, though in the interests of psychological realism, it should actually read "differences-similarities," in the opposite order.

The symmetric difference is no stranger to psychology. Frank Restle (1961J appears to have been the first to introduce the operation in his analysis of choice phenomena. Tversky (1977J and his colleagues (Tversky and Gati 1978J have also made use of a weighted form of the symmetric difference in their extensive investigation of paradigms for scaling of similarities. Bart (1971J generalized Piaget's stage of formal operations in terms of the symmetric difference of the algebra of n-ary operations. Rosenblatt's Perceptron, a simple adaptable learning precursor of neural networks, came a cropper when Minsky and Papert (1969J showed that the Perceptron Learning Rule could not perform the XOR (symmetric differenceJ operation.

As will be demonstrated below in detail, we therefore have:

Principle 2: *To the extent that dialectical psychology applies, cognitive processes are governed by the symmetric difference and its complement.*

2. Consciousness

Consciousness is a word of many meanings. Is it simple awareness, self-awareness, perception, cognition, "experience," decision-making, or whatever? Gazzinaga (1995: 1392J defines it as "that subjective state that we all possess when awake." However, it is not experience as such but how we regard that experience. Thus "consciousness" refers primarily to the decision-making faculties of the mind. A zombie would have experience but no decision- making capability.

2.1. *Awareness*

Though there may be some residual subconscious sensory or memory features present when one is asleep or under anesthesia, these latter phenomena are characterized by the absence of input from the Ascending Reticular Activating System (ARAFJ to the brain (Lindsley 1960J, (Scheibel and Scheibel 1967J. This is the fundamental awareness level of consciousness. It is only when input from the ARAF is present that we experience awareness. The ARAF input floods the brain (thus presenting a "blackboard"J to provide awareness of our specific surroundings via cancellation (inhibitionJ, like a blackboard "chalk mark." This view has much in common with Baars and Newman's (1994J ERTAS (Extended Reticular-Thalamic Activating SystemJ, which generates the basic brain structure for the cognitive functional global workspace (GWS) for both conscious and unconscious experience.

2.2. *Self-awareness*

The next higher consciousness level is that of *self*-awareness. Neuropsychologists make much of "motion primacy" — moving stimuli are sensed much more strongly than stationary ones. In the fine neurons of the Central Nervous System, nerve impulses propagate at about 2 to 5 meters/second. The result, according to a basic theorem of Special Relativity Theory (Rindler 1991: 8–10; Sachs and Wu 1977: secs. 1.4 and 2.4), is that this finite velocity of signal propagation forces the relation between the outside moving frame of reference and the internal egocentered frame of reference to be Lorentzian rather than Galilean (Caelli, Hoffman, and Lindman 1978).[3] The result is to force a relativistic distinction between the egocentered self and the external world (Hoffman 1990). This is basic self-awareness, upon which the other later developing self-directed cognitive and emotional faculties superimpose.

2.3. *Perception*

At the next higher level of consciousness, one finds the phenomena of perception, generated by what Pribram has termed the posterior intrinsic systems of the brain, which "mediate invariant properties of specific sensory modalities." Perception is of form, whether visual, auditory, sensorimotor, or of taste and smell. Associated with these forms is form memory — "pattern recognition" — and the psychological "constancies." The latter comprise the visual constancies of shape, size, motion, and color; the auditory constancies of pitch, loudness, and binaural localization; the "roman soldier leggings" of haptic perception (Werner and Whitsel 1968); and the regularly repetitive phenomena of taste and smell. Without the constancies, we would, as von Fieandt has observed, always be moving through a surrealistic world of perpetually deforming rubbery objects. The penalty for memory storage is obvious, for every form would have to be remembered in all of the countless distortions which viewing conditions might impose.

The visual field of view is a manifold in the technical, mathematical sense (Hoffman 1989). It is covered at both the retinal and post-retinal level by "charts" (local patches) of a "center-surround" nature that, in the aggregate, provide a mosaic-like covering, an "atlas," for the perceptual manifold. At the level of the "cortical retina" something new makes its appearance: superimposed upon the "center-surround" patches are the "orientation responses" of Hubel and Wiesel, Mountcastle and others. These constitute vector fields upon the visual manifold. Integration of these vector fields by flows and/or parallel transport gives rise to visual contours (Hoffman 1977, 1978, 1984, 1989, 1994b). According to the Figure-Ground Relation, visual contours are the basic states of visual perception.

Not every neuropsychologist will accept the Figure-Ground Relation, but all will accept the existence of visual contours. Visual contours are the "states" for the perception of visual form. The topological requirement of transversality (Hoffman and Dodwell 1985) imposes texture upon the contours. The invariance of these visual contours under the *continuous transformations* worked by viewing conditions constitutes the basis for the psychological constancies (Hoffman 1966, 1968, 1970, 1971b, 1977, 1978, 1984, 1985, 1986). These "continuous symmetries" of perceived forms constitute the essence of perception (Teuber 1960; Rock 1980).[4] The continuous transformation (Lie) group that generates these perceptual invariances is the conformal group CO(1,3) in one time dimension and three spatial dimensions (Hoffman 1994a).

2.4. *Cognition*

At the next higher stage of consciousness, meanings are imparted to the perceived forms via associated cognitions. Percepts can occur independently of associated meanings. The latter occur at this subsequent cognitive stage of consciousness. The maiden lady and her cat see, according to the psychological doctrine of veridicality, the same "mouse" form, but the meaning of that form to each is vastly different. And there are numerous cases in the medical literature of damage to the frontal or inferotemporal regions of the brain wherein perception of shapes persists undiminished, but the subject loses the ability to recognize the meanings of "life objects," (e.g., trucks).

Neuronal flows occur in both directions in the Central Nervous System — up to the cortex and back down again (Livingston 1967). These percepts also have emotional qualities imparted to them (Donovan 1985), (Levin, Decker, and Butcher 1992) by the limbic system and are projected down to the seat of consciousness in the subcortical regions of the brain. Mathematically speaking, such phenomena constitute a *fibration* in the sense of algebraic topology (Hoffman 1985). The brain model postulated here is shown in a sagittal section in Figure 2.

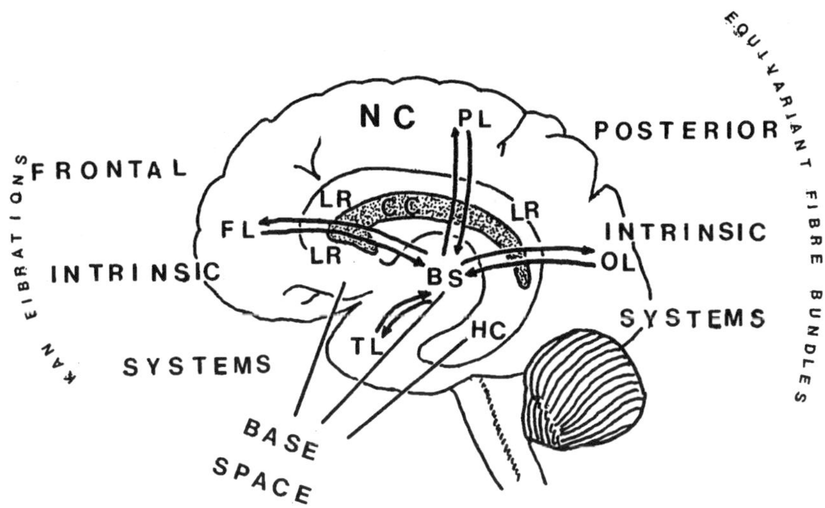

Figure 2. The brain model postulated here. The figure shows (diagrammatically) a sagittal section through the human brain. On the right are the posterior intrinsic systems which mediate invariant properties of specific sensory modalities. These invariants are determined by the transformation groups of the psychological constancies and their prolongations that generate form recognition. Since group actions and mappings must at all times be consistent, these constitute equivariant fibre bundles. OL is the occipital lobe; PL the parietal lobe. On the left are the frontal intrinsic systems, consisting of prefrontal cortex FL and inferotemporal cortex TL, which mediate cognitions and long trains of thought. According to information processing psychology and dialectical psychology, these frontal brain regions have the mathematical character of Kan fibrations. Both forebrain and hind brain have projections back and forth to the hippocampus HC and subcortical base space BS (the brain stem), where the seat of consciousness lies. These projections are colored by the limbic system LR by emotion and are mediated by various neurotransmitters. NC denotes the neocortex of the entire brain, and CC is the corpus callosum, which connects the two hemispheres of the brain.

Pribram's posterior intrinsic systems, which "mediate invariant properties of specific sensory modalities," are, in mathematical terms, fibre bundles associated with Lie transformation groups and contact transformations (Hoffman 1989). The latter are equivariant in the sense that the group actions and mappings commute. The cognitive systems residing mainly in prefrontal and inferotemporal cortex are Kan fibrations[5] consisting of simplicial objects of thought that have projections both ways — up to cortex from subcortical regions and the posterior intrinsic systems ("cross sections," in mathematical terms) and back and down to these same structures (projections). These are the "loops" or "circuits" of the brain anatomy literature.

2.5. *Subconscious Mental Processes*

As a lead-in to the discussion next of dialectical psychology, consider the operation of the symmetric difference process on conscious and subconscious brain states. By their very nature these are disjoint; yet they inevitably have also certain aspects in common. A stimulus pattern can result in an automatic or nonautomatic response. A word-list task unfamiliar to the subject activates a brain pathway that, as seen in a PET brain scan, consists of the anterior cingulate, the temporal and frontal lobes, and right cerebellum (Posner and Raichle 1994: 127). Once the subject becomes practiced in the task, however, the performance becomes automatic and no activation of these areas is seen. However, a new word list undoes the practice-induced non-activation, and the pathway again becomes evident in a PET scan.

3. Dialectical Psychology

In this writer's opinion, Riegel's dialectical psychology (Riegel 1973) offers the best avenue to understanding the nature of cognition. It is isomorphic to information processing psychology, and encompasses a whole host of cognitive phenomena. Dialectical psychology is a life-long developmental psychology that emphasizes individual development and the influence of the social environment. It postulates that one's mental processes move freely back and forth among the four Piagetian stages meanwhile "transforming contradictory experience into momentary stable structures."

As indicated in Principle 2 above, it will now be demonstrated that the symmetric difference operation of set-theoretic topology and its complement can be used to represent the fundamental operations of dialectical psychology. Thesis and antithesis are governed by the symmetric difference; synthesis and context are generated by the complement of the symmetric difference operation. The isomorphisms of the symmetric difference with simplicial k-chains (Henle 1979: 148), (Hocking and Young 1961) and the cyclic group of order 2 (Budden 1974) relate dialectical psychology to information processing psychology. The oft decried "binary fallacy" (Selfridge 1990) involved in modeling cognitive processes by the 0,1 elements of computer logic is thus replaced by a disjoint union of simplices from the category of simplicial objects representative of information processing psychology. The simplicial structure thus provides a bridge between information processing psychology and geometric psychology (Hoffman 1980a).

One of the most compelling features of Riegel's theory was the resolution of the contretemps surrounding Piaget's fourth developmental stage, formal operations. The negative results of Lovell (1961), the fact that 37% of college students and fully half of all adults fail to demonstrate formal operations in their thinking, as well as the prevalence of double negatives and other grammatical errors in everyday speech all cast doubt upon "logical thought" as such in the "if ... then ..." style of Piagetian scientific method. According to Bart (1971), the human mind has only limited capability for consciously exploring a complete lattice of 2^n logical possibilities. Instead, according to cognitive research (Bruner and Olver 1963: 434), the mind acts to reduce the number of possible alternatives and select those which are of primary interest.

The essence of dialectic lies in the apposition of opposites — thesis and antithesis — in order to arrive, through removal of sensed contradictions, at a synthesis, at perhaps a higher level. Dialectical thinking thus consists of an exploration of contradictory possibilities which results in cognition that reduces cognitive conflict. According to Kahneman and Miller (1986) all perceived events are compared to counterfactual alternatives, counterfactual in that they constitute alternative realities to the one experienced. Johnson-Laird (1995), in his study of mental models for deductive reasoning, also notes the importance of counterfactuals. Knight and Grabowecky (1995) state that counterfactual alternatives are omnipresent in normal human cognition. Whenever a figure has been perceived, it has emerged from (back)ground. For every experiment there is a control. To every agonist, there is an antagonist. Evaluation of such a set of internally generated alternatives provide the basis for judgments and decisions. Revision of any document (such as this one) is a classic instance of dialectical thinking.

Curiosity, attention, belief, and interest — "intentionality" — are inherent in such a dialectical structure though not in such a closed system as logical deduction. Ever since the term was introduced, theories of consciousness and cognition have relied heavily on "intentionality." Intention/al/ity can have three meanings (Marcel and Bisiach 1988):

1. the sum total of attributes or objects comprehended in a concept or set. This is the opposite of "extension;" it is always spelled "intension."
2. A goal or purpose that is explicity represented — an *intention* — which always requires a referential: "aboutness."

3. Content, reference, or indication — what something is about — spelled intention or "intention" — a goal.

The first of these — intension — is the meaning employed here. We shall have to do with psychological sets of attributes or percepts. These are such things as consciousness (C) or cognitions (C_i), percepts (P) or quasi-percepts (Q) (such as imagery or phonological rehearsal), expectations (E), meanings (M), etc.

Following Hegel's lead, Riegel laid down the following three "laws" for dialectical psychology:

I. The unity and struggle of opposites.
II. The transformation of quantitative into qualitative change.
III. The negation of the negation.

These three "laws" may be well expressed in the context of both dialectical logic and dialectical psychology by means of the set-theoretic operation of symmetric difference, the "exclusive or," XOR, or "disjoint union" — "one or the other, but not both."

The symmetric difference operation \$ upon two cognitive sets C_1 and C_2 is thus defined by the formula:

$$C_1 \,\$\, C_2 = (C_1 \cap \sim C_2) \cup (\sim C_1 \cap C_2) = (C_1 \cup C_2) \setminus (C_1 \cap C_2). \qquad (1)$$

Exemplified in this expression are the first two of the Hegel-Riegel "laws" above. Also apparent is the "law of the excluded middle." The symmetric difference and its complement are equivalent to the Sheffer stroke, which is enough in itself to generate first order logic. Formal, "logical" thought is thus accessible via the symmetric difference even though intuitive thought processes are more fundamental (Hoffman 1980b, 1985; Riegel 1973: 363).

Synthesis of C_1 and C_2 — the generation of their commonality — comes about by taking "the negation of the negation" ("law" number 3 above). Even though this "law" is traditionally regarded as complementation of a set, in the interests of cognitive reality the complement here is taken to apply to the symmetric difference rather than the sets themselves. The result is to restore the commonality of C_1 and C_2 and also to generate the context — "everything else" other than C_1 and C_2. That is:

$$\sim(C_1 \,\$\, C_2) = (C_1 \cap C_2) \cup \sim(C_1 \cup C_2). \qquad (2)$$

It is postulated here that the complement of $C_1 \,\$\, C_2$ is what properly constitutes the formal parallel of the third Hegel-Riegel "law" embodied in dialectical psychology. The first term on the right in Eq. (2) represents the synthesis, the commonality, of C_1 and C_2. The second provides their context within the universe of discourse D.

3.1. *Metric and Probabilistic Aspects*

A fundamental postulate of dialectical psychology is that quantitative change is transformed into qualitative change. Yet any psychological theory that does not admit quantitative description is obviously lacking. The bridge from the qualitative set-theoretic formulation above to metric and probabilistic phenomena is provided by the following theorem (Moran 1968: 211): A family of sets $\{C_i\}$ on which a measure $\mu(C_i)$ is defined can be made into a quasi-metric space by defining as a distance function, $d(C_i,C_j)$, between any two members C_i and C_j the expression:

$$d(C_i,C_j) = \mu(C_i \,\$\, C_j). \tag{3}$$

The symmetric difference thus provides a scale for such a family of sets. In the present context, C_i and C_j can be any cognitive elements.

Metric properties would thus be provided by psychometric measures involving scaling on such sets. For probability measure $p(C_i)$, the above definition leads immediately to the phase-one formula:

$$p(C_i \,\$\, C_j) = p(C_i \cup C_j) - p(C_i \cap C_j) = p(C_i) + p(C_j) - 2p(C_i \cap C_j), \tag{4}$$

while for the second-phase complement:

$$p[\sim(C_i \,\$\, C_j)] = p(C_i \cap C_j) + p[(\sim(C_i \cup C_j)] = 1 + p(C_i \cap C_j) - p(C_i \cup C_j). \tag{5}$$

It follows that these two formulas would govern the subjective statistical decision process involved in cognitive decisionmaking. As noted above, we are all, subjectively at least, "statisticians" (Hoffman 1980b).

3.2. *Memory in Dialectical Psychology*

Before going on, it is necessary to define three varieties of memory for use in the present context: Attentional Perception, Working Memory, and Long Term Memory. These may differ here from their usage in psychology,[6] but are required to make a proper distinction between a percept P or imagery Q, current expectations W at an attentional conscious level, and the unconscious Long Term Memory L residing, to paraphrase Robertson Davies, in that "fibrous darkness below consciousness." In addition, there is all knowledge K, past, present, and future, known to you personally or, what is more likely, not known by you at the present time.

- The Perceptual Field P. P denotes awareness of the entire perceptual field. It is non-attentional. P corresponds to the sensory registers of the Atkinson-Shiffrin memory model or the visuospatial sketchpad/ phonological loop of the Baddeley-Hitch model. In the case of vision, it would represent the entirety of what one sees within the two-sided field of view (right and left eyes). It is continuous in time and space and merges smoothly into the preceding and following perceptual fields.
- The Perceived Field $\mathcal{P} = P/A$. This is the perceptual field P "factored" (filtered) by attention A. It corresponds to efferent blocking of elements or objects of the perceptual field. $\mathcal{P}$ represents the passage from the sensory system to the STM of the Atkinson-Shiffrin model or the Central Executive of the Baddeley-Hitch model of working memory.
- Working Memory (WM or W). WM is conscious memory dredged up from the Long Term Memory store by attention (Baddeley 1993). It is enhanced by rehearsal and/or repetition but vulnerable to masking by new percepts or thoughts. It, too, is short term but can be maintained continuously conscious over a small time interval by phonological rehearsal. Working memory is perhaps a better measure of intelligence than IQ.
- Long Term Memory (LTM or L). This is the sum total of your mentally stored experience, procedural, categorical, or episodic. It is ordinarily below the conscious level and requires retrieval by interest, attention, or relevant STM or WM. It is a long term memory store LTS. Though apparently forgotten, elements of LTM can still be recalled under hypnosis or appropriate neuroactive drugs like sodium amytal.
- All knowledge, present and future (K). There is much information not currently presently stored in your LTM, and there will be more in the future. K represents this potential between what you know and what you don't know. Long Term Memory is inevitably a subset of all knowledge.

The symmetric difference operation can be used to identify a novel pattern. Suppose that X denotes such an unknown percept within perceived field $\mathcal{P}$. The problem is how to identify such an unknown entity X from the comparison:

$$W \$ X = \mathcal{P}. \tag{6}$$

It is known (Budden 1972: 63) that the set of all subsets {C} with \$ as operation generates a group, the null set ∅ being the identity element. Further the inverse to C is C itself, i.e., the group is idempotent. To solve Eq. (6), first of all group-multiply the left hand side by W:

$$W \$ W \$ X = \varnothing \$ X = X.$$

On the other hand, it also follows from Eq. (6) that:

$$W \$ W \$ X = W \$ \mathcal{P},$$

so that:

$$X = W \$ \mathcal{P}. \tag{7}$$

The symmetric difference operation thus admits the identification of an unknown pattern in the context of $\mathcal{P}$ and W. Budden (1972: 112) gives a graphic demonstration of how a successive approximation of this kind to the solution takes place. This recursive solution appears to have much in common with the multistage processes involved in problem solving that are described by Frederiksen (1984). We therefore have:

Principle 3. *Application of the symmetric difference operation to the perceived field $\mathcal{P}$ and working memory W identifies, by means of the classification and discrimination inherent in the symmetric difference operation, an unknown entity X.*

But suppose that the contents of $\mathcal{P}$ are not to be found within either WM or LTM. However, they are part of knowledge, currently known or not. We then have the more general principle:

Principle 4. *The symmetric difference among L, $\mathcal{P}$, and W leads not only to (i) the commonality of $\mathcal{P}$ and W but also leads "memory outside of itself by generating (ii) those elements in LTM not presently in $\mathcal{P}$ or W WM and (iii) new knowledge common to $\mathcal{P}$ and K but not already in LTM.*

The proof goes as follows:

$$L \$ (\mathcal{P} \$ W) = (L \$ W) \$ \mathcal{P} = \mathcal{P} \$ (L \$ W), \text{ by associativity and commutativity.}$$

But always W is contained within L, so that:

$$L \,\$\, W = (L \cup W) \setminus (L \cap W) = L \setminus W = L \cap {\sim}W.$$

Therefore:

$$\begin{aligned} \mathcal{P}\$\,(L \,\$\, W) &= [\mathcal{P} \cap {\sim}(L \cap {\sim}W] \cup [{\sim}\mathcal{P} \cap (L \cap {\sim}W)] \\ &= [(\mathcal{P} \cap {\sim}L) \cup (\mathcal{P} \cap W)] \cup [({\sim}P \cup {\sim}W) \cup L] \\ &= [\mathcal{P} \cap W] \cup [L \cap {\sim}(\mathcal{P} \cup W)] \cup [(\mathcal{P} \cap K) \setminus L]. \end{aligned} \tag{8}$$

The first term on the right of Eq. (8) is identifiable as the commonality of WM and $\mathcal{P}$. The second term represents the residual content of L that is not then common to WM and $\mathcal{P}$. And finally the third term comprises whatever portion of $\mathcal{P}$ and K that is not already stored in LTM.

3.3. *Learning*

Shulman (1986) has classified the kinds of knowledge as being propositional, case, and strategic. Propositional knowledge abstracts a situation, strips it down to its essentials, discarding in the process detail and context, and comes up with principles, maxims, and norms. Case knowledge is “common sense” knowledge. It arises out of detected regularities and leads to a representative rule by analogical reasoning. The highest form, strategic knowledge, enters when case rules contradict or principles collide. It consists of analytically comparing and contrasting cases or principles and their implications for practice. It occurs in the presence of the other forms of knowledge and is the primary means for testing, amending, and extending them. The connection with the symmetric-difference-and-its-complement model is not far to seek and has been discussed at length in (Hoffman 1995).

3.4. *Information Processing Psychology*

As discussed above, information processing psychology views memory as organized into nodes (“chunks” of information) and edges that interconnect such nodes in meaningful associations at an unconscious level. Forms in an incoming stimulus pattern are recognized during perceptual processing in the posterior intrinsic systems, then classified and assessed by processes of dialectical psychology, stored for the moment in WM, and perhaps transmitted to LTM the long term memory store. The nodes can stand for either a single item or a cluster ("chunk") of related items. Retrieval of some elements of the cluster generally recalls the entire "chunk." Nodes can represent sensory-perceptual knowledge; or such semantic information as facts, concepts, word meanings, beliefs, theories, emotional casts attached to such concepts; or procedural information required for motor or cognitive skills. LTM consists of countless such simplicial structures, all interconnected in complex ways. At any given instant, the great majority of the contents of LTM are at the subconscious level; only those activated at the moment comprise WM. The information processing resides in the retrieval of information from LTM and control of the flow into and out of WM. WM thus provides a subjective representation of the environment of the moment but is constrained by the number of distinct "chunks" (Miller's 7 ± 2) available for processing the information. Indeed, the capacity of WM has been found to be better related to problem solving ability than IQ.

The nodal network of information processing psychology thus constitutes, at least in the case of paired comparisons, a *k-chain* of k-simplices $\Delta(C_k)$, $(k = 0,1,2)$ of a simplicial complex Σ. The cognitive elements previously denoted by C_k now become $\Delta(C_k)$ to emphasize the simplicial nature of the cognitions. The following discussion is therefore limited to paired comparisons.

The sum of two such k-chains $\Delta(C_1)$ and $\Delta(C_2)$ consists of the set of k-simplexes contained in either C_1 or C_2 but not both (Henle 1979: p. 148); in other words, the symmetric difference of $\Delta(C_1)$ and $\Delta(C_2)$:

$$\Delta(C_1) \$ \Delta(C_2).$$

The set $C_k(\Sigma)$ of k-chains constitutes a group, where the identity element is the empty set $\varnothing$ of k-simplices and each element is idempotent, i.e., its own inverse: $\Delta(C_k) \$ \Delta(C_k) = \varnothing$. We thus have via the symmetric difference operation an isomorphism between information processing psychology and dialectical psychology.

Principle 5. *With respect to paired comparisons at least, information processing psychology and dialectical psychology are isomorphic.*

3.5. *Problem solving*

The majority of current theories of problem solving rest on information processing psychology (Frederiksen 1984) and are best suited to well structured problems which have a complete problem statement and well defined procedural knowledge such as an algorithm or automatic processing for solution. Frederiksen (1984) believes that there actually exist three types of problems: "well-structured" problems, which are clearly formulated, have an associated algorithm, and for which there are well defined criteria for testing the correctness of a solution; "structured problems requiring productive thinking," which are similar in the main to well structured problems but have some element in the problem solving process that must be created by the problem solver; and "ill structured" problems which lack both clear formulation and a known solution procedure plus solution-evaluation criteria. Most real world problems are of this third kind.

Problem solving has much in common with Resnick's paradigm for learning by removing contradictions from "naive theories." Such a hypothesize-and-test strategy for problem solving is especially appropriate for ill structured problems and ordinarily leads to a multistage process. Newell and Simon have found that a heuristic "means-end analysis," wherein the problem solver repeatedly compares the present state of the solution with an anticipated solution state, is often used. The question put during the process is, "What is the difference between where I now am and where I hope to end up? What can be done to reduce this difference?" This is strongly reminiscent of formula (3) above. It also has much in common with Nemhauser's (1966: 20) description of multistage strategies for mathematical proof. To determine the truth or falsity of a hypothesized mathematical theorem m_0, look for statements related to m_0 whose truth value (true or false) is known. Suppose m_1 is such a statement. Starting then from m_0, derive (if possible) m_1 from m_0. Then m_0 implies m_1 ($m_0 \Rightarrow m_1$). If m_1 is known to be false, then m_0 is false. However, if the truth value of m_1 is unknown, continue on in this way to obtain a chain of implications:

$$m_0 \Rightarrow m_1 \Rightarrow m_2 \Rightarrow \ldots \Rightarrow m_{n-1} \Rightarrow m_n.$$

Suppose that m_n is false. Then, working backwards through the intermediate statements to m_0, the conclusion is that m_0 itself is false.

The truth of m_0 also can be proved in this way. Take the contradictory of m_0, i.e.,$\sim m_0$. Working through a similar chain for $\sim m_0$, suppose we are led to the falsity of $\sim m_0$. We can then conclude that m_0 is true. The backward multistage approach comprises this and the preceding chain. It frequently leads to easier problem solving, as in the "juggling jugs" problem where exactly 7 ounces of fluid are to be measured out using only a 5–ounce jug and an 8–ounce jug (Nemhauser 1966).

There is also a forward multistage strategy, wherein one begins with some true statement m_n (or false, $\sim m_n$) and derives a chain $m_n \Rightarrow m_{n-1} \Rightarrow \ldots \Rightarrow m_1$ leading to m_0 or $\sim m_0$. If the result is m_0, it then follows that m_0 is true. On the other hand, if the result turns out to be $\sim m_0$, then one would conclude that m_0 is false.

Such strategies as these have a close connection with Boolean algebras and their unit elements. The representation of such a Boolean algebra requires the assignment to each class C a real number $\mu(C)$, similar to Eq. (3), having the properties:

$$\mu(C) = 0 \text{ or } 1,$$
$$\mu(D) = 1, \; \mu(\emptyset) = 0,$$
$$\mu(C_1 \$ C_2) = \mu(C_1) + \mu(C_2)(\text{mod } 2) = \mu(C_1 \cup C_2) - 2\,\mu(C_1 \cap C_2).$$

For an algebra of logical propositions, as here, the thing of main interest to Boole was the truth function t(C). The preceding requirements then oblige t(C) to take the value 0 or the value 1:

$$t(C) = 1 \text{ (true)}$$
or
$$t(C) = 0 \text{ (false)}.$$

Thus $\mu(m_{n-1} \$ m_n)$ and $\mu(m_1 \$ m_0)$ correspond respectively to backward processing and forward processing in the problem solving strategies above.

A strategy frequently employed in problem solving is to recall analogous problems whose solution is already known, determine their differences from the problem at hand, and try to resolve those differences. If A_i denotes such an analogous solution, and S_j is the current j^{th} stage of the solution process, then the first stage symmetric-difference-model assesses the difference: $S_j \$ A_i$, while the second stage, $\sim(S_j \$ A_i) = (S_j \cap A_i) \cup \sim(S_j \cup A_i)$ finds what is common to S_j and A_i and extends the process into their context.

Such structural approaches to working memory and problem solving have a long history in theoretical psychology (Jonassen 1993) and artificial intelligence (Chang 1985; Shapiro *et al.* 1987). Problem reduction methodology (Shapiro *et al.* 1987) exhibits many aspects that would seemingly benefit from use of the symmetric difference operation. Semantic networks, transition networks, petri nets, discrimination nets, frames and scripts, top down and bottom up parsing (Chang 1985) all have a simplicial structure that appears to be amenable to formulation in terms of the symmetric difference. Johnson-Laird (1995: 1005) states that:

> The psychological problem of deduction is to keep track of alternative possibilities.One way in which performance can be strikingly improved is to use diagrams rather than verbal premises. Not any sort of diagram will do, however. The evidence suggests that the diagram must use graphical means to make the alternative possibilities more explicit.

4. Conclusion

The key elements of dialectical psychology, information processing psychology, and consciousness have been analyzed in terms of the elements of the geometry of systems that correspond to brain structure and function. These elements consist of flows, dynamical systems, Lie transformation groups, control theory, fibre bundles, and fibrations (the archetype of parallel processing — serial along the fibres and parallel among the fibres). If "computation" is to play a role in brain and mind, it must have the character of Beck's finite information process. Consciousness (together with its complement, subconsciousness) is a fundamental fact for psychology. Consciousness is a term of many meanings, but here represented in an ascending hierarchy of awareness (as opposed to unconsciousness), self-awareness (arising from subjective special relativity), perception ("what" shape and "where"), cognition (meanings attached to percepts), and emotion (mediating cognitive decision making). Cognition depends strongly upon memory — short term, working, and long term — and classification of cognitions into categories. Information processing psychology, which expresses cognition in terms of concept nodes and edges for interrelationships among these concepts, imparts the structure of simplicial sets to such psychological categories. Classification and discrimination are fundamental to both perception and cognition. This leads naturally into representation of dialectical psychology in terms of the symmetric difference operation and its complement, and it was shown that dialectical psychology is isomorphic with information processing psychology (at least for paired comparisons). Problem solving and the structure of memory and learning were analyzed in terms of the symmetric difference operation and its complement.

Notes

1. Note: These are mathematical *fibre* bundles as distinguished from anatomical *fiber* bundles of heavily myelinated nerve fibers. *Fibre bundles* consist of a projection mapping $\pi: E \rightarrow B$ from total space E to base space B and a cross-section mapping from B to E.
2. A flow is an integral curve generated by a succession of arrows of a vector field.
3. The "laboratory frame" in the parlance of Special Relativity Theory.
4. Teuber (1960): The constancies are "absent only in the absence of perception of pattern, depth, and motion, with which they are indissolubly linked." Rock (1980): "... the perceptual constancies ... are at the very core of, and ... virtually synonymous with the field of perception." Rock has since somewhat modified this position,

however, with the view (Rock 1983) that perception results from thoughtlike processes. But even there, constancy and unconscious mental processes play an important role.

5. A simplicial map $\pi: E \rightarrow B$ is a Kan fibration if for every collection of n+1 n-simplices $x_0, \ldots, x_{k-1}, x_{k+1}, \ldots, x_{n+1}$ of E which satisfy the compactibility condition $\partial_i x_j = \partial_{j-1} x_i$, $i < j$, $i \neq k$, $j \neq k$, and for every (n+1)-simplex y of B such that $\partial_i y = \pi(x_i)$, $i \neq k$, there exists an (n+1)-simplex x of E such that $\partial_i x = x_i$, $i \neq k$, and $\pi(x) = y$. E is called the total complex, B the base complex, and (E, π, B) is called the fibre space. If ϕ denotes the complex generated by a vertex of B, $F = \pi^{-1}(\phi)$ is called the fibre over ϕ. If ψ is the complex generated by a vertex of F, then the sequence $(F, \psi) \rightarrow (E, \psi) \rightarrow (B, \phi)$ is called a fibre sequence (May 1967).
6. Which is not uniform either.

References

Abraham, R. and Marsden, J.E. 1978. *Foundations of Mechanics*. 2nd edition. Reading, MA: Benjamin Cummings.

Baars, B.J. 1988. *A Cognitive Theory of Consciousness*. New York: Cambridge University Press.

Baars, B.J. and Newman, J. 1994. A neurobiological interpretation of global workspace theory. In *Consciousness in Philosophy and Cognitive Neuroscience*, A. Revonsuo and M. Kamppinen (eds), 211–226. Hillsdale, NJ: Erlbaum Associates.

Baddeley, A. 1993. Working memory or working attention? In *Attention: Selection, Awareness, and Control, A Tribute to Donald Broadbent*, A. Baddeley and L. Weiskrantz (eds), 165–168. Oxford: Clarendon Press.

Barinaga, M. 1995. Researchers get a sharper image of the human brain. *Science* 268, 803–804.

Bart, W.M. 1971. A generalization of Piaget's logical-mathematical model for the stage of Formal Operations. *Jour. of Math. Psychol.* 8, 539–553.

Beck, J.M. 1977. Simplicial sets and the foundations of analysis. In *Applications of Sheaves*, M.P. Fourman *et al.* (eds). New York: Springer-Verlag.

Bruner, J.S. and Olver, R.R. 1963. Development of equivalence transformations in children. *Monographs of the Soc. for Res. in Child Dev.* 28, 125–141.

Budden, F.J. 1972. *The Fascination of Groups*. Cambridge, UK: Cambridge University Press.

Caelli, T., Hoffman, W.C., and Lindman, H. 1978. Subjective Lorentz transformations and the perception of motion. *Jour. Optical Soc. Am.* 68, 402–411.

Cassirer, E. 1944. The concept of group and the theory of perception. *Philos. and Phenom. Res.* 5, 1–35.

Chang, C.-L. 1985. *Intro. to Artificial Intelligence Techniques*. Austin, TX: JMA Press.

Donovan, B.T. 1985. *Humor, Hormones, and the Mind: An Approach to the Understanding of Behavior*. Cambridge, UK: Cambridge University Press.

Frederiksen, N. 1984. Implications of cognitive theory for instruction in problem solving. *Review of Educational Res.* 54, 363–407.

Gazzinaga, M.S. and Bizzi, E. (eds). 1995. *The Cognitive Neurosciences*. Cambridge, MA: MIT Press.

Henle, M. 1979. *A Combinatorial Introduction to Topology*. San Francisco: Freeman.

Herlihy, M. and Shavit, N. 1994. Applications of algebraic topology to concurrent computation. *SIAM News*, December 1994: 10–12.

Herlihy, M. and Shavit. N. 1995. Set consensus using arbitrary objects. In press: *Proc. 13th Annual ACM on Principles of Distributed Computing*.

Hillyard, S.A., Picton, T.W., and Regan, D. 1978. Sensation, perception, and attention: Analysis using ERP's. In *Event-Related Brain Potentials in Man*, E. Callaway *et al.* (eds), 223–321. New York: Academic Press.

Hocking, J.G. and Young, G.S. 1961. *Topology*. Reading, MA: Addison-Wesley.

Hoffman, W.C. 1966. The Lie algebra of visual perception. *Jour. Math. Psychol.* 3, 65–98.

Hoffman, W.C. 1968. The neuron as a Lie group germ and a Lie product. *Quart. Applied Math.* 25, 423–441.

Hoffman, W.C. 1970. Higher visual perception as prolongation of the basic Lie transformation group. *Math. Biosciences* 6, 437–471.

Hoffman, W.C. 1971a. Memory grows. *Kybernetik* 8, 151–157.

Hoffman, W.C. 1971b. Visual illusions of angle as an application of Lie transformation groups. *S.I.A.M. Review* 13, 169–184.

Hoffman, W.C. 1977. An informal, historical description (with bibliography) of the L.T.G./N.P. *Cahiers de Psychologie, Université de Provence* 20, 219–231.

Hoffman, W.C. 1978. The Lie transformation group approach to visual neuropsychology. In *Formal Theories of Visual Perception*, E.L.J. Leeuwenberg and H. Buffart (eds), 27–66. Chichester, UK: Halsted Press.

Hoffman, W.C. 1980a. Subjective geometry and Geometric Psychology. *Internat. Jour. Math. Modeling* 1, 349–367.

Hoffman, W.C. 1980b. Mathematical models of Piagetian Psychology. In *Toward a Theory of Psychological Development*, Chap. 13, S. and C. Modgil (eds). Windsor, UK: National Foundation for Educational Research.

Hoffman, W.C. 1981. Environmental Psychology, Ecological Psychology, and the Geometry of Systems. In *Proc. Internat. Congress on Applied Sys. Res. and Cybernetics*, Vol. IV, G.E. Lasker (ed). London: Pergamon Press.

Hoffman, W.C. 1984. Figural synthesis by vectorfields: Geometric Psychology. In *Figural Synthesis*, P.C. Dodwell and T. Caelli (eds). Hillsdale, NJ; Erlbaum

Hoffman, W.C. 1985. Some reasons why Algebraic Topology is important in neuropsychology: Perceptual and cognitive systems as fibrations. *Internat. Jour. Man-Machine Studies* 22, 613–650.

Hoffman, W.C. 1986. Invariant and programmable neuropsychological systems are fibrations. *Behavioral and Brain Sciences* 9, 99–100.

Hoffman, W.C. 1989. The visual cortex is a contact bundle. *Applied Math. and Computation* 32, 137–167.

Hoffman, W.C. 1990. The conformal group CO(1,3) as basis for both Nature and perception cum self-reference. In *Proc. 8th Internat. Congress of Cybernetics and Systems*, Vol. I, C. N. Manikopoulos (ed). Newark, NJ: New Jersey Inst. Technology Press.

Hoffman, W.C. 1994a. Conformal structures in perceptual psychology. *Spatial Vision* 8, 19–31.

Hoffman, W.C. 1994b. Equivariant dynamical systems: A formal model for generation of arbitrary shapes. In *Shape in Picture: Mathematical Description of Shape in Grey-Level Images*, O. Ying-Lie *et al.* (eds). Berlin: Springer-Verlag.

Hoffman, W.C. 1995. The dialectics of giftedness: Cognitive conflict and creativity. *Roeper Review* 17, 201–206.

Hoffman, W.C. and Dodwell, P.C. 1985. Geometric Psychology generates the visual Gestalt. *Canad. Jour. Psychol.* 39, 491–528.

Jacobson, M. 1971. *Developmental Neurobiology*. New York: Holt, Rinehart, and Winston.

Johnson-Laird, P.N. 1995. Mental models, deductive reasoning, and the brain. In *The Cognitive Neurosciences*, M.S. Gazzinaga (ed). Cambridge, MA: MIT Press.

Jonassen, D.H., Beissner, K., and Yacci, M. 1993. *Structural Knowledge: Techniques for Representing, Conveying, and Acquiring Structural Knowledge*. Hillsdale, NJ: Erlbaum.

Kahneman, D. and Miller, D.T. 1986. Norm theory: Comparing reality to its alternatives. *Psychol. Rev.* 93, 136–153.

Knight, R.T. and Grabowecky, M. 1995. Escape from linear time: Prefrontal cortex and conscious experience. In *The Cognitive Neurosciences*, M.S. Gazzinaga (ed), 1357–1371. Cambridge, MA: MIT Press.

Levin, E.D., Decker, M.W., and Butcher, L.L. (eds). 1992. *Neurotransmitter Interactions and Cognitive Function*. Boston: Birkhauser.

Lindsley, D.B. 1960. Attention, consciousness, sleep and wakefulness. In *Handbook of Physiology: Neurophysiology*, Sec. 1, Vol. III, J. Field and H.W. Magoun (eds). Washington, DC: Am. Physiol. Soc.

Livingston, R.B. 1967. Brain circuitry relating to complex behavior. In *The Neurosciences*, G.C. Quarton, T. Melnechuk, and F.O. Schmitt (eds). New York: Rockefeller University Press.

Lovell, K. 1961. A follow-up study of Inhelder and Piaget's The Growth of Logical Thinking. *British Jour. Psychol.* 52, 143–154.

Marcel, A.J. and Bisiach, E. 1988. *Consciousness in Contemporary Science*. Oxford: Clarendon Press.

May, J.P. 1967. *Simplicial Objects in Algebraic Topology*. Princeton, NJ: Van Nostrand Co.

Mayne, D.Q. and Brockett, R.W. (eds). 1973. *Geometric Methods in Systems Theory*. Dordrecht/Boston: D. Reidel.

Minsky, M. and Papert, S. 1969. *Perceptrons: An introduction to computational geometry*. Cambridge, MA: MIT Press.

Moran, P.A.P. 1968. *Intro. to Probability Theory*. Oxford: Clarendon Press.

Nemhauser, G.L. 1966. *Intro. to Dynamic Programming*. New York: John Wiley.

Palmer, S.E. 1983. The psychology of perceptual organization: A transformational approach. In *Human and Machine Vision*, J. Beck *et al.* (eds). San Diego, CA: Academic Press.

Pitts, W. and McCulloch, W.S. 1947. How we know universals: The perception of auditory and visual forms. *Bull. Math. Biophys.* 9, 127–147.

Posner, M.I. and Raichle, M.E. 1994. *Images of Mind*. New York: W.H. Freeman and Co.

Pribram, K.H. 1971. *The Languages of the Brain*. Englewood Cliffs, NJ: Prentice-Hall.

Restle, F.A. 1961. *Psychology of Judgment and Choice*. New York: John Wiley.

Riegel, K. 1973. Dialectic operations: The final period of cognitive development. *Human Development* 16, 346–370.

Rindler, W. 1991. *Intro. to Special Relativity*, 2nd edition. Oxford: Clarendon Press.

Rock, I. 1980. Difficulties with a direct theory of perception. *Behavioral and Brain Sciences* 3, 398–399.

Rock, I. 1983. *The Logic of Perception*. Cambridge, MA: MIT Press.

Rosch, E. and Lloyd, B.B. 1978. *Cognition and Categorization*. Hillsdale, NJ: Erlbaum.

Sachs, R.K. and Wu, H. 1977. General relativity and cosmology. *Bull. Am. Math. Soc.* 83, 1101–1164.

Scheibel, M.E. and Scheibel, A.B. 1967. Anatomical basis of attention mechanisms in vertebrate brains. In *The Neurosciences*, G.C. Quarton, T. Melenchuk, and F.O. Schmitt (eds). New York: Rockefeller University Press.

Schutz, J.W. 1973. *Foundations of Special Relativity: Kinematic Axioms for Minkowski Space-Time*. New York: Springer-Verlag.

Selfridge, O. 1990. In mockery of mankind: The Binary Fallacy. Paper presented at the Plenary Session of the 8th Internat. Congress of Cybernetics and Systems, New York.

Sereno, M.I., Dale, A.M., Reppas, J.B., Kwong, K.K., Belliveau, J.W., Brady, T.J., Rosen, B.R., and Tootell, R.B.H. 1995. Borders of multiple visual areas in humans revealed by functional magnetic resonance imaging. *Science* 269, 889–893.

Shapiro, S.C. (ed). 1987. *Encyclopedia of Artificial Intelligence*. New York: Wiley-Interscience.

Shulman, L.S. 1986. Those who understand: Knowledge growth in teaching. *Educational Researcher* 15, 4–14.

Smith, E.E. and Medin, D.L. 1981. *Categories and Concepts.* Cambridge, MA: Harvard University Press.

Talbot, S.A. and Marshall, W.H. 1941. Physiological studies on neural mechanisms of visual localization and discrimination. *Am. Jour. Ophthalmol.* 24, 1255–1263.

Teuber, H.-L. 1960. Perception. In *Handbook of Physiology, Sec. l: Neurophysiology*, Vol. III. Washington, DC: American Physiological Society.

Tversky, A. 1977. Features of similarity. *Psychological Review* 84, 327–352.

Tversky, A. and Gati, I. 1978. Studies of similarity. In *Cognition and Categorization*, E. Rosch and B.B. Lloyd (eds). Hillsdale, NJ: Erlbaum Associates.

Werner, G. and Whitsel, B.L. 1968. The topology of the body representation in the somatosensory area I of primates. *Jour. Physiol. (London). Neurophysiol.* 31, 856–869.

Subject index

Name index

In the series ADVANCES IN CONSCIOUSNESS RESEARCH (AiCR) the following titles have been published thus far or are scheduled for publication:

1. GLOBUS, Gordon G.: *The Postmodern Brain.* 1995.
2. ELLIS, Ralph D.: *Questioning Consciousness. The interplay of imagery, cognition, and emotion in the human brain.* 1995.
3. JIBU, Mari and Kunio YASUE: *Quantum Brain Dynamics and Consciousness. An introduction.* 1995.
4. HARDCASTLE, Valerie Gray: *Locating Consciousness.* 1995.
5. STUBENBERG, Leopold: *Consciousness and Qualia.* n.y.p.
6. GENNARO, Rocco J.: *Consciousness and Self-Consciousness. A defense of the higher-order thought theory of consciousness.* 1996.
7. MAC CORMAC, Earl and Maxim I. STAMENOV (eds): *Fractals of Brain, Fractals of Mind. In search of a symmetry bond.* 1996.
8. GROSSENBACHER, Peter G. (ed.): *Finding Consciousness in the Brain. A neuro-cognitive approach.* n.y.p.
9. Ó NUALLÁIN, Seán, Paul MC KEVITT and Eoghan MAC AOGÁIN (eds): *Two Sciences of Mind. Readings in cognitive science and consciousness.* 1997.
10. NEWTON, Natika: *Foundations of Understanding.* 1996.
11. PYLKKO, Pauli: *The Aconceptual Mind. Heideggerian themes in holistic naturalism.* n.y.p.
12. STAMENOV, Maxim I. (ed.): *Language Structure, Discourse and the Access to Consciousness.* 1997.